PRIVACY AND TECHNOLOGIES OF IDENTITY

PRIVACY AND TECHNOLOGIES OF IDENTITY

A CROSS-DISCIPLINARY CONVERSATION

Edited by

KATHERINE J. STRANDBURG
DePaul University, College of Law
Chicago, Illinois, USA

DANIELA STAN RAICU
DePaul University, School of Computer Science, Telecommunications, and Information Systems, Chicago, Illinois, USA

Editors:

Katherine J. Strandburg
DePaul University
College of Law
Chicago, IL
USA

Daniela Stan Raicu
DePaul University
School of Computer Science,
Telecommunications and Information Systems
Chicago, IL USA

Library of Congress Cataloging-in-Publication Data

Privacy and technologies of identity : a cross-disciplinary conversation / edited by
Katherine J. Strandburg, Daniela Stan Raicu.
p. cm.
Book stemming from a symposium held Oct. 14-15, 2004, at DePaul University.
Includes bibliographical references and index.

ISBN- 13: 978-1-4419-3858-9
ISBN- 13: 978-0-387-28222-0 (e-ISBN)
ISBN- 10: 0-387-28222-X (e-ISBN)

1. Privacy, Right of. 2. Information technology--Social aspects. I. Strandburg, Katherine, Jo, 1957- II. Raicu, Daniela Stan, 1969-

JC596.P743 2006
323.44'8--dc22

2005052307

Softcover reprint of the hardcover 1st edition 2006

springeronline.com

Contents

Contributing Authors

Alessandro Acquisti
H. John Heinz III School of Public Policy and Management, Carnegie Mellon University, Pittsburgh, Pennsylvania

Daniel Barth-Jones
Center for Healthcare Effectiveness, Wayne State University, Detroit, Michigan

Alex Cameron
LL.D. (Law & Technology) Candidate, University of Ottawa, Ottawa, Ontario, Canada

Christopher Clifton
Department of Computer Sciences, Purdue University, West Lafayette, Michigan

Yuval Elovici
Department of Information Systems Engineering, Ben-Gurion University of the Negev, Beer-Sheva, Israel

Farshad Fatouhi
Department of Computer Science, Wayne State University, Detroit, Michigan

Eric Goldman
Marquette University Law School, Milwaukee, Wisconsin

Jens Grossklags
School of Information Management and Systems, University of California, Berkeley, California

Dilek Z. Hakkani-Tur
AT&T Labs-Research, Florham Park, New Jersey

Joseph Lorenzo Hall
School of Information Management and Systems, University of California, Berkeley, California

Dennis Hirsch
Capital University Law School, Columbus, Ohio

Ari Juels
RSA Laboratories, Bedford, Massachusetts

Arthur Keller
Computer Science Department, University of California, Santa Cruz and Open Voting Consortium, California

Ian Kerr
Faculty of Law, University of Ottawa, Ottawa, Ontario, Canada

Dongge Li
Motorola Labs, Schaumburg, Illinois

David Mertz
Gnosis Software, Inc.

Mark Monmonier
Department of Geography, Maxwell School of Citizenship and Public Affairs, Syracuse University, Syracuse, New York

Deirdre Mulligan
School of Law (Boalt Hall), University of California, Berkeley, California

Lisa Nelson
Graduate School of Public and International Affairs, University of Pittsburgh, Pittsburgh, Pennsylvania

Raghu Ramakrishnan
Computer Sciences Department, University of Wisconsin, Madison, Wisconsin

Yucel Saygin
Sabanci University, Istanbul, Turkey

Paul M. Schwartz
Brooklyn Law School, Brooklyn, New York and School of Law (Boalt Hall), University of California, Berkeley, California (Visiting 2005)

Ishwar K. Sethi
Intelligent Information Engineering Lab, Department of Computer Science and Engineering, Oakland University, Rochester, Michigan

Bracha Shapira
Department of Information Systems Engineering, Ben-Gurion University of the Negev, Beer-Sheva, Israel

Daniel J. Solove
George Washington University Law School, Washington, D.C.

Yael Spanglet
Department of Information Systems Engineering, Ben-Gurion University of the Negev, Beer-Sheva, Israel

John A. Stefani
J.D., 2004, DePaul University College of Law, Chicago, Illinois

Daniel J. Steinbock
University of Toledo College of Law, Toledo, Ohio

Katherine J. Strandburg
DePaul University College of Law, Chicago, Illinois

Min Tang
Center for Spoken Language Research, University of Colorado, Boulder, Colorado

Traian Marius Truta
Department of Mathematics and Computer Science, Northern Kentucky University, Highland Heights, Kentucky

Gokhan Tur
AT&T Labs-Research, Florham Park, New Jersey

Arnold Urken
Stevens Institute of Technology, Castle Point on Hudson, Hoboken, New Jersey

Gang Wei
Accenture Technology Labs, Chicago, Illinois

Tal Z. Zarsky
Faculty of Law, University of Haifa, Haifa, Israel and Information Society Project, Yale Law School, New Haven, Connecticut

Preface

It is by now nearly a cliché that advances in technology, especially in the areas of computer science and data processing, have led to social changes that pose difficult challenges – some would say threats – to individual privacy. Computer scientists, legal scholars, policymakers, and others struggle to find ways to address these challenges so as to preserve essential aspects of individual privacy, while leaving society free to reap the benefits of technological progress. Depending upon our backgrounds and training, we approach these issues differently and bring different skills to bear on them. Thus, interdisciplinary dialogue and collaboration will be essential to resolving the issues of privacy, identity, and technology that confront our age. The CIPLIT® 2004 Symposium, Privacy and Identity: The Promise and Perils of a Technological Age, was organized jointly by computer scientists and legal scholars to promote such dialogue and to help us to continue to educate one another.

This book seeks to enlarge the audience for those interdisciplinary dialogue efforts. While specialists will find much of interest in the book, many of the chapters are devoted to tutorials and overviews for the non-specialist that are intended to reach across the disciplinary divides.

The book is organized into five parts. The first part contains three chapters which address general issues at the intersection of privacy and technology. Each of the following four parts is devoted to a particular aspect of modern technology that raises privacy concerns. Each part includes chapters by technologists and chapters by legal scholars to provide a multi-faceted view of the issues raised by the technology it addresses. Part II addresses the privacy implications of Radio Frequency Identification (RFID) and other means of locational tracking. Part III focuses on biometric technologies. Part IV considers the implications of data mining and knowledge discovery. Part V contains chapters devoted to the interactions between technology and anonymity.

As organizers of the 2004 CIPLIT® Symposium and editors of this volume, we hope that readers of this volume will find it useful in their own efforts to reach across disciplinary boundaries to find ways to promote a successful marriage of privacy and technology in this new millennium.

Katherine J. Strandburg, J.D., Ph.D.
Associate Professor
DePaul University College of Law

Daniela Stan Raicu, Ph.D.
Assistant Professor
DePaul University School of Computer Science, Telecommunications, and Information Systems

Acknowledgments

This book grows out of the 2004 CIPLIT® Symposium, which was jointly organized by the Center for Intellectual Property Law and Information Technology of the DePaul University College of Law (CIPLIT®) and the Computer Information and Network Security Center (CINS) of the DePaul University School of Computer Science, Telecommunications, and Information Technology (CTI). The participation of all the speakers and attendees at the Symposium was indispensable to the success of the Symposium and contributed substantially, though in some cases indirectly, to this conference book.

The Symposium was made possible by the generous support of the law firm of McGuire Woods LLP and particularly by the efforts of partner Thomas R. Mulroy. Financial support for the symposium from the DePaul University Research Council and from CIPLIT® is also gratefully acknowledged.

Professors Strandburg and Raicu are very appreciative of support from College of Law and CTI deans Glen Weissenberger and Helmut Epp. They are especially grateful to CIPLIT® Director, Professor Roberta Kwall, for her enthusiasm for the concept of this Symposium and her continuing advice and direction in its execution. The administrative and organizational talents of CIPLIT® associate director, Ellen Gutiontov, and administrative assistant, Vadim Shifrin, were absolutely essential to the success of the Symposium.

Professor Strandburg also wishes to acknowledge the excellent assistance of her research assistants, Nicholas Lucius and Cole Dunnick, and the indefatigable efforts of her faculty clerk, Lawrence Arendt, in the editing and preparation of this manuscript.

I

Introductory Issues in Privacy and Technology

Chapter 1

THE DIGITAL PERSON AND THE FUTURE OF PRIVACY

Daniel J. Solove
Associate Professor, George Washington University Law School; J.D. Yale, 1997. This chapter is based on an article that was printed in The Chronicle of Higher Education (Dec. 10, 2004), which was adapted from Professor Solove's book, *The Digital Person: Technology and Privacy in the Information Age* (NYU Press 2004).

Abstract: This chapter, based on Professor Solove's book, The Digital Person: Technology and Privacy in the Information Age (NYU Press 2004), explores the social, political, and legal implications of the collection and use of personal information in computer databases. In the Information Age, our lives are documented in digital dossiers maintained by a multitude of businesses and government agencies. These dossiers are composed of bits of our personal information, which when assembled together begin to paint a portrait of our personalities. The dossiers are increasingly used to make decisions about our lives – whether we get a loan, a mortgage, a license, or a job; whether we are investigated or arrested; and whether we are permitted to fly on an airplane. In this chapter, Solove explores the implications of these developments and sets forth a new understanding of privacy, one that is appropriate for the challenges of the Information Age.

Key words: privacy, information, data, Kafka, Orwell, technology, database

We are in the midst of an information revolution, and we are only beginning to understand its implications. The past few decades have witnessed a dramatic transformation in the way we shop, bank, and go about our daily business – changes that have resulted in an unprecedented proliferation of records and data. Small details that were once captured in dim memories or fading scraps of paper are now constantly sifted, sorted, rearranged, and combined in hundreds of ways, then preserved forever in vast, fertile fields of data. The minutiae of our everyday comings and goings, of our likes and dislikes, of who we are and what we own, illuminate and shadow us. They are our digital selves.

Think of them as "digital dossiers." There are hundreds of companies that are constructing gigantic databases of psychological profiles, amassing information about an individual's race, gender, income, hobbies, and purchases. Shards of data from our daily existence are assembled to investigate backgrounds, check credit, market products, and make a wide variety of decisions affecting our lives.

A new breed of company is emerging whose primary business is the collection of personal information. Catalina Marketing Corporation, based in Florida, maintains supermarket-buying-history databases on 30 million households from more than 5,000 stores.[1] Those data contain a complete inventory of one's groceries, over-the-counter medications, hygiene supplies, and contraceptive devices, among other purchases. Aristotle International markets a database of 150 million registered voters, recording voters' names, addresses, phone numbers, party affiliations, and voting frequencies. Aristotle combines those data with about 25 other categories of information, such as one's race, income, and employer – even the make and model of one's car – to market a list of wealthy campaign donors called Fat Cat. Aristotle boasts: "Hit your opponent in the Wallet! Using Fat Cats, you can ferret out your adversary's contributors and slam them with a mail piece explaining why they shouldn't donate money to the other side."[2] Another company manufactures software called GeoVoter, which combines about 5,000 categories of information about each voter to calculate how that individual is likely to vote.[3]

The most powerful database builders construct empires of information on more than half of the American population. For example, Donnelly Marketing Information Services keeps track of 125 million people. Wiland Services has constructed a database containing over 1,000 elements, from demographic information to behavioral data, on more than 215 million people. About five database compilers have information on almost every household in the United States.[4]

Beyond marketers, hundreds of companies keep data about us in their record systems. The complete benefits of the information age do not simply come to us – we must "plug in" to join in. In other words, we must establish

[1] Robert O'Harrow, Jr., *Behind the Instant Coupons, a Data-Crunching Powerhouse*, WASH. POST, Dec. 31, 1998, at A20.

[2] Leslie Wayne, *Voter Profiles Selling Briskly as Privacy Issues Are Raised*, N.Y. TIMES, Sept. 9, 2000, at A1.

[3] Marcia Stepanek, *How the Data-Miners Want to Influence Your Vote*, BUSINESS WEEK (Oct. 26, 2000), http://www.businessweek.com/bwdaily/dnflash/oct2000/nf20001026_969.htm.

[4] Arthur M. Hughes, THE COMPLETE DATABASE MARKETER 354 (2d ed. 1996).

relationships with Internet-service providers, cable companies, phone companies, insurance companies, and so on.

Increasingly, companies rely on various records and documents to assess individuals' financial reputations. Creditors rely on credit-reporting agencies to obtain information about a person's credit history. Credit reports reveal a person's consistency in paying back debts as well as the risk of a person's defaulting on a loan. People are assigned a credit score, which influences whether they will be extended credit, and, if so, what rate of interest will be charged. Credit reports contain detailed financial histories, financial-account information, outstanding debts, bankruptcy filings, judgments, liens, and mortgage foreclosures. Today there are three major credit-reporting agencies: Equifax, Experian, and TransUnion. Each one has compiled extensive dossiers on almost every adult American citizen.[5] Credit reports have become essential to securing a loan, obtaining a job, purchasing a home or a car, applying for a license, or even renting an apartment.

Founded in 2002, Regulatory DataCorp has created a massive database to investigate people opening bank accounts. RDC, as it is known, was created by many of the world's largest financial companies. Its Global Regulatory Information Database gathers information from over 20,000 sources around the world.[6] RDC's purpose is to help financial companies conduct background checks of potential customers for fraud, money laundering, terrorism, and other criminal activity. Although some information in the database may be incorrect, people lack the ability to correct errors about themselves. The company's CEO and president responds: "There are no guarantees. Is the public information wrong? We don't have enough information to say it's wrong."[7]

The Internet is rapidly becoming the hub of the personal-information market, for it has made the peddling and purchasing of data much easier. Focus USA's Web site boasts that it has detailed information on 203 million people.[8] Among its more than 100 targeted mailing lists are "Affluent Hispanics," "Big-Spending Parents," "First Time Credit Card Holders," "Grown but Still at Home," "Hi-Tech Seniors," "New Homeowners,"

[5] For example, Experian has information on 205 million Americans. See http://www.experian.com/corporate/factsheet.html.

[6] Regulatory DataCorp, *RDC Services*, http://www.regulatorydatacorp.com/ourServices.html.

[7] Tyler Hamilton, *Getting to Know You: Opening a Bank Account Gives Regulatory DataCorp a Window on Your Life*, TORONTO STAR, June 18, 2003.

[8] Focus USA, http://www.focus-usa-1.com/lists_az.html.

"Status Spenders," "Big Spending Vitamin Shoppers," and "Waist Watchers."[9]

There's a list of "Mr. Twentysomethings," which contains mostly college-educated men who, the company believes, are eager to spend money on electronic equipment. And there are lists of pet lovers, fitness-conscious people, cat and dog owners, motorcycle enthusiasts, casino gamblers, and subprime prospects (people who are risky to lend to).[10] Dunhill International also markets a variety of lists, including "America's Wealthiest Families," which includes 9.6 million records "appended with demographic and psychographic data."[11] Other databases cover disabled people, consumers who recently applied for a credit card, cruise-ship passengers, teachers, and couples who just had a baby. Hippo Direct markets lists of people suffering from "medical maladies" like constipation, cancer, diabetes, heart disease, impotence, migraines, and enlarged prostates.[12] Another company markets a list of five million elderly incontinent women.[13]

Because of the interactive nature of the Internet, marketers can learn how we respond to what we hear and see. A Web site will collect information about the way a user interacts with the site and will store the information in its database.

Another information-collection device, known as a "Web bug," can be embedded in a Web page or even an e-mail message. The Web bug is a hidden snippet of code that can gather data about a person.[14] For example, a company can send a spam e-mail with a Web bug that will report back when the message is opened. The bug can also record when the message is forwarded to others. Web bugs also can collect information about people as they explore a Web site. Some of the nastier versions of Web bugs can even access a person's computer files.[15]

[9] *Id.*

[10] *Id.*

[11] Dunhill International List Co., *Residential Database Info.*, *available at* http://www.datacardcentral.com/dataindex-r.cfm.

[12] Hippo Direct, *available at* http://www.hippodirect.com/ListSubjectsN_1.asp?1Subject=37. For a terrific discussion of various databases such as the ones discussed in this section, see Electronic Privacy Information Center, *Privacy and Consumer Profiling, available at* http://www.epic.org/privacy/profiling.

[13] *See* Standards for Privacy of Individually Identifiable Health Information, Preamble, 65 Fed. Reg. 82,461 (Dec. 28, 2000).

[14] See Robert O'Harrow, Jr., *Fearing a Plague of "Web Bugs": Invisible Fact-Gathering Code Raises Privacy Concerns*, WASH. POST, Nov. 13, 1999, at E1.

[15] *See* Leslie Walker, *Bugs That Go through Computer Screens*, WASH. POST, Mar. 15, 2001, at E1. *See also* Richard M. Smith, *FAQ: Web Bugs, available at* http://www.privacyfoundation.org/resources/webbug.asp.

Companies also use what has become known as "spyware," which is software that is often deceptively and secretly installed into people's computers. Spyware can gather information about every move one makes when surfing the Internet. Those data are then used by spyware companies to target pop-up ads and other forms of advertising.[16]

As we stand at the threshold of an age structured around information, we are only beginning to realize the extent to which our lives can be encompassed within its architecture. "The time will come," predicts one marketer, "when we are well known for our inclinations, our predilections, our proclivities, and our wants. We will be classified, profiled, categorized, and our every click will be watched."[17]

All these developments certainly suggest a threat to privacy, but what specifically is the problem? The way that question is answered has profound implications for the way the law will grapple with the problem.

Journalists, politicians, and jurists often describe the problem created by databases with the metaphor of Big Brother – the harrowing, totalitarian government portrayed in George Orwell's *1984*.[18] A metaphor, as the legal scholar Steven L. Winter of Wayne State University Law School aptly defines it, "is the imaginative capacity by which we relate one thing to another."[19] Much of our thinking about a problem involves the metaphors we use. Metaphors are tools of shared cultural understanding.[20] Privacy involves the type of society we are creating, and we often use metaphors to envision different possible worlds, ones that we want to live in and ones that

[16] James R. Hagerty & Dennis K. Berman, *Caught in the Net: New Battleground over Web Privacy: Ads That Snoop*, WALL ST. J., Aug. 27, 2003.

[17] Jim Sterne, WHAT MAKES PEOPLE CLICK: ADVERTISING ON THE WEB 255 (1997).

[18] *See, e.g.*, William Branigin, *Employment Database Proposal Raises Cries of "Big Brother"*, WASH. POST, Oct. 3, 1995, at A17; James Gleick, *Big Brother Is Us: Our Privacy Is Disappearing, But Not by Force. We're Selling It, Even Giving It Away*, N.Y. TIMES MAG., Sept. 29, 1996, at 130; Priscilla M. Regan, LEGISLATING PRIVACY 93 (1995) (a House committee held hearings called "1984 and the National Security State" to examine the growing computerization of records); 140 CONG. REC. H9797, H9810 (statement of Rep. Kennedy) ("the promise of the information highway has given way to an Orwellian nightmare of erroneous and unknowingly disseminated credit reports"); J. Roderick MacArthur Found. v. FBI, 102 F.3d 600, 608 (D.C. Cir. 1996) (Tatel, J., dissenting) ("Congress passed the Privacy Act to give individuals some defenses against governmental tendencies towards secrecy and 'Big Brother' surveillance"); McVeigh v. Cohen, 983 F. Supp. 215, 220 (D.D.C. 1998) ("In these days of 'big brother,' where through technology and otherwise the privacy interests of individuals from all walks of life are being ignored or marginalized, it is imperative that statutes explicitly protecting these rights be strictly observed").

[19] Steven L. Winter, A CLEARING IN THE FOREST: LAW, LIFE, AND MIND 65 (2001).

[20] J.M. Balkin, CULTURAL SOFTWARE: A THEORY OF IDEOLOGY 141 (1998).

we don't. Orwell's Big Brother is an example of this type of metaphor; it is a shared cultural narrative, one that people can readily comprehend and react to.

Big Brother is an all-knowing, constantly vigilant government that regulates every aspect of one's existence. It demands complete obedience from its citizens and controls all aspects of their lives. Big Brother is constantly monitoring and spying; uniformed patrols linger on street corners; helicopters hover in the skies, poised to peer into windows. The primary surveillance tool is a device called a "telescreen" that is installed in each house and apartment. The telescreen is a bilateral television – individuals can watch it, but it also enables Big Brother to watch them.[21]

The metaphor of Big Brother dominates the discourse of information privacy.[22] Internet "surveillance" can be readily compared to Orwell's telescreen. Commentators view databases as having many of the same

[21] George Orwell, 1984 3-20 (1949).

[22] *See, e.g.*, William G. Staples, THE CULTURE OF SURVEILLANCE: DISCIPLINE AND SOCIAL CONTROL IN THE UNITED STATES 129-34 (1997) (noting that we have become a Big Brother culture); Abbe Mowshowitz, *Social Control and the Network Marketplace*, 79 COMPUTERS, SURVEILLANCE, AND PRIVACY 95-96 (David Lyon & Elia Zureik eds., 1996) ("The specter of Big Brother has haunted computerization from the beginning"); Dorothy Glancy, *At the Intersection of Visible and Invisible Worlds: United States Privacy Law and the Internet*, 16 SANTA CLARA COMPUTER & HIGH TECH. L.J. 357, 377 (2000) (describing privacy problem created by the private sector as the "little brother" problem); Marsha Morrow McLauglin & Suzanne Vaupel, *Constitutional Right of Privacy and Investigative Consumer Reports: Little Brother Is Watching You*, 2 HASTING CONST. L.Q. 773 (1975); Hon. Ben F. Overton & Katherine E. Giddings, *The Right of Privacy in Florida in the Age of Technology and the Twenty-First Century: A Need for Protection from Private and Commercial Intrusion*, 25 FLA. ST. U. L REV. 25, 27 (1997) ("In his book, 1984, we were warned by George Orwell to watch out for 'Big Brother.' Today, we are cautioned to look out for 'little brother' and 'little sister.'"); Thomas L. Friedman, *Foreign Affairs: Little Brother*, N.Y. TIMES, Sept. 26, 1999, Sec. 4 at 17; Wendy R. Leibowitz, *Personal Privacy and High Tech: Little Brothers Are Watching You*, NAT'L L.J., Apr. 7, 1997, at B16; Charles N. Faerber, *Book versus Byte: The Prospects and Desirability of a Paperless Society*, 17 J. MARSHALL J. COMPUTER & INFO. L. 797, 798 (1999) ("Many are terrified of an Orwellian linkage of databases."); Brian S. Schultz, *Electronic Money, Internet Commerce, and the Right to Financial Privacy: A Call for New Federal Guidelines*, 67 U. CIN. L. REV. 779, 797 (1999) ("As technology propels America toward a cashless marketplace...society inches closer to fulfilling George Orwell's startling vision."); Alan F. Westin, *Privacy in the Workplace: How Well Does American Law Reflect American Values?*, 72 CHI.-KENT L. REV. 271, 273 (1996) (stating that Americans would view the idea of government data protection boards to regulate private sector databases as "calling on 'Big Brother' to protect citizens from 'Big Brother.'"); Wendy Wuchek, *Conspiracy Theory: Big Brother Enters the Brave New World of Health Care Reform*, 3 DEPAUL J. HEALTH CARE L. 293, 303 (2000).

purposes (social control, suppression of individuality) and employing many of the same techniques (surveillance and monitoring) as Big Brother.[23] According to this view, the problem with databases is that they are a form of surveillance that curtails individual freedom.

Although the metaphor has proved quite useful for a number of privacy problems, it only partially captures the problems of digital dossiers. Big Brother envisions a centralized authoritarian power that aims for absolute control, but the digital dossiers constructed by businesses aren't controlled by a central power, and their goal is not to oppress us but to get us to buy products and services. Even our government is a far cry from Big Brother, for most government officials don't act out of malicious intent or a desire for domination. Moreover, Big Brother achieves its control by brutally punishing people for disobedience and making people fear that they are constantly being watched. But businesses don't punish us so long as we keep on buying, and they don't make us feel as though we are being watched. To the contrary, they try to gather information as inconspicuously as possible. Making us feel threatened would undermine rather than advance the goal of unencumbered information collection. Finally, while Big Brother aims to control the most intimate details of a citizen's life, much of the information in digital dossiers is not intimate or unusual.

Franz Kafka's harrowing depiction of bureaucracy in *The Trial* captures dimensions of the digital-dossier problem that the Big Brother metaphor does not.[24] *The Trial* opens with the protagonist, Joseph K., awakening one morning to find a group of officials in his apartment, who inform him that he is under arrest. He is bewildered as to why he has been placed under arrest, and the officials give him no explanation. Throughout the rest of the novel, Joseph K. undertakes a frustrating quest to discover why he has been arrested and how his case will be resolved. A vast bureaucratic court has apparently scrutinized his life and assembled a dossier on him. However, no matter how hard he tries, Joseph K. cannot discover what information is in his dossier or what will be done with it.

Kafka depicts an indifferent bureaucracy, where individuals are pawns, not knowing what is happening, having no say or ability to exercise meaningful control over the process. This lack of control allows the trial to completely take over Joseph K.'s life. The story captures the sense of helplessness, frustration, and vulnerability one experiences when a large

[23] Jerry Kang, *Information Privacy in Cyberspace Transactions*, 50 STAN. L. REV. 1193, 1261 (1998); Paul M. Schwartz, *Privacy and Participation: Personal Information and Public Sector Regulation in the United States*, 80 IOWA L. REV. 553, 560 (1995); Oscar H. Gandy, Jr., THE PANOPTIC SORT: A POLITICAL ECONOMY OF PERSONAL INFORMATION 10 (1993).

[24] Frank Kafka, THE TRIAL 16-17 (Willa & Edwin Muir trans., 1937).

bureaucratic organization has control over a vast dossier of details about one's life. Decisions are made based on Joseph K.'s data, and he has no say, no knowledge, and no ability to fight back. He is completely at the mercy of the bureaucracy.

Understood in light of the Kafka metaphor, the primary problem with databases stems from the way the bureaucratic process treats individuals and their information. Although bureaucratic organization is an essential and beneficial feature of modern society, bureaucracy also presents numerous problems. It often cannot adequately attend to the needs of particular individuals – not because bureaucrats are malicious, but because they must act within strict time constraints, have limited training, and are frequently not able to respond to unusual situations in unique or creative ways.

Additionally, decisions within bureaucratic organizations are often hidden from public view, decreasing accountability. As Max Weber noted, "bureaucratic administration always tends to exclude the public, to hide its knowledge and action from criticism as well as it can."[25] Frequently, bureaucracies fail to train employees adequately and may employ subpar security measures over personal data.

In this view, the problem with databases and the practices currently associated with them is that they disempower people. They make people vulnerable by stripping them of control over their personal information. There is no diabolical motive or secret plan for domination; rather, there is a web of thoughtless decisions made by low-level bureaucrats, standardized policies, rigid routines, and a way of relating to individuals and their information that often becomes indifferent to their welfare.

Identity theft is an example of the kind of Kafkaesque problems that are occurring because of the collection and use of people's personal data. In an identity theft, the culprit obtains personal information and uses it in a variety of fraudulent ways to impersonate the victim. Identity theft is the most rapidly growing type of white-collar criminal activity.[26] According to a Federal Trade Commission estimate in September 2003, "almost 10 million Americans have discovered that they were the victim of some form of ID theft within the past year."[27]

Identity theft plunges people into a bureaucratic nightmare. Victims experience profound difficulty in dealing with credit-reporting agencies, and often find recurring wrong entries on their credit reports even after

[25] Max Weber, ECONOMY AND SOCIETY 992 (Guenther Roth & Claus Wittich eds., 1978).

[26] *See* Jennifer 8. Lee, *Fighting Back When Someone Steals Your Name*, N.Y. TIMES, Apr. 8, 2001.

[27] Federal Trade Commission, Identity Theft Survey Report 4 (Sept. 2003).

contacting the agencies.[28] Identity theft creates those problems because our digital dossiers are becoming so critical to our ability to function in modern life. The identity thief not only pilfers victims' personal information, but also pollutes their dossiers by adding unpaid debts, traffic violations, parking tickets, and arrests.[29]

Identity theft is caused in large part by the private sector's inadequate security measures in handling personal information. Companies lack adequate ways of controlling access to records and accounts in a person's name, and numerous companies use Social Security numbers, mothers' maiden names, and addresses for access to account information.[30] Creditors give out credit and establish new accounts if the applicant supplies simply a name, a Social Security number, and an address.

Social Security numbers are relatively easy for the identity thief to obtain. They are harvested by database firms from a number of public and nonpublic sources, such as court records and credit reports.[31] It is currently legal for private firms to sell or disclose the numbers, and identity thieves can readily purchase them.[32]

The law has attempted to respond to problems of information privacy in a number of ways, but it has often proved ineffective. Since the early 1970s, Congress has passed over 20 laws pertaining to privacy.[33] Federal regulation covers records from federal agencies, educational institutions, cable-television and video-rental companies, and state motor-vehicle agencies, but it does not cover most records maintained by state and local officials, libraries, and charities, and by supermarkets, department stores, mail-order catalogs, bookstores, and other merchants. Moreover, many of Congress's

[28] *See* Janine Benner *et al.*, NOWHERE TO TURN: VICTIMS SPEAK OUT ON IDENTITY THEFT: A CALPIRG/PRIVACY RIGHTS CLEARINGHOUSE REPORT (May 2000), *available at* http://www.privacyrights.org/ar/idtheft2000.htm.

[29] Privacy Rights Clearinghouse & Identity Theft Resource Center, *Criminal Identity Theft* (May 2002), *available at* http://www.privacyrights.org/fs/fs11g-CrimIdTheft.htm; Stephen Mihm, *Dumpster Driving for Your Identity*, N.Y. TIMES MAG., Dec. 21, 2003.

[30] Lynn M. LoPucki, *Human Identification Theory and the Identity Theft Problem*, 80 TEX. L. REV. 89 (2001).

[31] *See* Robert O'Harrow Jr., *Identity Thieves Thrive in Information Age: Rise of Online Data Brokers Makes Criminal Impersonation Easier*, WASH. POST, May 31, 2001, at A1; Jennifer Lee, *Dirty Laundry for All to See: By Posting Court Records, Cincinnati Opens a Pandora's Box of Privacy Issues*, N.Y. TIMES, Sept. 5, 2002, at G1.

[32] For example, an identity thief purchased the SSNs of several top corporate executives from Internet database companies. The thief then used the SSNs to obtain more personal information about the victims. Benjamin Weiser, *Identity Theft, and These Were Big Identities*, N.Y. TIMES, May 29, 2002.

[33] For a survey of these laws, see Daniel J. Solove & Marc Rotenberg, INFORMATION PRIVACY LAW (2003).

privacy statutes are hard to enforce. It is often difficult, if not impossible, for an individual to find out if information has been disclosed. A person who begins receiving unsolicited marketing mail and e-mail messages may have a clue that some entity has disclosed her personal information, but she often will not be able to discover which entity was the culprit.

The Kafka metaphor illustrates that our personal information escapes us into an undisciplined bureaucratic environment. Understood from the standpoint of *The Trial,* the solution to regulating information flow is not to radically curtail the collection of information, but to regulate its uses.

For example, Amazon.com's book-recommendation service collects extensive information about a customer's tastes. If the problem is surveillance, then the most obvious solution would be to provide strict limits on Amazon.com's collection of such information. But that would curtail much data gathering that is necessary for business in today's society and is put to beneficial uses. Indeed, many Amazon.com customers, including me, find its book recommendations to be very helpful.

From *The Trial*'s perspective, the problem is not that Amazon is "spying" on its users or that it can use personal data to induce its customers to buy more books. What's troubling is its unfettered ability to do whatever it wants with that information. That was underscored when Amazon.com abruptly changed its privacy policy in 2000 to allow the transfer of personal information to third parties in the event that the company sold any of its assets or went bankrupt.[34] As a customer, I had no say in that change of policy, no ability to change it or bargain for additional privacy protection, and no sense as to whether it would apply retroactively to the purchases I had already made. And what's to prevent Amazon.com from changing its policy again, perhaps retroactively?

The law's first step should be to redefine the nature of our relationships to businesses and government entities that gather and use our personal information. Consider our relationships to our doctors and lawyers. The law imposes a number of obligations on them to focus on our welfare. They must keep our information confidential and must disclose to us any financial or other interest that may affect their relationships with us.[35] The law should

[34] *See* Richard Hunter, WORLD WITHOUT SECRETS: BUSINESS, CRIME, AND PRIVACY IN THE AGE OF UBIQUITOUS COMPUTING 7 (2002); *Amazon Draws Fire for DVD-Pricing Test, Privacy-Policy Change*, WALL ST. J., Sept. 14, 2000, at B4.

[35] Doctors and lawyers are said to stand in a fiduciary relationship with their patients and clients. As one court has aptly explained the relationship: "A fiduciary relationship is one founded on trust or confidence reposed by one person in the integrity and fidelity of another. Out of such a relation, the laws raise the rule that neither party may exert influence or pressure upon the other, take selfish advantage of his trust[,] or deal with the

stipulate that the organizations that collect and use our personal data have similar obligations.

Regarding identity theft, for example, the primary legislative response has been to enact greater criminal penalties for wrongdoers. So far, that approach has failed. Law-enforcement agencies have not devoted adequate resources toward investigating and prosecuting identity-theft cases, most of which remain unsolved.[36] The research firm Gartner Inc. estimates that fewer than one in 700 instances of identity theft results in a conviction.[37]

Instead, the law should recognize that the collection and use of personal information are activities that carry duties and responsibilities. Minimum security practices must be established for handling people's personal information or accounts. Use of a Social Security number, mother's maiden name, or birth date as the means of gaining access to accounts should be prohibited. Identity theft can be curtailed by employing alternative means of identification, such as passwords. Although passwords are far from foolproof, they can readily be changed if a thief discovers them. Social Security numbers are very difficult to change, and other personal information can never be modified.

When the government discloses personal information in public records, it can prevent those data from being amassed by companies for commercial purposes, from being sold to other companies, and from being combined with other information and sold back to the government.

There are some who glibly claim that protecting privacy is impossible in the information age. To the contrary, the law can go quite far in safeguarding privacy. But to do so, we need to think about the nature of our privacy problems in a way that underscores their most serious dangers.

subject matter of the trust in such a way as to benefit himself or prejudice the other except in the exercise of utmost good faith." Mobile Oil Corp. v. Rubenfeld, 339 N.Y.S.2d 623, 632 (1972). Courts have imposed liability on doctors who breached patient confidentiality based on fiduciary duty theory. Hammonds v. Aetna Casualty & Surety Co., 243 F. Supp. 793 (D. Ohio 1965); McCormick v. England, 494 S.E. 2d 431 (S.C. Ct. App. 1997).

[36] GAO Identity Theft Report, supra notes 17-18.

[37] Stephen Mihm, *Dumpster Diving for Your Identity*, N.Y. TIMES MAG., Dec. 21, 2003.

Chapter 2

PRIVACY AND RATIONALITY

A Survey

Alessandro Acquisti[1] and Jens Grossklags[2]
[1]H. John Heinz III School of Public Policy and Management, Carnegie Mellon University, Hamburg Hall, 5000 Forbes Avenue, Pittsburgh, PA 15213; [2]School of Information Management and Systems, University of California at Berkeley, 102 South Hall, Berkeley, CA 94720

Abstract: We present preliminary evidence from a survey of individual privacy attitudes, privacy behavior, and economic rationality. We discuss the theoretical approach that drives our analysis, the survey design, the empirical hypotheses, and our initial results. In particular, we present evidence of overconfidence in privacy assessments, lack of information prior to privacy-sensitive decisions, misconceptions about one's own exposure to privacy risks, bounded rationality, and hyperbolic discounting.

Key words: privacy, rationality, experimental economics, consumer behavior

1. INTRODUCTION[1]

From their early days[2] to more recent incarnations,[3] economic models of

[1] A previous version of this chapter appeared in 3(1) IEEE SECURITY AND PRIVACY 26 (2005).

[2] *See* R.A. Posner, *An Economic Theory of Privacy*, 2(3) REGULATION 19 (1978); R.A. Posner, *The Economics of Privacy*, 71 AM. ECON. REV. 405 (1981); and G.J. Stigler, *An Introduction to Privacy in Economics and Politics*, 9 J. LEGAL STUDIES 623 (1980).

[3] A. Acquisti and H.R. Varian, *Conditioning Prices on Purchase History*, MARKETING SCIENCE (forthcoming 2005); G. Calzolari and A. Pavan, *On the Optimality of Privacy in Special Contracting,* (2004) (working paper), *available at* http://faculty.econ.nwu.edu/faculty/pavan/privacy.pdf; C.R. Taylor, *Consumer Privacy and the Market for Customer Information*, 35 RAND J. ECON. 631 (2004).

privacy have viewed individuals as rational economic agents who rationally decide how to protect or divulge their personal information. According to this view, individuals are forward lookers, utility maximizers, Bayesian updaters who are fully informed or who base their decisions on probabilities coming from known random distributions. (Some recent works[4] have contrasted myopic and fully rational consumers, but focus on the latter.) This approach also permeates the policy debate, in which some believe not only that individuals should be given the right to manage their own privacy trade-offs without regulative intervention, but that individuals can, in fact, use that right in their own best interest.

While several empirical studies have reported growing privacy concerns across the U.S. population,[5] recent surveys, anecdotal evidence, and experiments[6] have highlighted an apparent dichotomy between privacy attitudes and actual behavior. First, individuals are willing to trade privacy for convenience or to bargain the release of personal information in exchange for relatively small rewards. Second, individuals are seldom willing to adopt privacy protective technologies.

Our research combines theoretical and empirical approaches to investigate the drivers and apparent inconsistencies of privacy decision making and behavior. We present the theoretical groundings to critique the assumption of rationality in privacy decision making. In addition, through an anonymous, online survey, we have started testing the rationality

[4] Acquisti and Varian, *supra* note 3; Taylor, *supra* note 3.

[5] M. Ackerman, L. Cranor and J. Reagle, *Privacy in E-Commerce: Examining User Scenarios and Privacy Preferences*, PROC. ACM CONFERENCE ON ELECTRONIC COMMERCE (1998); A.F. Westin, *Harris-Equifax Consumer Privacy Survey 1991, available at* http://privacyexchange.org/iss/surveys/eqfx.execsum.1991.html;
Federal Trade Commission, Privacy On-line: Fair Information Practices in the Electronic Marketplace (2000), *available at* http://www.ftc.gov/reports/privacy2000/privacy.pdf; B. Huberman, E. Adar, and L.R. Fine, *Valuating Privacy*, forthcoming in IEEE SECURITY AND PRIVACY (on file with author); Jupiter Research, *Online Privacy: Managing Complexity to Realize Marketing Benefits* (2002), *available at* http://jupiterdirect.com/bin/toc.pl/89141/951.

[6] Jupiter Research, *supra* note 5; Harris Interactive, *Most People are Privacy Pragmatists Who, While Concerned with Privacy, Will Sometimes Trade It Off for Other Benefits* (2003), *available at* http://www.harrisinteractive.com/harris_poll/index.asp?PID=365; R.K. Chellappa and R. Sin, *Personalization Versus Privacy: An Empirical Examination of the Online Consumers Dilemma*, 6(2-3) INFORMATION TECH. AND MGM'T 181 (2005); Il-Horn Hann *et al.*, *Online Information Privacy: Measuring the Cost-Benefit Trade-Off*, PROC. 23RD INT'L CONF. ON INF. SYS. (ICIS 02) (2002), *available at* http://comp.nus.edu.sg/~ipng/research/privacy_icis.pdf; S. Spiekermann, J. Grossklags, and B. Berendt, *E-Privacy in Second Generation E-Commerce: Privacy Preferences versus Actual Behavior*, PROC. ACM CONF. ON ELECTRONIC COMMERCE (EC 01) (2001).

assumption by analyzing individual knowledge, economic behavior, and psychological deviations from rationality when making privacy-sensitive decisions.

2. CHALLENGES IN PRIVACY DECISION MAKING

The individual decision process with respect to privacy is affected and hampered by multiple factors. Among those factors, incomplete information, bounded rationality, and systematic psychological deviations from rationality suggest that the assumption of perfect rationality might not adequately capture the nuances of an individual's privacy sensitive behavior.[7]

First, incomplete information affects privacy decision making because of externalities (when third parties share personal information about an individual, they may affect that individual without his being part of the transaction between those parties);[8] information asymmetries (information relevant to the privacy decision process – for example, how personal information will be used – may be known only to a subset of the parties making decisions); risk (most privacy related payoffs are not deterministic); and uncertainties (payoffs may not only be stochastic, but dependent on unknown random distributions). Benefits and costs associated with privacy intrusions and protection are complex, multifaceted, and context-specific. They are frequently bundled with other products and services (for example, a search engine query can prompt the desired result but can also give observers information about the searcher's interests), and they are often recognized only after privacy violations have taken place. They can be monetary but also immaterial, and thus difficult to quantify.

Second, even if individuals had access to complete information, they would be unable to process and act optimally on such vast amounts of data. Especially in the presence of complex, ramified consequences associated with the protection or release of personal information, our innate bounded

[7] A. Acquisti, *Privacy in Electronic Commerce and the Economics of Immediate Gratification, in* PROC. ACM CONF. ON ELECTRONIC COMMERCE (EC 04) (2004).

[8] H.R. Varian, *Economic Aspects of Personal Privacy*, in U.S. Dept. of Commerce, PRIVACY AND SELF-REGULATION IN THE INFORMATION AGE (1997), *available at* http://www.ntia.doc.gov/reports/privacy/privacy_rpt.htm.

rationality[9] limits our ability to acquire, memorize and process all relevant information, and makes us rely on simplified mental models, approximate strategies, and heuristics. These strategies replace theoretical quantitative approaches with qualitative evaluations and "aspirational" solutions that stop short of perfect (numerical) optimization. Bounded problem solving is usually neither unreasonable nor irrational, and it need not be inferior to rational utility maximization. However, even marginal deviations by several individuals from their optimal strategies can lead to substantial impacts on the market outcome.[10]

Third, even if individuals had access to complete information and could successfully calculate optimization strategies for their privacy sensitive decisions, they might still deviate from the rational strategy. A vast body of economic and psychological literature reveals several forms of systematic psychological deviations from rationality that affect individual decision making.[11] For example, in addition to their cognitive and computational bounds, individuals are influenced by motivational limitations and misrepresentations of personal utility. Experiments have shown an idiosyncrasy between losses and gains (in general, losses are weighted heavier than gains of the same absolute value), and documented a diminishing sensitivity for higher absolute deviations from the *status quo*. Psychological research also documents how individuals mispredict their own future preferences or draw inaccurate conclusions from past choices. In addition, individuals often suffer from self-control problems – particularly, the tendency to trade off costs and benefits in ways that damage future utility in favor of immediate gratification. Individuals' behavior can also be guided by social preferences or norms, such as fairness or altruism. Many of these deviations apply naturally to privacy sensitive scenarios.[12]

Any of these factors may influence decision-making behavior inside and outside the privacy domain, though not all factors need to always be present. Empirical evidence of their influence on privacy decision making would not necessarily imply that individuals act recklessly or make choices against their own best interest.[13] It would, however, imply bias and limitations in the

[9] H.A. Simon, MODELS OF BOUNDED RATIONALITY, (1982); R. Selten, *What is Bounded Rationality? in* BOUNDED RATIONALITY: THE ADAPTIVE TOOLBOX (G. Gigerenzer and R. Selten, eds., 2001).

[10] N. Christin, J. Grossklags and J. Chuang, *Near Rationality and Competitive Equilibria in Networked Systems, in* PROC. SIGCOMM WORKSHOP ON PRACTICE AND THEORY OF INCENTIVES IN NETWORKED SYSTEMS (2004).

[11] D. Kahneman and A. Tversky, CHOICES, VALUES AND FRAMES (2000).

[12] A. Acquisti, *supra* note 7.

[13] P. Syverson and A. Shostack, *What Price Privacy? in* THE ECONOMICS OF INFORMATION SECURITY (L.J. Camp and S. Lewis, eds., 2004)

individual decision process that should be considered in the design of privacy public policy and privacy-enhancing technologies.

3. THE SURVEY

In May 2004 we contacted potential subjects who had shown interest in participating in economic studies at Carnegie Mellon University. We offered participants a lump sum payment of $16 to fill out an online, anonymous survey about "ECommerce preferences" and we gathered 119 responses (the title for the survey was chosen to mitigate self selection bias from pre-existing privacy beliefs). The survey contained several questions organized around various categories: demographics, a set of behavioral economic characteristics (such as risk and discounting attitudes), past behavior with respect to protection or release of personal information, knowledge of privacy risks and protection against them, and attitudes toward privacy.[14] This survey was the second round of a research project funded by the Berkman Faculty Development Fund. The first round was a pilot survey we conducted in January, and in the third round, forthcoming, we will further investigate the findings of this article.

Participants ranged from 19 to 55 years old (with the mean age of 24). Eighty-three percent were US citizens, with the remainder having heterogeneous backgrounds. More than half of our subjects worked full or part time or were unemployed at the time of the survey, although students represented the largest group (41.3 percent). All participants had studied or were studying at a higher education institution. Hence, our population of relatively sophisticated individuals is not an accurate sample of the US population, but this makes our results even more surprising.

Most participants had personal and household incomes below $60,000. Approximately 16.5 percent reported household incomes above that level, including 6.6 percent with an income greater than $120,000. Most respondents are also frequent computer users (62.0 percent spend more than 20 hours per week) and Internet browsers (69.4 percent spend more than 10 hours per week) and access computers both at home and work (76.0 percent). Our respondents predominantly use computers running Windows (81.6 percent), 9.6 percent primarily use Macintosh, and 8.8 percent rely on Linux or Unix systems.

[14] We discuss only a subset of questions in this article; the full survey is available online at http://www.heinz.cmu.edu/~acquisti/survey/page1.htm.

4. ATTITUDES

A large portion of our sample (89.3 percent) reported they were either moderately or very concerned about privacy (see Table 1).[15] Our subjects provided answers compatible with patterns observed in previous surveys. For example, when asked, "Do you think you have enough privacy in today's society?" 73.1 percent answered that they did not. And, when asked, "How do you personally value the importance of the following issues for your own life on a day-to-day basis?," 37.2 percent answered that information privacy policy was "very important" – less than the fraction that held education policy (47.9 percent) and economic policy (38.0 percent) to be very important, but more than the fraction of people who believed that the threat of terrorism (35.5 percent), environmental policy (22.3 percent), or same-sex marriage (16.5 percent) were very important.

Privacy attitudes appear to correlate with income; the lowest personal income group (less than $15,000 a year) tended to be less concerned about privacy than all other income groups, with a statistically significant difference in the distributions of concerns by income grouping ($\chi^2 = 17.5$, $p = 0.008$).

Table 2- 1. Generic Privacy Attitudes (in percent)

	General Privacy Concern	Data about offline identity	Data about online identity	Data about personal profile	Data about profession al profile	Data about sexual and political identity
High Concern	53.7	39.6	25.2	0.9	11.9	12.1
Medium Concern	35.6	48.3	41.2	16.8	50.8	25.8
Low Concern	10.7	12.1	33.6	82.3	37.3	62.1

Table 2-1 also shows that requests for identifying information (such as the subject's name or email address) lead to higher concerns than requests for profiling information (such as age; weight; or professional, sexual, and

[15] Several survey questions on privacy attitudes were framed using the 5- or 7-points Likert Scale (for example, from "I strongly agree" to "I strongly disagree" with a certain statement). In this paper we summarize the results by clustering answers in broader categories.

political profile). When asked for isolated pieces of personal information, subjects were not particularly concerned if the information was not connected to their identifiers. Sensitivity to such data collection practices is generally below the reported general level of concern.

However, subjects were more sensitive to data bundled into meaningful groups. A correlation of data from subjects' offline and online identities caused strong resistance in 58.3 percent of the sample.

We employed *k*-means multivariate clustering techniques[16] to classify subjects according to their privacy attitudes, extracting base variables used for clustering from several questions related to privacy attitudes. Hierarchical clustering (average linkage) outsets the data analysis. We selected the best partitioning using the Calinski-Harabasz criterion.[17] We derived four distinct clusters: privacy fundamentalists with high concern toward all collection categories (26.1 percent), two medium groups with concerns either focused on the accumulation of data belonging to online or offline identity (23.5 percent and 20.2 percent, respectively), and a group with low concerns in all fields (27.7 percent).

Not surprisingly, concerns for privacy were found to be correlated to how important an individual regards privacy to be. However, by contrasting privacy *importance* and privacy *concerns*, we found that for those who regard privacy as most important, concerns were not always equally intense; 46.5 percent of those who declared privacy to be very important expressed lower levels of privacy concerns.

A vast majority of respondents strongly agrees with the definition of privacy as ability to control disclosure and access to personal information (90.1 percent and 91.7 percent, respectively). However, a significant number of subjects also care about certain aspects of privacy that do not have immediate informational or monetary interpretation, such as privacy as personal dignity (61.2 percent) and freedom to develop (50.4 percent). In fact, only 26.4 percent strongly agreed with a definition of privacy as the "ability to assign monetary values to each flow of personal information." Our subjects seemed to care for privacy issues even beyond their potential financial implications.

These results paint a picture of multifaceted attitudes. Respondents distinguish types of information bundles and associated risks, discern between the importance of privacy in general and their personal concerns, and care for privacy also for nonmonetary reasons.

[16] M. Berry and G. Linoff, DATA MINING TECHNIQUES FOR MARKETING, SALES AND CUSTOMER SUPPORT (1997).

[17] R.B. Calinski and J. Harabasz, *A Dendrite Method for Cluster Analysis*, 3 COMM. STATISTICS 1 (1974).

5. BEHAVIOR

We investigated two forms of privacy-related behavior: self-reported adoption of privacy preserving strategies and self-reported past release of personal information.

We investigated the use of several privacy technologies or strategies and found a nuanced picture. Usage of specific technologies was consistently low – for example, 67.0 percent of our sample never encrypted their emails, 82.3 percent never put a credit alert on their credit report, and 82.7 percent never removed their phone numbers from public directories. However, aggregating, at least 75 percent did adopt at least one strategy or technology, or otherwise took some action to protect their privacy (such as interrupting purchases before entering personal information or providing fake information in forms).

These results indicate a multifaceted behavior: because privacy is a personal concept, not all individuals protect it all the time. Nor do they have the same strategies or motivations. But most *do* act.

Several questions investigated the subjects' reported release of various types of personal information (ranging from name and home address to email content, social security numbers, or political views) in different contexts (such as interaction with merchants, raffles, and so forth). For example, 21.8 percent of our sample admitted having revealed their social security numbers for discounts or better services or recommendations, and 28.6 percent gave their phone numbers. A cluster analysis of the relevant variables revealed two groups, one with a substantially higher degree of information revelation and risk exposure along all measured dimensions (64.7 percent) than the other (35.3 percent). We observed significant differences between the two clusters in past behavior regarding the release of social security numbers and descriptions of professional occupation, and the least difference for name and nonprofessional interests.

When comparing privacy attitudes with reported behavior, individuals' generic attitudes may often appear to contradict the frequent and voluntary release of personal information in specific situations.[18] However, from a methodological perspective, we should investigate how psychological attitudes relate to behavior under the same scenario conditions (or frames), since a person's generic attitude may be affected by different factors than those influencing her conduct in a specific situation.[19] Under homogeneous

[18] Jupiter Research, *supra* note 5; Harris Interactive, *supra* note 6; Chellappa and Sin, *supra* note 6; Spiekermann *et al.*, *supra* note 6.

[19] M. Fishbein and I. Azjen, BELIEF, ATTITUDE, INTENTION AND BEHAVIOR: AN INTRODUCTION TO THEORY AND RESEARCH (1975).

frames, we found supporting evidence for an attitude/behavior dichotomy. For example, we compared stated privacy concerns to ownership of supermarket loyalty cards. In our sample 87.5 percent of individuals with high concerns toward the collection of offline identifying information (such as name and address) signed up for a loyalty card using their real identifying information. Furthermore, we asked individuals about specific privacy concerns they have (participants could provide answers in a free text format) and found that of those who were particularly concerned about credit card fraud and identity theft only 25.9 percent used credit alert features. In addition, of those respondents that suggested elsewhere in the survey that privacy should be protected by each individual with the help of technology, 62.5 percent never used encryption, 43.7 percent do not use email filtering technologies, and 50.0 percent do not use shredders for documents to avoid leaking sensitive information.

6. ANALYSIS

These dichotomies do not imply irrationality or reckless behavior. Individuals make privacy sensitive decisions based on multiple factors, including what they know, how much they care, and how effective they believe their actions can be. Although our respondents displayed sophisticated privacy attitudes and a certain level of privacy-consistent behavior, their decision processes seem affected also by incomplete information, bounded rationality, and systematic psychological deviations from rationality.

6.1 Constrained knowledge and incomplete information

Survey questions about respondents' knowledge of privacy risks and modes of protection (from identity theft and third-party monitoring to privacy-enhancing technologies and legal means for privacy protection) produced nuanced results. As in prior studies,[20] our evidence points to an alternation of awareness and unawareness from one scenario to the other (though a cluster of 31.9 percent of respondents displayed high unawareness

[20] S. Fox, *et al.*, *Trust and Privacy Online: Why Americans Want to Rewrite the Rules*, (working paper) The Pew Internet and American Life Project (2000), *available at* http://www.pewinternet.org/pdfs/PIP_Trust_Privacy_Report.pdf; J. Turow, *Americans and Online Privacy: The System is Broken*, Annenberg Center Public Policy Report (2003), *available at* http://www.annenbergpublicpolicycenter.org/04_info_society/2003_online_privacy_version_09.pdf.

of simple forms of risks across most scenarios).

On the one hand, 83.5 percent of respondents believe that it is most or very likely that information revealed during an e-commerce transaction would be used for marketing purposes, 76.0 percent believe that it is very or quite likely that a third party can monitor some details of usage of a file sharing client, and 26.4 percent believe that it is very or quite likely that personal information will be used to vary prices during future purchases. On the other hand, most of our subjects attributed incorrect values to the likelihood and magnitude of privacy abuses. In a calibration study, we asked subjects several factual questions about values associated with security and privacy scenarios. Participants had to provide a 95-percent confidence interval (that is a low and high estimate so that they are 95 percent certain that the true value will fall within these limits) for specific privacy-related questions. Most answers greatly under or overestimated the likelihood and consequences of privacy issues. For example, when we compared estimates for the number of people affected by identity theft (specifically for the US in 2003) to data from public sources (such as US Federal Trade Commission), we found that 63.8 percent of our sample set their confidence intervals too narrowly – an indication of overconfidence.[21] Of those individuals, 73.1 percent underestimated the risk of becoming a victim of identity theft.

Similarly, although respondents realized the risks associated with links between different pieces of personal data, they are not fully aware of how revealing those links are. For example, when asked, "Imagine that somebody does not know you but knows your date of birth, sex, and zip code. What do you think the probability is that this person can uniquely identify you based on those data?" 68.6 percent answered that the probability was 50 percent or less (and 45.5 percent of respondents believed that probability to be less than 25 percent). According to Carnegie Mellon University researcher Latanya Sweeney,[22] 87 percent of the US population can be uniquely identified with a 5-digit zip code, birth date, and sex.

In addition, 87.5 percent of our sample claimed not to know what Echelon (an alleged network of government surveillance) is; 73.1 percent claimed not to know about the FBI's Carnivore system; and 82.5 percent claimed not to know what the Total Information Awareness program is.

Our sample also showed a lack of knowledge about technological or legal forms of privacy protection. Even in our technologically savvy and educated sample, many respondents could not name or describe an activity or technology to browse the Internet anonymously so that nobody can

[21] S. Oskamp, *Overconfidence in Case Study Judgments*, 29 J. CONSULTING PSYCH. 261 (1965).

[22] L. Sweeney, *K-Anonymity: A Model for Protecting Privacy*, 10 INT'L. J. UNCERTAINTY, FUZZINESS AND KNOWLEDGE-BASED SYSTEMS 557 (2002).

identify one's IP address (over 70 percent), be warned if a Web site's privacy policy was incompatible with their privacy preferences (over 75 percent), remain anonymous when completing online payments (over 80 percent), and protect emails so that only the intended recipient can read them (over 65 percent). Fifty-four percent of respondents could not cite or describe any law that influenced or impacted privacy. Respondents had a fuzzy knowledge about general privacy guidelines. For example, when asked to identify the OECD Fair Information Principles,[23] some incorrectly stated that they include litigation against wrongful behavior and remuneration for personal data (34.2 percent and 14.2 percent, respectively).

6.2 Bounded Rationality

Even if individuals have access to complete information about their privacy risks and modes of protection, they might not be able to process vast amounts of data to formulate a rational privacy-sensitive decision. Human beings' rationality is bounded, which limits our ability to acquire and then apply information.[24]

First, even individuals who claim to be very concerned about their privacy do not necessarily take steps to become informed about privacy risks when information is available. For example, we observed discrepancies when comparing whether subjects were informed about the policy regarding monitoring activities of employees and students in their organization with their reported level of privacy concern. Only 46 percent of those individuals with high privacy concerns claimed to have informed themselves about the existence and content of an organizational monitoring policy. Similarly, from the group of respondents with high privacy concerns, 41 percent freely admit that they rarely read privacy policies.[25]

In addition, in an unframed (that is, not specific to privacy) test of bounded rationality, we asked our respondents to play the beauty contest game, that behavioral economists sometimes utilize to understand individuals' strategizing behavior.[26] While some of our subjects (less than

[23] Organization for Economic Cooperation and Development (OECD), *The Recommendation Concerning Guidelines Governing the Protection of Privacy and Transborder Flows of Personal Data, available at* http://europa.eu.int/comm/internal_market/privacy/instruments/odecguideline_en.htm.

[24] Simon, *supra* note 9.

[25] T. Vila, R. Greenstadt and D. Molnar, *Why We Can't be Bothered to Read Privacy Policies: Models of Privacy Economies as a Lemons Market, in* THE ECONOMICS OF INFORMATION SECURITY (L.J. Camp and S. Lewis, eds., 2004).

[26] In the most popular form of the beauty contest game, experimenters ask subjects to respond to the following question: "Suppose you are in a room with 10 other people and

10 percent) followed the perfectly rational strategy, most seemed to be limited to a low number of clearly identifiable reasoning steps.

This result does not imply bounded rationality in privacy-relevant scenarios; it simply demonstrates the subjects' difficulties navigating in complex environments. However, we found evidence of simplified mental models also in specific privacy scenarios. For example, when asked the open-ended question, "You completed a credit-card purchase with an online merchant. Besides you and the merchant Web site, who else has data about parts of your transaction?" 34.5 percent of our sample answered "nobody," 21.9 percent indicated "my credit card company or bank," and 19.3 percent answered "hackers or distributors of spyware." How is it possible that 34.5 percent of our respondents forget to think of their own bank or other financial intermediaries when asked to list which parties would see their credit-card transactions? When cued, obviously most people would include those parties too. Without such cues, however, many respondents did not consider obvious options. The information is somehow known to the respondents but not available to them during the survey – as it might not be at decision-making time in the real world. In other words, the respondents considered a simplified mental model of credit-card transactions.[27] (We found similar results in questions related to email and browsing monitoring.)

Further evidence of simplified mental models comes from comments that expanded respondents' answers. For example, some commented that if a transaction with the merchant was secure, nobody else would be able to see data about the transaction. However, the security of a transaction does not imply its privacy. Yet, security and privacy seem to be synonyms in simplified mental models of certain individuals. Similar misconceptions were found related to the ability to browse anonymously by deleting browser cookies, or to send emails that only the intended recipient can open by using free email accounts such as Yahoo mail.

Similarly, a small number of subjects that reported to have joined loyalty programs and to have revealed accurate identifying information also claimed elsewhere in the survey that they had never given away personal information for monetary or other rewards, showing misconceptions about their own behavior and exposure to privacy risks. (We tested for information items

you all play a game. You write down a number between 0 and 100. The numbers are collected, and the average is calculated. The person who wrote the number closest to two-thirds of the average wins the game. What number would you write?" The beauty contest is dominance solvable through iterated elimination of weakly dominated strategies; it leads to the game's unique equilibrium where everybody chooses zero – this is what a rational agent would do if it believed that all other agents are rational. *See* R. Nagel, *Unraveling in Guessing Games: An Experimental Study*, 85 AMER. ECON. REV. 1313 (1995).

[27] Simon, *supra* note 9.

commonly asked for during sign-up processes for loyalty programs, such as name [4.2 percent exhibited such misconceptions], address [10.1 percent], and phone number [12.6 percent].)

6.3 Psychology and Systematic Deviations from Rationality

Even with access to complete information and unbounded ability to process it, human beings are subject to numerous psychological deviations from rationality that a vast body of economic and psychological literature has highlighted: from hyperbolic discounting to underinsurance, optimism bias, and others.[28] (In previous works,[29] we discussed which deviations are particularly relevant to privacy decision making.) Corroborating those theories with evidence generally requires experimental tests rather than surveys. Here we comment on indirect, preliminary evidence in our data.

First, we have already discussed overconfidence in risk assessment and misconception about an individual's information exposing behavior.

Discounting[30] might also affect privacy behavior. Traditionally, economists model people as discounting future utilities exponentially, yielding the intertemporal utility function[31] where payoffs in later periods, t, are discounted by δ^t, with δ being a constant discount factor.[32] Time-inconsistent (hyperbolic) discounting[33] suggests instead that people have a systematic bias to overrate the present over the future. This notion is captured with a parameter $\beta < 1$ that discounts later periods in addition to δ. Intuitively, an individual with a $\beta < 1$ will propose to act in the future in a certain way ("I will work on my paper this weekend.") but, when the date arrives, might change her mind ("I can start working on my paper on

[28] D. Kahneman and A. Tversky, *supra* note 11.

[29] A. Acquisti, *supra* note 7; A. Acquisti and J. Grossklags, *Losses, Gains and Hyperbolic Discounting: An Experimental Approach to Information Security and Behavior*, *in* THE ECONOMICS OF INFORMATION SECURITY (L.J. Camp and S. Lewis, eds., 2004).

[30] Acquisti, *supra* note 7.

[31] Indifference curve analysis was first applied to intertemporal choice by I. Fisher, THE THEORY OF INTEREST (1930).

[32] The constant rate discounted utility model with positive time preferences was first introduced by P. Samuelson, *A Note on Measurement of Utility*, 4 REV. ECON. STUDS. 155 (1937).

[33] D. Laibson, *Golden Eggs and Hyperbolic Discounting*, 112 QUARTERLY J. ECON. 443 (1997); T. O'Donoghue and M. Rabin, *The Economics of Immediate Gratification*, 13 J. BEHAV. DECISION MAKING 233 (2000). See also an early model of time-inconsistent preferences by E.S. Phelps and R. A. Pollak, *On Second-Best National Saving and Game-Equilibrium Growth*, 35 REV. ECON. STUD. 185 (1968).

Monday."). If individuals have such time inconsistencies, they might easily fall for marketing offers that offer low rewards now and a possibly permanent negative annuity in the future. Moreover, although they might suffer in every future time period from their earlier mistake, they might decide against incurring the immediate cost of adopting a privacy technology (for example, paying for an anonymous browsing service or a credit alert) even when they originally planned to do so.[34] In an unframed test of our sample, 39.6 percent acted time consistently according to the classical economic perception ($\beta = 1$). However, 44.0 percent acted time inconsistently by discounting later periods at a higher rate (16.4 percent could not be assigned to any of these two categories).

Although the discounting results we discuss are unframed and might not relate to privacy behavior, preliminary evidence about the use of protective technologies was compatible with the theory of immediate gratification. The share of users of a privacy-related technology seems to decrease with the length of time before one would expect to incur the penalty from the privacy intrusions against which that technology is supposed to protect. For example, 52.0 percent of our respondents regularly use their answering machine or caller-ID to screen calls, 54.2 percent have registered their number in a do-not-call list, and 37.5 percent have often demanded to be removed from specific calling lists (when a marketer calls them). However, as we noted earlier, 82.3 percent never put a credit alert on their credit report (of those, however, at least 34.2 percent are not aware of this possibility at all): the negative consequences of not using this kind of protection could be much more damaging than nuisances associated with unwanted phone calls, but are also postponed in time and uncertain, while the activation costs are immediate and certain. (From our calibration study, we know that 17.4 percent of those individuals that did not use the credit alert option of their credit card company even overestimated the risk of becoming a victim of identity theft.) We will subject these findings to further study to differentiate between alternative explanations that might be valid, such as lack of knowledge or trust in the accuracy of a technology or a service.

7. DISCUSSION AND CONCLUDING REMARKS

Based on theoretical principles and empirical findings, we are working towards the development of models of individual's privacy decision-making

[34] Acquisti, *supra* note 7.

that recognize the impact of incomplete information, bounded rationality, and various forms of psychological deviations from rationality.

Many factors affect privacy decision making, including personal attitudes, knowledge of risks and protection, trust in other parties, faith in the ability to protect information, and monetary considerations. Our preliminary data show that privacy attitudes and behavior are complex but are also compatible with the explanation that time inconsistencies in discounting could lead to under-protection and over-release of personal information. In conclusion, we do not support a model of strict rationality to describe individual privacy behavior. We plan further work on understanding and modeling these behavioral alternatives and on their experimental validation.

Even our preliminary data has implications for public policy and technology design. The current public debate on privacy seems anchored on two prominent positions: either consumers should be granted the right to manage their own privacy trade-offs or the government should step in to protect consumers. Our observations suggest that several difficulties might obstruct even concerned and motivated individuals in their attempts to protect their privacy.

While respondents' actual knowledge about law and legislative recommendations was weak, many favored governmental legislation and intervention as a means for privacy protection (53.7 percent). Our test population also supported group protection through behavioral norms (30.6 percent) and self-protection through technology (14.9 percent). Nobody favored the absence of any kind of protection; only one subject suggested self-regulation by the private sector. This is a striking result, contrasting the traditional assumption that US citizens are skeptical toward government intervention and favor industry-led solutions.

ACKNOWLEDGMENTS

We thank Nicolas Christin, Lorrie Cranor, Allan Friedman, Rachel Greenstadt, Adam Shostack, Paul Syverson, Cristina Fong, the participants at the WEIS 2004 Workshop, Blackhat 2004 Conference, CIPLIT® 2004 Symposium, CACR 2004 Conference, and seminar participants at Carnegie Mellon University, the Helsinki Institute for Information Technology (HIIT), the Information Technology department at HEC Montréal, and the University of Pittsburgh for many helpful suggestions. Alessandro Acquisti's work is supported in part by the Carnegie Mellon Berkman Faculty Development Fund. Jens Grossklags' work is supported in part by the National Science Foundation under grant number ANI-0331659.

Chapter 3

SOCIAL NORMS, SELF CONTROL, AND PRIVACY IN THE ONLINE WORLD

Katherine J. Strandburg
Assistant Professor of Law, DePaul University College of Law. An extended version of this chapter will be published as Privacy, Rationality, Temptation, and the Implications of Willpower Norms (forthcoming Rutgers Law Review, 2005).

Abstract: This chapter explores ways in which human limitations of rationality and susceptibility to temptation might affect the flow of personal information in the online environment. It relies on the concept of "willpower norms" to understand how the online environment might undermine the effectiveness of social norms that may have developed to regulate the flow of personal information in the offline world. Finally, the chapter discusses whether legal regulation of information privacy is an appropriate response to this issue and how such regulation should be formulated in light of tensions between concerns about self-control and paternalism.

Key words: willpower, temptation, self control, data mining, online communities, privacy torts, social norms

1. INTRODUCTION

Everyone over the age of four or five knows that certain personal information is appropriately discussed in some social contexts, but not in others. Teenagers have an acronym for inappropriate personal disclosures; "TMI!," they say, "Too much information!" Social norms delineate the types of personal information that are appropriately disclosed in particular circumstances and those who do not follow these personal information norms are looked upon with disfavor. This reluctance to receive certain personal information is surprising in light of the important role that accurate information plays in making decisions and assessing options, in interpersonal as well as commercial contexts. In a more extended version of this chapter,[1] I argue that the complicated and nuanced social norms that surround the flow of personal information are examples of "willpower norms" which can arise in response to the interplay between self-control temptation and human cognitive limitations.

Here I focus on the implications of the analysis of social norms about personal information dissemination for the online context. Section 2 argues that the disclosure, dissemination, and processing of personal information are rife with issues of self control. Section 3 analyzes personal information norms in light of these issues. In particular, Section 3 discusses why socially beneficial personal information norms will probably not develop to deal with widespread computerized data processing and argues that people will likely disclose more personal information online than is consistent with their long-term preferences. Failures of the "market" for personal information, which are mitigated by social norms in the interpersonal context, thus must be addressed by other means. Section 4 discusses whether and how legal regulation should address these norm failures. Care must be taken in devising such regulations in light of the personal autonomy questions that are inevitably raised by regulations that respond to issues of self-control.

2. PERSONAL INFORMATION AND SELF CONTROL

The disclosure and dissemination of personal information about individuals within a community has traditionally been regulated by a nuanced system of social norms. Social norms regulate prying and gossip and also the "inappropriate" disclosure of certain kinds of personal information in certain contexts. By and large the function of these personal information norms has yet to be analyzed in depth from a social norms theory perspective.[2] Yet the question is important for at least two reasons.

[1] Katherine J. Strandburg, Privacy, Rationality, and Temptation: A Theory of Willpower Norms, RUTGERS L. REV. (forthcoming 2005).

[2] Richard H. McAdams, *The Origin, Development and Regulation of Norms*, 96 MICH. L. REV. 338, 424-33 (1997), has analyzed the effects that restricting the dissemination of personal information through privacy might have on the evolution of social norms. Paul M. Schwartz, *Internet Privacy and the State*, 32 CONN. L. REV. 815, 838-43 (2000), has discussed how information privacy might mitigate some of the harmful effects of overly zealous enforcement of social norms. Steven A. Hetcher, NORMS IN A WIRED WORLD 243-305 (2004), has devoted considerable attention to the emergence and function of social norms that govern the behavior of online website owners, but has not applied the analysis to interpersonal dissemination and disclosure of personal information.

First, since social norms often arise out of underlying conflicts between individual preferences and collective benefits,[3] they provide clues to those underlying conflicts. Studying them may bring to light social issues that need to be addressed in contexts in which norms are ineffective. Second, many have mourned a perceived decline in social norms about personal information disclosure[4] and predicted that modern data processing and other technological means for obtaining personal information will lead to major and undesirable social changes.[5] To understand whether these social forces will result in a breakdown of personal information norms – and to determine whether such a breakdown is a problem that should be addressed by legal regulation – it is important to try to understand these norms and to see what underlying social purposes they have served.

In this chapter, I argue that many personal information norms are best understood as "willpower norms," which are social norms aimed at compensating for human failures of rationality and self-control. Willpower norms arise in the personal information context because both disclosing information and obtaining information are subject to problems of self control and temptation. Individuals regularly both disclose information they wish they had not disclosed and obtain information they are unable to consider rationally, with resulting social disutility. Personal information norms can curb these tendencies by reinforcing the self-control of both disclosers and recipients of the information.[6]

Though a large fraction of social discourse consists of discussion of the activities – including the follies and foibles – of friends, colleagues, and acquaintances, there are well-recognized boundaries to the types of personal information that can "appropriately" be discussed. The boundaries depend on the identities of the participants in the conversation. Most interesting – and puzzling – are social norms opposing disclosure of one's own personal information to others. Even in today's "tell-all" society (or perhaps especially so), there is widespread aversion to and disapproval of the disclosure of personal information at the wrong time and place. It is not

[3] *See, e.g.*, Hetcher, *supra* note 2 at 38-78, for a review of social norm theory.

[4] *See, e.g.*, Rochelle Gurstein, THE REPEAL OF RETICENCE: A HISTORY OF AMERICA'S CULTURAL AND LEGAL STRUGGLES OVER FREE SPEECH, OBSCENITY, SEXUAL LIBERATION, AND MODERN ART (1996); Anita L. Allen, *The Wanted Gaze: Accountability for Interpersonal Conduct at Work*, 89 GEO. L. REV. 2013 (2001); Anita L. Allen, *Coercing Privacy*, 40 WM. AND MARY L. REV. 723 (1999) for arguments that the widespread "waiver" of privacy by individuals is a social problem.

[5] *See, e.g.*, Daniel J. Solove, THE DIGITAL PERSON (2004), and references therein.

[6] Unfortunately, personal information norms can also serve as socially pernicious and inefficient "silencing norms," as exemplified by the "don't ask, don't tell" policy against gays in the military. This possibility is analyzed in Strandburg, *supra* note 1.

simply that individuals do not want to share everything with everyone, but also that *individuals do not want to know everything about everyone else.*[7]

These social norms against providing information are puzzling when one stops to think about it, especially in light of the high value usually placed by society on the free flow of information. The availability of information is an assumed prerequisite to a well-functioning and efficient marketplace. In addition, transparency is a widely held democratic value and the ability to obtain and discuss information about others is considered by many to be an important aspect of free speech.[8] If personal information disclosures were always valuable to the recipients of the information, it is difficult to see how social norms against disclosing personal information could arise.

2.1 Self-Control Issues in Personal Information Disclosure and Dissemination

Both disclosing one's own personal information and obtaining personal information about other people can raise self control issues. Moreover, there can be social costs when people yield to the temptation to disclose or to obtain personal information.

2.1.1 Temptations to Disclose Personal Information

People often disclose personal information despite having indicated a preference for keeping such information private. This tendency has been documented particularly well in the context of online disclosures. In the past few years, a large number of surveys and a few experimental studies have

[7] *See* Daniel J. Solove, *The Virtues of Knowing Less: Justifying Privacy Protections against Disclosure*, 53 DUKE L. J. 967, 1035-44 (2003), noting that more information can lead to misjudgment because information is taken out of context and because of irrational judgments based on stigmas; and Julie E. Cohen, *Examined Lives: Informational Privacy and the Subject as Object*, 52 STAN. L. REV. 1373 at 1403-04, arguing against the contention that more information always leads to more knowledge.

[8] For example, Eugene Volokh has characterized the right to information privacy as a "right to stop you from speaking about me." Eugene Volokh, *Freedom of Speech and Information Privacy: The Troubling Implications of a Right to Stop People From Speaking About You,* 52 STAN. L. REV. 1049, 1117 (2000). Volokh analyzes various rationales for permitting the subject of personal information to control its dissemination and argues all could be expanded easily to justify restricting other types of discourse. *See also* Diane L. Zimmerman, 68 CORNELL L. REV. 291, 326-36 (1983), discussing tensions between the privacy torts and constitutional values and also arguing that gossip serves a positive function.

probed public attitudes about online disclosure of personal information.[9] Respondents generally report a high degree of concern about the privacy of their personal information[10] and strongly believe that they should have control over its use. However, as noted by Stan Karas, "despite warnings against personality-warping self-censorship, there exists little evidence that consumers actually alter their behavior because of ongoing data collection."[11] And few individuals take affirmative steps to protect their online privacy.[12] In other words, despite having indicated a preference not to disclose personal information online, individuals in fact disclose such information frequently.

One possible conclusion from these studies would be that people are simply not as concerned about revealing their personal information in the course of real transactions as they report themselves to be when considering a disclosure in the abstract. However, a very interesting study by Sarah Spiekermann and collaborators suggests another possibility. The study probed the correspondence between subjects' reported privacy concerns and their behavior during an online shopping trip involving real purchases.[13] The study involved an online "bot" named Luci which asked the shoppers a variety of personal questions to guide their product selection. The subjects' behavior during the shopping trip seemed inconsistent with reported privacy preferences. All participants, including those who reported strong privacy preferences, answered a high number of the "Luci's" personal questions.

A closer inspection of the participants' behavior brought to light correlations between shopping behavior and reported privacy preferences, however. Those study participants who had expressed greater privacy concerns delayed answering the bot's questions, checking frequently to see whether an information search would provide a satisfactory recommendation for their purchase, such that answering the question could be avoided. Though many of them eventually revealed nearly as much personal information as those ostensibly less concerned with privacy, they thus

[9] For a very useful bibliography of surveys of the American public regarding privacy issues, see Bibliography of Surveys of the U.S. Public, 1970-2003 *at* http://www.privacyexchange.org/iss/surveys/surveybibliography603.pdf.

[10] *See, e.g.*, Susannah Fox, *Trust and Privacy Online: Why Americans Want to Rewrite the Rule* (2000).

[11] Stan Karas, *Privacy, Identity, and Databases*, 52 AM. U. L. REV. 393, 414 (2002).

[12] A. Acquisti and J. Grossklags, *Losses, Gains and Hyperbolic Discounting: An Experimental Approach to Information Security and Behavior*, *in* THE ECONOMICS OF INFORMATION SECURITY (L.J. Camp and S. Lewis, eds., 2004) at 165-178, reviewing and citing surveys.

[13] Sarah Spiekermann, Jens Grossklags, and Bertina Berendt, *E-Privacy in 2nd Generation E-Commerce: Privacy Preferences v. Actual Behavior*, *in* 3RD ACM CONFERENCE ON ELECTRONIC COMMERCE - EC '01 (2002).

appeared to be much more conflicted about providing the personal information that was requested, providing it only after a period of delay.

These experiments suggest that the more privacy-conscious individuals engaged in an internal struggle over whether to reveal the information. If this speculation is correct, the study may provide evidence that the inconsistency between individuals' stated attitudes about disclosing personal information and their behavior is a result of struggles pitting long term desires for privacy against short term temptations to disclose information in exchange for relatively minor, but immediately attractive, savings or conveniences. As recognized in a recent economic treatment of information privacy online, "the protection against one's own future lack of willpower could be a crucial aspect [of] providing a link between information security attitudes and actual behavior."[14]

Temptations to disclose personal information may arise not only from the opportunity to trade privacy for immediate gain, as in the usual online context, but from a taste for the disclosure itself. Of course, much of the personal information that provokes concern when it is disclosed in commercial transactions seems unlikely to be the subject of such a taste for disclosure. Likely few people get an inherent kick out of disclosing their credit card numbers or social security numbers. However, the issue of personal information disclosure is significant in many arenas other than commercial transactions. Disclosure of personal information is at issue in social, employment, health care and other contexts in which an inherent appetite for expression is much more likely to come into play.

In summary, individuals demonstrate ambivalent attitudes and behavior with respect to disclosing their own personal information, suggesting that time-inconsistent preferences and self-control may have important ramifications for personal information disclosure.[15]

[14] Acquisti and Grossklags, *supra* note 12.

[15] These issues of temptation and willpower arise in addition to other related concerns. Individuals may not make good decisions about whether to disclose personal information for a variety of reasons beyond lack of self-control. *See, e.g.*, Alessandro Acquisti and Jens Grossklags, *Privacy and Rationality: Preliminary Evidence from Pilot Data*, 3RD ANNUAL WORKSHOP ON ECONOMICS AND INFORMATION SECURITY at 3 (WEIS 2004); Acquisti and Grossklags, *supra* note 12; Schwartz, *supra* note 2. *See generally,* Daniel Kahneman and Amos Tversky, eds., CHOICES, VALUES, AND FRAMES (2000) for a review of "bounded rationality."

2.1.2 Self-Control Issues with Obtaining and Handling Personal Information about Others

It is relatively easy to understand why people might wish to obtain personal information about others. Such information may be directly useful in evaluating the advisability of transacting with them, might even provide opportunities to exploit or defraud them, might be indirectly useful as a kind of "morality tale" about the costs and benefits of particular lifestyle choices, and might satisfy a simple curiosity or taste for information about others.

Despite the obvious potential to benefit from obtaining personal information about others, it is well known that individuals sometimes obtain information against their better judgment. Many have experienced regret at having yielded to the temptation to read a letter carelessly left out on a desk or an unintentionally forwarded email. As noted by Solove, "Similarly, people may recognize the value of being restrained from learning certain details about others, even if they crave gossip and would gain much pleasure from hearing it."[16] Elsewhere, I analyze in detail why this might be the case.[17] The analysis has three steps: (1) explaining that people might make better decisions if they avoid certain information because the accuracy of decisionmaking does not always increase with increasing information availability and because people exhibit systematic irrationalities in information processing, including a tendency to over-emphasize certain distracting information; (2) arguing that people experience self-control problems about obtaining and considering such information; and (3) arguing that personal information is particularly likely to be the object of such problems because it is often highly contextual and tends to be distracting.

The conclusion from this analysis is that we should not be surprised if people have complex preferences about learning personal information about others. Such information is likely to be useful and interesting in some contexts and distracting and confusing in others. Frequently, personal information about others is interesting in the short term but distracting or confusing in the long run, leading to the time-inconsistent preferences about nosing into other people's affairs that are familiar from daily life. These complicated time-inconsistent preferences about personal information may underlie the complex social regulation of personal information dissemination by personal information norms.

[16] Daniel J. Solove, *The Virtues of Knowing Less: Justifying Privacy Protections against Disclosure*, 53 DUKE L. J. 967, 1050 (2003).

[17] Strandburg, *supra* note 1.

3. PERSONAL INFORMATION NORMS AS WILLPOWER NORMS

Because the short-term and long-term costs and benefits of disclosing and obtaining personal information are highly dependent on the context of the disclosure, it is not surprising that personal information norms are highly contextual. While a detailed analysis of such norms is beyond the empirical foundation of this chapter, it is possible to make some general arguments about these norms based on the concept of "willpower norms." I also give a more detailed analysis of the online disclosure of personal information, a context in which personal information norms can be ineffective.

3.1 Norms against Disclosing Personal Information

Under the traditional view that more information is better, one would expect self-disclosures to be welcomed and reticence to be suspect. As explained by Professor Richard Murphy: "in grossly oversimplified terms, the consensus of the law and economics literature is this: more information is better, and restrictions on the flow of information in the name of privacy are generally not social wealth maximizing, because they inhibit decision-making, increase transaction costs, and encourage fraud."[18] Those who wish to keep others from obtaining personal information about them are suspected of seeking to exploit third party misapprehensions.[19]

If one accepts that both disclosing and receiving personal information are subject to self control failures, however, it is much easier to understand the phenomenon of norms against disclosing information. Individuals' long-term incentives both as disclosers and recipients of information could then motivate them to participate in penalizing the disclosure of "inappropriate" personal information. Norms against disclosing certain types of personal information in certain contexts can compensate for common cognitive and self-control problems related to the taste for disclosure and for inquiry, the likely irrelevance of personal information, and the difficulty in properly processing – or ignoring – distracting information. Because the extent to which information is "more prejudicial than probative" will vary according to the context, the norms of disclosure can be expected to vary accordingly. Information that is considered "inappropriate" in one context might be considered a perfectly acceptable subject of disclosure in another. For

[18] Richard S. Murphy, *Property Rights in Personal Information: An Economic Defense of Privacy*, 84 GEO. L.J. 2381, 2382 (1996).

[19] *See* Richard A. Posner, *The Right of Privacy*, 12 GA. L. REV. 393, 399-400 (1977-78).

example, norms cordoning off the disclosure and dissemination of certain types of personal information in environments such as the workplace may be explicable on these grounds. If personal information is "floating around" the office it may be imposed on those who would prefer not to know it.

To summarize, the willpower norms concept is helpful in understanding why personal information norms prohibit both disclosing and prying into personal information in some circumstances. When disclosers and hearers are more or less similarly situated members of a close-knit community, these personal information norms will often be socially beneficial and efficient as ways of coping with common human weaknesses.[20] Deviations from these norms will frequently be detectable and can be punished with social disfavor. The next sub-section asks whether and how personal information norms will function in the context of online disclosures.

3.2 Personal Information Norms and the Disclosure of Personal Information Online

Disclosure of personal information online can be analyzed as a willpower issue.[21] Individuals may disclose personal information in transactions with websites – even, for example, when those websites do not have effective privacy controls – because the temptation of immediate gratification overcomes a longer-term preference not to disclose such information. Will social norms arise to compensate for such self-control failures? The development of social norms generally requires a repeatedly-interacting group, which is able to detect and penalize deviations from the norm. These conditions may not be satisfied in the online context. To explore this question, it is helpful to distinguish two contexts for personal information disclosure: the online community and the online commercial transaction.

3.2.1 Personal Information Norms and Online Communities

Online communities are groups of individuals who interact on an interpersonal level over the internet. These interactions take place in various venues, including chat rooms, listserves, bulletin boards, and so forth. There

[20] In Strandburg, *supra* note 1, I discuss how personal information norms may go awry, leading to pernicious effects on minority groups and impeding the evolution of norms about other types of behavior.

[21] *See* Alessandro Acquisti, Privacy in Electronic Commerce and the Economics of Immediate Gratification, *in* PROCEEDINGS OF ACM ELECTRONIC COMMERCE CONFERENCE (EC 04) (2004), and Acquisti and Grossklags, *supra* note 12, for explorations of this possibility from a behavioral economics perspective.

may be some question about whether online communities can develop personal information norms, since participants are often identified only be screen names and the social displeasure of the online group cannot easily penetrate into the "real world" context. However, online communities often do develop informal norms about disclosure of personal information.[22] Moreover, if norms fail to develop, such communities often promulgate rules of behavior, enforced by a central authority, to regulate personal information disclosure, among other things. Such codes of "netiquette" are increasingly common. To the extent that group members value their online interactions, they are susceptible to reputational and shunning penalties for "inappropriate" behavior just as people are in the offline world.

Even if netiquettes develop, however, two factors make personal information norms less effective in such online communities than in real world communities. First, if online conversations are maintained in a searchable archive (as they may be either intentionally or unintentionally by the website owner), personal information may "escape" from the community in which it was disclosed to contexts in which it will not be correctly understood. The community's norms may not account for this possibility, especially if the archiving is invisible to participants. Second, such online communities may have corporate sponsors, who are silent observers of the interpersonal exchange and may also communicate the information beyond the community. These corporate sponsors may not be constrained by community norms about disclosure if their norm violations are not visible to the group. Moreover, they may in some instances have access to information that maps an online identity to a real world identity. For these reasons, personal information norms may fail to deter some undesirable secondary disclosures of personal information.

3.2.2 Personal Information Norms and Commercial Transactions

In the commercial online context, norms governing personal information disclosures are difficult to establish. One consumer's disclosure of or failure to disclose personal information in an online transaction is of no direct interest to other consumers since they are not the recipients of the information.[23] I argue elsewhere that a willpower norm might develop even around behavior that does not affect other people because a repeatedly

[22] *See, e.g.*, Uta Pankoke-Babatz and Phillip Jeffrey, *Documented Norms and Conventions on the Internet*, 14 INT'L J. HUMAN-COMPUTER INTERACTION 219 (2002).

[23] This is not strictly true, since consumers share a common interest, as explored by Hetcher, *supra* note 2, at 258-59, and discussed below, in pressuring websites to change their data collection policies.

interacting group can assume reciprocal obligations to punish common lapses of self-control.[24] Online commercial disclosures cannot easily support such a norm, however, both because failures of will are not observed by other online consumers and because online consumers do not form a repeatedly interacting group.

While consumers are unable to develop a norm of penalizing one another for excessive online disclosures of personal information, they do have ways to penalize "tempters" – in this case websites without protective privacy policies which offer immediate commercial gratification in exchange for personal information disclosures which may be regretted later. For example, online consumers could punish such websites for failing to protect privacy by boycotting them. Is a norm against dealing with such websites likely to arise? Professor Steven Hetcher argues that consumers face a collective action problem in attempting to influence website companies to adopt more privacy-protective policies.[25] Consumers as a whole might be better off refusing to deal with sites that do not offer a high level of privacy protection, but for each individual consumer it is rational to attempt to free ride on the boycott efforts of others. Unlike consumers in the offline world, who can enforce boycotts by observing when people deal with "forbidden" retailers, online consumers are unable to enforce a boycott based on a website's privacy policies because they cannot detect and penalize defectors. This analysis suggests that consumers will not be able to sustain a "norm against tempters" that penalizes such websites.

However, as discussed in detail elsewhere,[26] the self control context provides a mechanism that may permit social norms to influence unobservable behavior. One personal mechanism for enhancing self control is to categorize or "bundle" behaviors using personal rules.[27] Willpower norms can reinforce such individual self control measures by operating to "bundle" undetectable behaviors with socially disfavored detectable behaviors. Because an individual may perceive that her *undetectable* violation of a rule signals that she is also likely to violate the rule *in public*, "bundling" provides a mechanism by which social norms can influence

[24] Strandburg, *supra* note 1.

[25] Hetcher also discusses in detail why a functioning market for privacy-protective policies may fail to develop. *Supra* note 2 at 243-60.

[26] Strandburg, *supra* note 1.

[27] *See, e.g.*, Drazen Prelec and Ronit Bodner, *Self-Signaling and Self-Control* in TIME AND DECISION (George Loewenstein *et al.*, eds., 2003), arguing that activities with low marginal efficacy, such as voting, may be explained as signals to the self of future propensity to resist temptation. Such a propensity increases the attractiveness of resisting temptation now. *See also* George Ainslie, BREAKDOWN OF WILL 90-104 (2001), at 90-104.

undetectable behavior. In the online context, for example, if a consumer were to categorize online personal information disclosures with inappropriate *public* disclosures of personal information as part of a personal rule, her ability to resist the temptation to disclose online could increase.

In fact, Hetcher's description of the activities of "norm entrepreneurs" in the context of website privacy policies may be evidence of just such a bundling willpower norm. Hetcher details how norm entrepreneurs created a consumer demand for online information privacy in part by moralizing the meaning of online data collection by re-categorizing data collection as a privacy invasion.[28] Norm entrepreneurs framed the behavior of websites that do not protect privacy as "disrespectful." It is not immediately clear from Hetcher's description, however, why framing behavior as "disrespectful" will help to solve the consumer collective action problem. Consumers can free ride on the efforts of others to boycott "disrespectful" websites just as easily as on efforts to boycott privacy-invasive websites. The answer may lie in the "bundling" self-control strategy just discussed. The moralizing of privacy policies has the effect of re-framing "dealing with a website that does not provide privacy protection" as "allowing oneself to be treated with disrespect." Dealing with such a website then becomes just one of many ways in which an individual might violate a personal rule that says "I do not allow myself to be treated with disrespect." Bundling the undetectable online behavior of "dealing with websites that do not protect privacy" with more detectable instances of being treated with disrespect has the potential to aid the individual in resisting the temptation to deal with those websites.[29]

3.3 Implications of Personal Information Norms for Computerized Data Processing

3.3.1 Computerized Data Processing and the Relevance of Personal Information Norms

Jange and Schwartz define "information privacy" as "the creation and maintenance of rules that structure and limit access to and use of personal data." They note that the rules are sometimes found in norms, such as those involving gossip, and sometimes in law.[30] Under this definition, an understanding of personal information norms would seem central to an

[28] Hetcher, *supra* note 2, at 268-72.
[29] Whether such a norm can be effective when websites have deceptive privacy policies is another question, of course.
[30] Edward J. Jange and Paul M. Schwartz, *The Gramm-Leach-Bliley Act, Information Privacy, and the Limits of Default Rules*, 86 MINN. L. REV. 1219, 1223 (2002).

understanding of information privacy. It may thus be a bit surprising that rather minimal attention has been paid to the theory of personal information norms in the information privacy debate. This relative inattention to the norms that govern the interpersonal flow of personal data is not as surprising, however, when one considers that much of the focus of recent discussions of information privacy rightly has been on analyzing the impact of computerized data processing and collection by corporate entities.

Social norms of the sort analyzed here do not govern the behavior of these commercial entities directly. One may thus ask whether social norm analysis has anything to add to the debate about the regulation of commercial applications of computerized data processing. However, the relevance of personal information norms to this debate stems in part from the very fact that these entities are not subject to social norms. In today's society, the flow of personal information is less and less subject to social norm mechanisms of enforcement: disclosures are often made in unobservable online contexts; information is disseminated to and by corporate entities; and information flow is no longer confined to a community of individuals who interact repeatedly. This social norm failure can be beneficial in some instances, by undermining silencing norms and permitting minority group discourse. However, it also seems likely that social norms will no longer be able to mitigate the effects of bounded rationality and willpower in cases where long-term preferences would counsel against disclosure. Moreover, where, as is usual in the online context, disclosures are made to corporate entities, there is the danger that those entities will develop their own norms that promote their private interests but are socially detrimental. For example, Hetcher has argued that websites are likely to develop a coordination norm of deceptive privacy practices, while individuals will have difficulty developing and enforcing norms of privacy-protective behavior.[31]

Unless other self-control mechanisms are available, one may thus expect individuals to disclose significantly more information in the online context than they would prefer to disclose in the long term. At the same time, computerized data processing permits far more extensive aggregation and dissemination of personal information than is possible in the interpersonal context. Roger Clarke coined the term "dataveillance" to describe the monitoring of individual behavior that is made possible by computerized data collection.[32] Professor Julie Cohen argues that data collection reduces

[31] Hetcher, supra note 2 at 255-58.
[32] *See* Roger Clarke, *Information Technology and Dataveillance*, 315 COMMUN. ACM 498 (1988).

autonomy because it provides a means of informational surveillance which generates "a 'picture' that, in some respects, is more detailed and intimate than that produced by visual observation."[33] Data may be aggregated over space and time and may be searched to discern patterns.

Data aggregation exacerbates the self-control issues that affect decisions to disclose personal information. Disclosure of each piece of information is particularly tempting because each disclosure forms an insignificant part of the picture. Personal information may also be disseminated far more rapidly and widely by computer than is likely in the interpersonal context, to parties that have no contextual knowledge of the individual involved. This widespread and rapid dissemination makes it difficult, if not impossible, for individuals to make rational predictions of the costs and benefits of each disclosure. As explained in a recent economic treatment, "[T]he negative utility coming from future potential misuses of somebody's personal information is a random shock whose probability and scope are extremely variable . . . [the] individual [] is facing risks whose amounts are distributed between zero and possibly large (but mostly uncertain) amounts according to mostly unknown functions."[34]

One might argue that concern about this increased circulation of personal information should be mitigated by the fact that the human failings of rationality and will that were central to our explanation of personal information norms are not at issue in computerized data processing. Thus, the argument would go, commercial entities will process information rationally and, though individuals may not like the outcome of that rational decision-making, society will not suffer when individuals disclose more personal information than they might in the long run prefer.

There are many reasons to criticize this hypothesis. I discuss two reasons here. First, it may be privately advantageous for online data collectors to use personal information in socially suboptimal ways and the market will fail to correct this tendency. Second, it is a mistake to assume that computerized data processing entities are immune from human failings in interpreting personal information. Inescapably, decisions about what data to collect and how to analyze it are made by human beings.

[33] *See* Julie E. Cohen, *Examined Lives: Informational Privacy and the Subject as Object*, 52 STAN. L. REV. 1373, 1425 (2000).

[34] Acquisti and Grossklags, *supra* note 12.

3.3.2 Information Externalities and Computerized Information Processing

As has become increasingly evident, commercial entities may prefer to collect more information and to be less careful about its security than is socially optimal for a variety of reasons.[35] Most importantly, the collecting entity does not bear the full cost of privacy losses to the subjects of the information, including any costs that arise from criminal misuse of the information by rogue employees or hackers or from errors in the data. Because of their own self-control failures and inability to take into account accurately the expected costs of disclosing personal information online, individuals will not force these entities to pay a high enough "price" for personal information. Thus, the personal information "market" will not solve this problem.[36] Recent high-profile demonstrations of insufficient information security may well lead to improved legal regulation aimed at curbing the possibility of malicious uses of personal information.[37] Such regulation raises few concerns about paternalism – not many individuals have long-term preferences for insecure or inaccurate data collection!

The self-control analysis highlights a further way in which rational uses of data processing may lead to socially harmful results, however. Targeted marketing, based on the analysis of consumer buying patterns, has the apparent potential for great social benefit. It promises to eliminate social and consumer waste associated with marketing that is aimed at those who have no interest in the services or products being offered. Despite these reasonable-sounding justifications, there is widespread uneasiness with the idea that commercial entities might base their marketing on personal profiles that are too detailed or complete. This uneasiness is somewhat mysterious if consumers are rational processors of marketing information. Presumably,

[35] For a more formal economic treatment of the incentives of private firms to obtain inefficient amounts of information, *see* Curtis R. Taylor, Privacy in competitive markets (Technical report, Department of Economics, Duke University, 2004), *available at* http://www.econ.duke.edu/Papers/Other/Taylor/privacy.pdf.

[36] *See* Jange and Schwartz, *supra* note 30, at 1241-46, making the point that because of bounded rationality the privacy market between financial institutions and consumers does not function well.

[37] *See, e.g., Some Sympathy for Paris Hilton,* New York Times Week in Review, February 27, 2005, discussing various personal information leaks, including the recent leak of personal information about 145,000 individuals by Choicepoint; *FTC Drops Probe into DoubleClick Privacy Practices*, CNET News.com, January 22, 2001 at http://news.com.com/2100-1023-251325.html?legacy=cnet, describing the "maelstrom of scrutiny from consumer advocates, media and legislators about data-collection practices on the Web" set off by an announcement of plans to combine online "clickstream" data with personally identifiable information for advertising purposes.

everyone could be subjected to less advertising if the advertising could be targeted accurately. If, however, one accounts for the influence of temptation and willpower on consumption decisions, the aversion to overly personalized marketing makes more sense. A more detailed personal profile permits a more targeted satisfaction of consumer preferences, but also a more targeted attack on consumer will. The possibility that targeted advertising can undermine long-term preferences may explain an intuition that such detailed profiling is a threat to personal autonomy. This analysis suggests that consumer antipathy to spam and pop-up ads, for example, may be premised on more than concerns about attention consumption.[38]

3.3.3 Rationality, Willpower, and Computerized Data Processing

Computerized data processing may also not be immune from the irrationalities that affect human processing and interpretation of personal information. Are individuals correct to fear that computerized processing of personal information will be used to make bad decisions about them? Or are they simply seeking to benefit from hiding negative information about themselves from their employers, insurance companies, and potential dates?

Computers, one may safely say, are not plagued by issues of temptation and self-control. They are neither curious nor confessional. They simply do what they are told. Thus, the relevance of bounded rationality and willpower to computerized data processing stems from three things: human choices affecting the input data; human choices about how to process the data; and human interpretation of the computerized analysis. Computerized uses of personal information can range from the simple collection of a mailing list to sophisticated and innovative techniques for "mining" patterns from large quantities of data. It is beyond the scope of this chapter to provide an extensive analysis of the ways in which these various uses of computer analysis might fall prey to bounded rationality and limited self-control. However, the potential pitfalls may be illustrated by a few specific points.

First, corporate entities may tend to over-collect information if the costs associated with processing excessive information are not imposed on the group within the entity that makes decisions about collection. Collecting more data may impose very minor immediate costs. Failing to collect and store the data may implicate loss aversion, resulting in a tendency to avoid the greater regret that might accompany an irretrievably lost opportunity to acquire information (particularly because information – especially in digital

[38] *See* Eric Goldman, *Data Mining and Attention Consumption*, *in* this volume, for an analysis of targeted marketing based on its costs in attention consumption.

form – seems very easy to "throw away later.")[39] Moreover, the individuals in charge of data collection may perceive (probably correctly) that they will suffer more dire personal consequences if they fail to collect data than if they collect an excess of data.

Second, where statistically based techniques are used to categorize individuals based on computerized data, the categorizations may be highly dependent upon the choice of input data and on judgment calls as to when a categorization is "sensible" and when the application of a particular technique (of which there are many) is "successful." Texts on data mining stress the importance of user input at many stages of the process, including selecting data to be used in the data mining computation and assessing the meaning and "interestingness" of patterns that are discovered by the data mining process.[40] Human cognitive limitations can enter at any of these points. Moreover, as a technical matter, computerized data analysis often depends on numerical optimization techniques that are not guaranteed to reach a global optimum. For such techniques, the inclusion of irrelevant information can lead the computation to become "stuck" in a local optimum. Often there are no objective methods for assessing whether a particular data mining result is even close to a global optimum.

Finally, when human decision-makers assess the output of any computerized data analysis, there are several cognitive biases that may come into play. There is a tendency to treat quantitative output as more certain and more objective than may be justified by its accuracy and the amount of subjective input on which it depends. The specificity of numerical output may also give it a salience that could lead to over-reliance on this information. Moreover, the results of computerized data analysis techniques are frequently statistical, giving probabilities that certain correlations or patterns of behavior apply to certain individuals. Even setting aside the normative issue of whether it is fair to make decisions about individuals based on statistical assessments, the well-established difficulties that people have in accurately reasoning based on uncertain, probabilistic information are likely to degrade the assessment of the output of computerized data analysis of personal information.

[39] *See, e.g.*, Russell Korobkin, *The Endowment Effect and Legal Analysis, Symposium on Empirical Legal Realism: A New Social Scientific Assessment of Law and Human Behavior*, 97 NW. U.L. REV. 1227 (2003). Of course, the low cost of data collection and disposal also figures into the rational calculus of how much information to collect and raises the rational limit.

[40] *See, e.g.*, Jiawei Han and Micheline Kamber, DATA MINING: CONCEPTS AND TECHNIQUES 1-34 (2001).

This subject clearly warrants more extensive treatment. However, the brief discussion here should serve as a warning against any facile assumption that computerized analysis is immune to the limitations of human information processing.

4. PERSONAL INFORMATION NORMS AND INFORMATION PRIVACY REGULATION

Underlying the debate about the social benefits of information privacy and information flow is the question of the extent to which information privacy is a matter for government regulation rather than individual action, social norms, and market forces. The fact that privacy has benefits and disclosure has costs is not, in and of itself, a justification for government action. Generally, individuals in American society are left to judge for themselves the desirability of engaging in activities that involve tradeoffs of costs and benefits. Government action may be justified, though, when pursuit of individual objectives leads to societal difficulties, including market failures of various sorts.

Here, the personal information norm analysis is instructive. The very existence of social norms governing personal information disclosure and dissemination is a red flag that indicates some underlying difficulty with purely individual decisions about when to disclose and disseminate personal information. Understanding the provenance of personal information norms thus provides clues as to whether legal regulation or other government intervention is likely to be appropriate. The analysis of personal information norms in this chapter thus has implications for the debate about information privacy regulation. In this section I provide a preliminary discussion of some implications for the debate about information privacy regulation in the context of computerized data processing.

First, the recognition that disclosing and obtaining personal information are subject to self control problems undermines the facile assumption that individual decisions to disclose or obtain such information reveal long-term preferences about information privacy. It also supports arguments that information privacy is not a simple tradeoff between privacy's benefits to subjects of the information and costs to others of being deprived of the information. Information privacy regulation has the potential for direct benefit both to subjects and to recipients of information. On the other hand, while providing additional justification for privacy regulation, the personal information norms analysis also highlights potential dangers of such regulation. As with other attempts to regulate behavior that is subject to temptation, the prospect of interfering with individual liberty must be taken seriously, since it is difficult to distinguish time-inconsistent preferences from either true minority preferences or long-term preference changes.

Second, where individual self-control measures fail and social norms are ineffective for some reason (such as where there is no close-knit community to enforce them or where violations are undetectable), legal regulation may promote the long-term social good. For example, as already discussed, modern technology may undermine the effectiveness of personal information norms in many instances. Yet the temptations associated with disclosing and disseminating personal information persist. A market approach to information privacy will not work in such circumstances. Regulation may be desirable to replace the social norms that have historically mitigated market failure in the interpersonal context.

There are serious hazards, however, to employing legal regulation to bolster self-control. As I argue in detail elsewhere,[41] the willpower norms analysis suggests that the use of the law in a paternalistic way to enforce self-control should be approached with care because of the possibility of imposing "silencing norms" on minority groups, of masking norm changes, and of other impositions on the liberty of those whose long-term preferences differ from those of the majority. The best approach thus may be to structure the law to support voluntary self-control measures whenever possible. Mechanisms for promoting information privacy should be designed as much as possible to permit individuals to distinguish for themselves between long term preferences and short term temptations. Usually this will mean enhancing – and perhaps mandating – the availability of self-control measures, while permitting individuals to "opt out" of those measures.

4.1 Personal Information Norms and the Legal Regulation of Online Communities

In the interpersonal context, legal strictures against disclosing one's own personal information to other individuals are likely to be unnecessary, intrusive, and insufficiently in tune with particular webs of interpersonal relationships. Section 3.2.1 identified two situations in which personal information norms may fail to function in online communities. Both concerned the behavior of those who obtain personal information "incidentally" rather than as participants in the online discussion.

One approach to these potential norm failures in online communities might be to require that websites providing interpersonal discussion fora

[41] Strandburg, *supra* note 1.

give clear notice of their policies about archiving the online discussion and about corporate use of information disclosed in the discussions. Because community members have repeated interactions with each other and with the website owners, notice may be enough to permit online communities to subject these owners to community norms about the treatment of personal information.

Another approach would be to subject website owners to tort liability to the extent that their behavior violates the social norms that pertain to other members of the online community. I argue elsewhere[42] that the interpretation of reasonable expectations of limited privacy in the tort law context should take into account the social norms of the group in which the disclosure occurred. This approach may be usefully applied in the context of online social groups. As already discussed, online interpersonal interactions often occur in the context of communities with well-defined social norms or rules about the treatment of personal information. The reasonableness of the behavior of individuals or corporate entities that "lurk" in the background of these discussions and then disseminate or record information disclosed by community members could be evaluated in light of the norms of those communities. Such an application of the privacy torts might also help control the uses of virtual "individuals" (what have been called "buddy-bots") which are increasingly capable of engaging in conversations that have the "look and feel" of friendly discourse. [43] Such recent advances in computer marketing technology seem likely to tap more and more into the kind of self-disclosure which seems to be inherently pleasurable, and thus potentially tempting. As online experiences begin more and more to resemble conversations with amiable acquaintances (wearing hidden microphones that feed into computer databases), it may be more and more reasonable to hold these virtual individuals to the standards set by appropriate social norms.

4.2 Legal Approaches in the Context of Computerized Data Processing

Some of the social problems resulting from computerized data processing, such as inadequate accuracy and security, may be best attacked

[42] Strandburg, *supra* note 1.

[43] *See* Ian R. Kerr and Marcus Bornfreund, *Buddy Bots: How Turing's Fast Friends Are Under-Mining Consumer Privacy, in* PRESENCE: TELEOPERATORS AND VIRTUAL ENVIRONMENTS (2005), arguing that virtual reality can now be used "to facilitate extensive, clandestine consumer profiling under the guise of harmless, friendly conversation between avatars and humans."

by regulating data handlers directly. Other issues, such as targeted marketing, raise more delicate questions of preserving autonomy while avoiding temptation. To avoid the hazards of imposing majority self-control solutions on individuals who may have a long term preference to disclose personal information, legal solutions to these kinds of problems might focus on providing self-control options for consumers. The law can also seek to regulate "the tempters" – for example by requiring data collectors to provide better notice about what will happen to the personal information that individuals disclose – to increase the likelihood that disclosures of personal information reflect long term preferences.

An example of a possible regulation of this sort in the context of disclosure of personal information might be a statute requiring opt-in policies for websites that collect and disseminate personal information. Such a rule would not prevent anyone from disclosing any personal information, but would assist individuals in resisting the temptation to disclose by aggregating most disclosure decisions under a non-disclosure category and setting that category as the default.[44] Individuals would no doubt still disclose information in many circumstances, but would be required to "justify" each disclosure to themselves as a reasonable "exception" to the default rule of non-disclosure. Perhaps even better would be to mandate that internet service providers provide a choice of "privacy plans" requiring each user to select a customized personal information disclosure "rule" that would apply to online transactions as a default, but could be over-ridden on a case by case basis. Thus, consumers with a long-term preference for disclosure could choose an "opt-in" rule and avoid the transaction costs of opting in in piecemeal fashion.[45] Just as a spam filter

[44] *See*, however, Jange and Schwartz, *supra* note 30 at 1245-59, questioning whether opt-in defaults are sufficient in light of consumer bounded rationality and asymmetrical information. Jange and Schwartz do not consider the role that opt-in defaults might play as self-commitment mechanisms, but their concerns are well-taken. The possibility of a voluntary technologically-implemented privacy rule that removes commercial entities from consideration if they do not comply with pre-set privacy choices may mitigate these concerns to some extent.

[45] The Platform for Privacy Preferences (P3P) provides a standard for machine-readable privacy policies that has been implemented on many websites. Lorrie Cranor et al., An Analysis of P3P Deployment on Commercial, Government, and Children's Web Sites as of May 2003, TECH. REP., FEDERAL TRADE COMMISSION WORKSHOP ON TECHNOLOGIES FOR PROTECTING PERSONAL INFORMATION (May 2003). Unfortunately, "P3P user agents available to date have focused on blocking cookies and on providing information about the privacy policy associated with a web page that a user is requesting. Even with these tools, it remains difficult for users to ferret out the web sites that have the best policies." Simon Byers, Lorrie Cranor, Dave Kormann, and Patrick McDaniel, *Searching for Privacy:*

takes away the need to consider (and perhaps even be tempted by) unsolicited email advertising, a privacy rule could permit individuals to choose not to be confronted with the possibility of transacting with websites with privacy policies that do not meet pre-selected standards or to choose to have website options for a particular transaction divided into compliant and non-compliant categories. While it is possible that a market for such privacy plans will develop – especially as privacy-protective technologies become more user-friendly – the possibility for a socially sub-optimal website coordination norm may counsel in favor of legal intervention.

The law might also increase the salience of the potential costs of disclosing personal information by requiring notice of these potential costs precisely at the point of disclosure (somewhat in the spirit of attaching warning labels to cigarettes). Issues of bounded rationality and self-control highlight the importance of the timing, framing, and placement of information about the potential uses of personal information, suggesting that legal regulation of notice should take these issues into account.

5. CONCLUSION

Self-control and temptation have played a significant role in human society throughout history. Personal information norms, which frown upon disclosing certain kinds of personal information and prying into personal information in certain contexts, are possibly explained as willpower norms that have arisen to deal with bounded rationality and self control problems related to disclosing, disseminating, and processing personal information. Because personal information norms frequently arise to compensate for collective action problems in the dissemination of personal information, they highlight the potential for information market failures that can lead to socially sub-optimal decision-making. Therefore, when situations arise in which personal information norms are expected to be ineffective, such as in the context of computerized data processing, alternative mechanisms, including legal regulation, for addressing these problems must be

Design and Implementation of a P3P-Enabled Search Engine, in PROCEEDINGS OF THE 2004 WORKSHOP ON PRIVACY ENHANCING TECHNOLOGIES (PET2004). Given the procrastinations and temptation issues discussed here, automated privacy rules are unlikely to provide significant social value unless they are easy to use. Most recently, a search engine has been developed that displays information about a website's privacy policy next to its entry in the search results. *Id.* If this feature is widely adopted by well-known search engines, it may dramatically improve the practical effectiveness of the technology. However, making choosing a privacy rule a standard aspect of obtaining internet service should drastically increase the effectiveness of such technologies.

considered. Potentially useful areas of legal regulation include the adaptation of the reasonable expectation of privacy analysis in the privacy torts to account for the personal information norms of a social group; and measures that increase consumer self-control in the context of online information processing – such as opt-in defaults, the ability to set across-the-board personal information rules, and increasing the salience of information about the hazards of personal information online. Norm entrepreneurship that categorizes online personal information disclosures with "real-space" behavior that is governed by social norms – such as the "moralization" of inadequate website privacy policies as "disrespectful" described by Hetcher – may also promote willpower through the bundling mechanism.

Further analysis, and especially further empirical work aimed at elucidating the role that temptation plays in information disclosure and dissemination, will be necessary to understand the interplay between personal information norms and information privacy law more completely. The concept of willpower norms should also find useful application outside of the information privacy context.

ACKNOWLEDGEMENTS

I am grateful to Alessandro Acquisti, Julie Cohen, Eric Goldman, Jay Kesan, Roberta Kwall, Richard McAdams, Joshua Sarnoff, Paul Schwartz, Daniel Solove, Lior Strahilevitz, Diane Zimmerman, the participants in the Works in Progress in Intellectual Property Conference at Boston University on September 10-11, 2004, the participants in the Intellectual Property Scholars Conference on August 3-4, 2004 at DePaul, the attendees at the panel on Privacy, Information, and Speech at the Law and Society Association 2004 Annual Meeting, and the attendees at faculty workshops at DePaul College of Law and Loyola Law School, Los Angeles for helpful comments on earlier versions of this work.

II

Privacy Implications of RFID and Location Tracking

Chapter 4

RFID PRIVACY

A Technical Primer for the Non-Technical Reader

Ari Juels

RSA Laboratories, Bedford, MA USA, ajuels@rsasecurity.com

Abstract: RFID (Radio-Frequency IDentification) is a wireless identification technology poised to sweep over the commercial world. A basic RFID device, often known as an "RFID tag," consists of a tiny, inexpensive chip that transmits a uniquely identifying number over a short distance to a reading device, and thereby permits rapid, automated tracking of objects. In this article, we provide an overview of the privacy issues raised by RFID. While technically slanted, our discussion aims primarily to educate the non-specialist.

We focus here on basic RFID tags of the type poised to supplant optical barcodes over the coming years, initially in industrial settings, and ultimately in consumer environments. We describe the challenges involved in simultaneously protecting the privacy of users and supporting the many beneficial functions of RFID. In particular, we suggest that straightforward approaches like "killing" and encryption will likely prove inadequate. We advance instead the notion of a "privacy bit," effectively an on/off data-privacy switch that supports several technical approaches to RFID privacy enforcement.

Key words: blocker, encryption, EPC, kill command, privacy, RFID

1. INTRODUCTION

RFID (Radio-Frequency IDentification) is a technology that facilitates the automated identification of objects. While people are generally skillful at visual identification of a range of objects, computers are not. The task of identifying a coffee mug as a coffee mug is one that many bleary-eyed people perform naturally and effectively every morning in a variety of contexts. For computing systems, this same task can pose a challenging exercise in artificial intelligence.

The simplest way to ease the process of automated identification is to equip objects with computer-readable tags. This is essentially what happens in a typical supermarket. Through a printed barcode on its packaging, a can of tomato soup identifies itself automatically to a checkout register. While a checkout clerk must manually position items to render them readable by a scanner, printed barcodes alleviate the overhead of human categorization and data entry. Over the course of well more than two decades, they have proven to be indispensable timesavers and productivity boosters.

An RFID chip, also referred to as an RFID tag, is in effect a wireless barcode. It comprises a silicon microprocessor and an antenna in a package that is generally in size and form like an ordinary adhesive label. An RFID tag can be as small, though, as a grain of sand, and can even be embedded in paper.[1] An RFID tag carries no internal source of power; rather, it is simultaneously powered and read by a radio-emitting scanner. Under ideal circumstances, an RFID tag is readable through obstructions at a distance of up to several meters.

RFID confers a powerful advantage lacking in the optical barcode: It largely eliminates the need for human positioning of objects during the scanning process. This feature promises a new order of automated object identification. For example, it could eventually render checkout clerks in supermarkets obsolete. Once RFID tagging is universal, a customer might be able to roll a shopping cart full of items by a point-of-sale scanner that would ring them up without human intervention – and automatically mediate payment as well. This vision extends to the factory and warehouse as well, where RFID could enable automated inventory-taking and ultimately even robot-guided item selection and assembly.

RFID tags have another advantage over optical barcodes. Today, every product of a given type – every 150-count box of Kleenex® tissues, for example – carries an identical barcode. With existing printing processes and scanning standards, it is impractical for individual boxes to bear unique serial numbers. In contrast, RFID tags do actually transmit unique serial numbers in addition to product information. An RFID scanner can distinguish one box of tissues from the many other millions of exactly the same type. RFID therefore permits much finer-grained data collection than optical barcodes do.

[1] RFID Journal, *Hitachi unveils smallest RFID chip*, March 14, 2003, *available at* http://www.rfidjournal.com/article/articleview/337/1/1.

RFID in some sense endows computing systems with the ability to "see" objects. By merit of their unique serial numbers and wireless transmission, RFID tags enable computing systems in certain respects to outstrip human beings. An RFID system can "see" visually obstructed objects, and can distinguish automatically between objects that are otherwise physically identical. The implications of such power for industrial automation and productivity are tremendous. Thanks to their role in streamlining inventory operations and thereby cutting costs, billions of RFID tags are likely to see use in the commercial world over the next few years. To name just a few examples: Wal-mart® and the United States Department of Defense, among others, are mandating that their major suppliers apply RFID tags to pallets of items by 2005[2] (although there has been some lag in compliance); the U.S. FDA is advocating use of RFID to secure pharmaceutical supplies;[3] tens of millions of pets have RFID tags implanted under their skin so that they can be traced to their owners in case of loss;[4] and a company called VeriChip is extending this concept to human beings by selling a human-implantable RFID tag.[5]

Our concern in this article is the effect on individual privacy of RFID-enabled computing systems that can automatically "see" everyday objects – the clothing on your person, the medical implants in your body, the prescription drugs you are carrying, the payment devices in your pocket, and perhaps even individual pieces of paper, like banknotes and airline tickets.

Computer perception of everyday objects would confer undoubted benefits: If you are lost in an airport or parking lot, an RFID-based system that can guide you to your gate or car would be appealing. So too would be the ability to return items to shops without receipts, either for refunds or warranty servicing, and RFID-enhanced medicine cabinets that ensure that you have remembered to take your medications. (In fact, a group at Intel has created prototypes of this idea.[6]) But RFID could engender many malicious activities and nuisances as well, including clandestine profiling and physical

[2] *Wal-Mart, DOD forcing RFID*, WIRED NEWS, November 3, 2003, *available at* http://www.wired.com/news/business/0,1367,61059,00.html.

[3] *Combatting counterfeit drugs: A report of the Food and Drug* Administration, United States Food and Drug Administration, February 18, 2004, *available at* http://www.fda.gov/oc/initiatives/counterfeit/report 02_04.html.

[4] C. Booth Thomas, *The See-it-all Chip*, TIME, September 22, 2003, *available at* http://time.com/time/globalbusiness/article/0,9171,1101030922-485764,00.html.

[5] Verichip Corporation Website (2004), *available at* http://www.4verichip.com.

[6] K.P. Fishkin, M. Wang, and G. Boriello, *A Ubiquitous System for Medication Monitoring*, PERVASIVE (2004), *available as* A Flexible, Low-Overhead Ubiquitous System for Medication Monitoring, Intel Research Seattle Technical Memo IRS-TR-03-011, October 25, 2003.

tracking. Articles in the popular press have tarred RFID with Orwellian catchwords and monikers like "spy-chips."[7] Privacy advocates have even mounted boycotts against companies using RFID.[8]

As we shall explain, both the utopian and dystopian visions surrounding RFID are largely hypothetical at this point. But privacy is and will be an important issue in RFID systems, one that we should take steps to address in the early stages of deployment, as standards and practices take shape that will persist for many years. This article will treat the question of RFID privacy from a technically focused perspective. In short, we shall consider the question: What technical options do we have for protecting privacy and simultaneously preserving the many benefits of RFID?

2. FOUR ESSENTIAL FACTS ABOUT RFID PRIVACY

Any meaningful discussion of RFID privacy must proceed in view of four essential facts.

1. "RFID" often serves as a catch-all term. Wireless barcodes are one manifestation of RFID. Other wireless devices may also be viewed as forms of RFID. Among these are the SpeedPass™ payment tokens now used by millions of consumers in the United States, as well as contactless building-access cards and toll-payment transponders in automobile windshields used worldwide. These different technologies have incommensurable technical properties. Toll-payment transponders, for instance, carry batteries to boost their transmission range, while SpeedPass™ and wireless barcodes are "passive," meaning that they have no internal power sources. SpeedPass™ executes a form of cryptographic challenge-response (anti-cloning) protocol, while wireless barcodes lack the circuitry to do so. Thus, while all of these wireless devices bear significantly on consumer privacy, they do not lend themselves to easy categorical discussion. Rather, "RFID" denotes a broad and fluid taxonomy of devices that share the characteristic of wireless transmission of identifying information. Loosely speaking, the term RFID may even apply to your mobile phone – a kind of

[7] D. McCullagh, *Are Spy Chips Set to Go Commercial?*, ZDNET, January 13, 2003, *available at* http://news.zdnet.com/2100-9595_22-980354.html.

[8] RFID JOURNAL, *Benetton to Tag 15 Million Items*, March 12, 2003, *available at* http://rfidjournal.com/article/articleview/344/1/1.

hypertrophied RFID tag. In this article, we use the term "RFID tag" to refer to the very basic and cheap (ultimately perhaps five-cent/unit) wireless barcode. Tags of this kind have only barebones computing power, and are essentially designed just to emit an identifier, *i.e.*, a string of numbers.

The major standard for RFID tags is under development by an entity known as EPCglobal – a joint venture of the UCC and EAN, the bodies regulating barcode use in the U.S. and Europe respectively. Tags defined by this standard are often referred to as Electronic Product Code (EPC) tags. These are the type we concern ourselves with principally here – particularly the most basic types, known as Class 0 and 1 tags. Up-to-date details on EPC tags may be found on the EPCglobal Web site.[9]

2. RFID tags – again, of the wireless-barcode variety – are unlikely to have a considerable presence in the hands of consumers for some years to come. The entities spearheading RFID tag development now through EPCglobal – including large corporations such as Wal-mart® and Proctor & Gamble® – are looking to RFID mainly to manage cases and pallets of items in the supply chain, not to tag individual consumer products. There are exceptions, of course. The U.K. retailer Marks and Spencer, for example, has initiated RFID tagging of individual items of apparel.[10] For several reasons, however, most notably tag cost and the persistence of existing data-management infrastructure, RFID tags will in all probability supplant product barcodes only gradually. Any discussion of the topic of RFID and consumer privacy in the year 2005 is necessarily futuristic. EPC-tag privacy may be a topic of immediate import for the year 2015 or 2020. This is not to discount the value of the debate now: The persistence of data-management infrastructure will not only mean gradual RFID deployment, but will also mean that once deployed, the RFID designs of 2005 – with all of their features and drawbacks – may be the predominant ones in 2020. Moreover, consumer use of barcode-type RFID is happening in a limited way already, as libraries, for

[9] EPCGlobal website, *available at* www.epcglobalinc.org, 2004.

[10] J. Collins, *Marks & Spencer Expands RFID Retail Trial*, RFID JOURNAL, February 10, 2004, *available at* http://www.rfidjournal.com/article/articleview/791/1/1.

instance, begin tagging books with RFID.[11]

3. RFID tags are unreliable – at least at present. The hypothetical scanning range of a passive RFID tag is on the order of some tens of meters. In practice, it is at best a few meters. RFID signals do propagate through obstructions. In practice, however, metals – such as the foil lining of a can of potato chips – can play havoc with RFID signals. Additionally, the type of passive RFID tag with the longest range, known as an ultra-high frequency (UHF) tag, is subject to interference in the presence of liquids. This factors significantly into the issue of consumer privacy, because human beings consist largely of water.[12] If you're worried about your RFID-tagged sweater being scanned, your best course of action may be to wear it!

   Even when RFID systems scan effectively, they do not achieve omniscient perception of their surroundings. The company NCR conducted a pilot involving automated shopping-cart inventorying at an RFID-based check-out register.[13] This exercise revealed that *good* scanning range could pose problems: Customers sometimes ended up paying for the purchases of those behind them in line!

   Of course, these are the technical obstacles of today. Improvements in reader and RFID-tag antenna technology, changes in packaging, different use of radio spectrum, and techniques yet to be conceived will no doubt lead to improved effectiveness. One should not wholly credit either the perception of RFID systems as unerring or the view that they are too shoddy to pose a threat to consumer privacy. It is hard to say exactly how they will evolve.

4. A final point: RFID privacy is not just a consumer issue. RFID tags on products could facilitate corporate espionage by offering an easy and clandestine avenue for harvesting inventory information.[14] Among leaders in the deployment of RFID is the United States Department of

[11] D. Molnar and D. Wagner, *Privacy and Security in Library RFID: Issues, Practices and Architectures*, PROC. ACM CONF. ON COMM. AND COMPUTER SECURITY 210-219 (B. Pfitzmann and P. McDaniel, eds., 2004).

[12] M. Reynolds, *The Physics of RFID*, Invited talk at MIT RFID Privacy Workshop, *slides available at* http://www.rfidprivacy.org/2003/papers/physicsofrfid.ppt.

[13] D. White, Video Presentation at MIT RFID Privacy Workshop, November 15, 2003.

[14] R. Stapleton-Gray, *Would Macy's Scan Gimbels? Competitive Intelligence and RFID*, 2003, *available at* http://www.stapleton-gray.com/papers/ci-20031027.PDF.

Defense. Battery-powered RFID tags played a significant role in management of materiel in the second Gulf War, for example.[15] (It is recounted that prior to this campaign, the supply chain was so poor that the only reliable way to procure a helmet, for example, was to order three helmets. RFID has purportedly remedied this situation to some extent.) RFID could create infringements of privacy that are uncomfortable or even dangerous for consumers. For the military, infringements of privacy could be lethal. The idea of RFID-sniffing munitions is illustration enough.

In its increasingly prevalent practice of using off-the-shelf technologies, the Department of Defense may use EPC tags – the same RFID tags serving the needs of industry. This places an extra burden of privacy enforcement on the developers of EPC tags.

Nor is it RFID tags alone that pose a threat of data compromise in an RFID system. RFID readers and their associated computing facilities in warehouses will harvest valuable business intelligence – valuable to its legitimate consumer as well as to industrial spies. Broadly speaking, RFID stretches the security perimeter of computing networks into the physical world.

3. THE NATURE OF THE THREAT

EPC tags will include several pieces of information, most notably a product type identifier, a manufacturer identifier, and a unique serial number. Thus, the RFID tag on a sneaker might indicate that it is a "year 2005 tennis shoe" that is "manufactured by Adidas®," and that it has been assigned the unique serial number "38976478623." (These pieces of information will be represented in the form of numerical codes.)

The threat of privacy infringement from RFID is twofold. First, the presence of a unique serial number in an RFID tag opens up the possibility of clandestine physical tracking. Suppose that Alice pays for her sneakers using a credit card. The shop she has patronized can make an association between the name "Alice" and the serial number "38976478623." Whenever Alice returns to the shop, her identity can be established automatically – a situation valuable for marketing. If this information is

[15] M. Roberti, *RFID Upgrade Gets Goods to Iraq*, RFID JOURNAL, July 23, 2004, *available at* http://rfidjournal.com/article/articleview/1061/1/1.

sold, then Alice's sneaker might betray her identity more widely. By further linking the sneaker serial number with Alice's credit history, shops might make decisions about the level of service that Alice should receive. And so forth.

In fact, the threat of physical tracking does not require a direct binding between names and serial numbers. If Alice participates in a political rally, for example, law enforcement officers might note her sneaker as belonging to a suspect individual. By using RFID readers deployed strategically around a city, officers might track and/or apprehend Alice.

A second threat arises from the presence of product information in RFID tags. This information in principle permits clandestine scanning of the objects on Alice's person. If Alice is carrying a painkilling drug with a high street value, she could be more vulnerable to mugging. The European Central Bank purportedly considered a plan a few years ago to embed RFID tags in banknotes.[16] These would probably have been very short-range tags designed for combating counterfeiting, but who knows what unanticipated abuses they might have engendered? Alice could also be subject to profiling of various types. If she is wearing a Rolex, she might receive preferential treatment in jewelry shops – and poor service if she is wearing a cheap digital watch. If she is carrying a book on anarchism, she might trigger law enforcement scrutiny when she walks by a police station. If she walks by a video screen carrying a bottle of Pepsi®, she might see a Coca-Cola® advertisement, and so on and so forth.

It is worth remarking that bold sallies into marketing-based exploitation of RFID as sketched above seem implausible. Corporations are too sensitive to their reputations among consumers. The threat in commercial settings probably stems more from gradual erosion of privacy. Clothing stores might, for instance, begin by offering discounts to customers who are wearing the stores' garments. (They could even do so with customer permission by retaining only scanned information present in a specially designated database.) Shops might offer automated RFID-based payment and RFID-based warranty fulfillment and returns. Habituation is a slippery slope. More aggressive RFID-based marketing and other infringements might increasingly assume an air of innocuousness to consumers that would open the door to abuses. (As a useful analogy, witness changes in the ethos of the entertainment industry. These have gradually led to common production of content that would have seemed outrageously inappropriate but a few decades ago.)

[16] *Security Technology: Where's the Smart Money?*, THE ECONOMIST, February 9, 2002, at 69-70.

Remark. In some circles there circulates a misconception that effective privacy protection may be achieved by storing the correspondence between EPC codes and their natural-language meanings in a "secure" database. Apart from the fact that such a database would need to be open to a large community to operate effectively, sophisticated resources are not required to ascertain that a number like "15872918" means "Coca-Cola Classic®." It suffices to scan a single bottle of Coca-Cola Classic® to learn this correspondence.

3.1 Why RFID privacy may matter

Irrespective of the contours that RFID technology ultimately assumes, it is the belief of this author that consumer privacy will be an important and psychologically evocative issue – more so than it is for other technologies that permit tracking of human behavior, such as credit cards and browser cookies. RFID has several properties of particular psychological potency.

To begin with, RFID tags are palpable, and physically present with their owners. This is true of other devices, such as mobile phones, but there is a key difference. A mobile phone transmits information that is accessible (without specialist eavesdropping equipment) only to a well-regulated service provider. In contrast, RFID tags will be readable by any off-the-shelf scanning device. Additionally, it is likely that consumers will make use of RFID tags in ways that will render them conscious of the technology's presence and function. When consumers perform item returns, when they are able to walk by scanners in clothing stores that read off their apparel sizes (for convenience), and so forth, they will perceive with strong immediacy that they are radiating personal information. Additionally, mobile phone models are available today that have (very short-range) RFID readers.[17] Thus consumers may come to scan RFID tags for their own purposes, such as comparison shopping and cataloging of personal possessions.

What could ultimately bring RFID privacy to the fore would be a few stark, baleful, and well publicized incidents of privacy infringement: Muggings that involve use of RFID scanners to locate bottles of prescription medication, for example. Passports are soon to be deployed with RFID tags;[18] these could cause an outcry if they betray personal information for use

[17] RFID JOURNAL, *Nokia Unveils RFID Phone Reader*, March 17, 2004, *available at* http://www.rfidjournal.com/article/view/834.

[18] M.L. Wald, *New High-Tech Passports Raise Concerns of Snooping*, NEW YORK TIMES, November 26, 2004, at 28.

in identity theft. Privacy infringement through RFID has the potential to draw attention to itself in striking ways.

4. PROPOSED REMEDIES TO THE RFID PRIVACY PROBLEM

A form of very basic radio-frequency technology is already familiar in retail shops today. Electronic Article Surveillance (EAS) systems rely on small plastic tags to detect article theft. Items that bear these tags trigger alarms at shop exits when improperly removed by customers. When EAS-tagged items are purchased, of course, their EAS tags are deactivated or removed. EAS systems have naturally pointed the way for RFID: Why not simply remove or deactivate RFID tags on purchased items to avoid privacy problems?

EPC tags support this approach by inclusion of a *kill* feature. When an EPC tag receives a special "kill" command from a reader (along with a tag-specific PIN for authorization), it permanently disables itself. Of course, "dead tags tell no tales." The presence of the kill command seems at first glance to kill the privacy debate.

Theft detection stops outside the shop door. The consumer benefits of RFID don't. We have mentioned the fact that consumers regularly carry RFID devices like SpeedPass™ and contactless building-access cards already, and have also described some of the useful applications of ubiquitous RFID tagging in the future, including "smart" medicine cabinets that monitor compliance with medication regimes,[19] and receipt-free consumer item returns. Many more such applications are envisaged. These include "smart" appliances, like refrigerators than can draw up shopping lists, suggest meals based on available ingredients, and detect expired foodstuffs; washing machines that can detect garments that may be harmed by a given temperature setting; and clothing closets that can provide fashion advice. A company called Merloni has already prototyped a range of RFID-enabled appliances.[20] For recycling – namely accurate identification of plastic types – RFID would be a boon.

There are countless other proposed examples – and examples not yet imagined – of how RFID can and undoubtedly will benefit ordinary people. The critical point here is that consumers will invariably want to have "live" RFID tags in their possession. Killing tags at the time of purchase will help

[19] *See supra* note 6.

[20] *See supra* note 4.

address privacy problems in the short term, but in the long term will prove unworkable as it undercuts too many of the benefits of RFID. The same remark applies even to partial information "killing," *e.g.*, the elimination of unique serial numbers at the point of sale, with the retention of item type information.

A kindred approach, advocated by EPCglobal,[21] is to make RFID tags easily visible to the consumer and easily removable. Indeed, Marks and Spencer adopted this tack; they incorporated RFID into price tags rather than directly into the garments they were tagging. In general, however, this approach has the same drawback as the killing of tags: It undercuts the consumer benefits. And reliance on tag removal carries the additional drawback of inconvenience. It is difficult to imagine consumers assiduously poring through their shopping bags peeling off RFID tags. It is likewise difficult to imagine a valetudinarian carefully peeling RFID tags off her collection of medication bottles at the exit to a pharmacy. She could remove tags at home instead, but by then may already have walked the streets broadcasting the presence of a bottle of painkillers with a high street value.

A supplementary remedy advocated by EPCglobal and by some policymakers is consumer notification. Signage or product packaging would notify consumers of the presence of RFID tags on store items. While this may result in somewhat more vigorous peeling of labels (the exercise of a right to opt out), it hardly offers consumers a convenient and effective avenue for privacy protection. It is indeed loosely analogous to signs that warn of video surveillance – except that RFID tags, unlike video cameras, will follow consumers home.

There is a simple physical means of enforcing RFID privacy protection. As mentioned above, metals interfere with RFID signals. It is possible to prevent radio signals from reaching an RFID device by enclosing it in a metal mesh or foil of an appropriate form, known as a Faraday cage. An agency of the State of California recently adopted this approach in offering mylar bags to shield toll-payment transponders from scanning when not in use. They bags offered a way to opt out of state-initiated programs that use such transponders to monitor traffic patterns. Faraday cages, however, are of limited utility. They not only prevent scanning of RFID tags on privately owned items, but also serve in evading EAS systems, *i.e.*, abetting in-store theft. For this reason, retail shops are unlikely to support their widespread use. Faraday cages are also likely to be of little utility when and if RFID

[21] Guidelines on EPC for Consumer Products, 2004, *available at* http://epcglobalinc.org/public_policy/public_policy_guidelines.html.

tags are embedded in a wide range of personal possessions, such as items of clothing.

4.1 The siren song of encryption

Encryption is a technique for shielding data from unauthorized access. Viewed in this light, its application to the problem of RFID privacy seems natural. But encryption does not provide a straightforward solution to the problems of privacy protection. If the product information and serial number on an RFID tag are encrypted, then they are readable only upon decryption under a valid secret key. This protects privacy, but introduces a new problem: How is the secret key to be managed?

A simple scenario is highly illustrative. Suppose that Alice purchases an RFID-tagged carton of milk at a supermarket. To protect her privacy as she walks home, the information on the carton is encrypted at the supermarket under some secret key k. Of course, Alice wants her refrigerator to be able to read the RFID tag on the carton of milk. Therefore her refrigerator must somehow be provisioned with the secret key k for use in decryption.

This key k might be printed on the milk carton, enabling Alice to enter it manually into her refrigerator by means of, *e.g.*, a numeric keypad. This would be laborious, though. Alternatively, the key might be stored in a special portion of RFID tag memory, and Alice might release it by making physical contact with the tag, *e.g.*, by touching it with a special-purpose wand. Physical intermediation of this kind would still be tedious.

A more convenient solution would be for Alice to make use of a supplementary device, a smartcard or a mobile phone, for instance, to manage the key k on her behalf. For example, Alice's mobile phone might furnish the key k to a supermarket point-of-sale device for encryption of her RFID tags at the time that she makes her purchase.

Suppose, however, that Alice is buying the carton of milk for her friend Bob. Alice would then either have to encrypt the milk information under a key belonging to Bob, or else transfer her secret key k to Bob. Yet Alice and Bob might not trust one another sufficiently to share keys.

As an alternative, the key k might be item-specific, *i.e.*, the carton of milk might have its own associated encryption key k that is not shared with any other item. That way, Alice could freely transfer it to Bob without compromising her own secrets. But then Alice assumes the burden of managing the secret key for a single carton of milk.

The problem of enabling users to protect and distribute secret keys is known to data security experts as the "key management problem." Key management has proven historically to be one of the great challenges in securing computing systems of any kind, even when each user possesses just a single cryptographic key. (Think of the headaches that passwords cause today.) Requiring ordinary consumers to manage keys for individual items would make the problem even more difficult. Suppose that the carton of milk is leaky, and Alice would like to return it to the supermarket, but her mobile phone battery is dead and she can't recover the key for the milk? When all of the natural scenarios involving Alice and her carton of milk are taken into account, encryption rapidly loses its appeal as a privacy-protection measure.

Moreover, straightforward encryption of tag data does not address the full range of basic privacy problems. Consider the problem of physical tracking. Suppose that the data D on the carton of milk are encrypted as a number E, and the RFID tag stores E instead of D. A malefactor that scans E may not know what data it represents, and therefore that he has scanned a carton of milk. Nonetheless, he can use E as a unique serial number for the purposes of physically tracking Alice. In other words, if D is a unique serial number for the carton of milk, then E is essentially a meta-serial-number!

Similar caveats vex related privacy-enhancing ideas, like that of putting tags to "sleep" upon purchase, and then "waking" them when they are ready for home use. If a tag can be awakened by just anyone, then the sleep function does not protect against surreptitious scanning of tags. Therefore, a sleeping tag must require a special accompanying key or PIN k to authorize waking. Management of the key k in this case presents many of the same challenges as those illustrated above in management of a decryption key k.

It is possible that as consumer devices like mobile phones evolve, reliable and convenient key-management systems will arise in support of RFID privacy. Prognostication on this score would be misguided. From today's perspective on RFID privacy, however, encryption does not adequately solve the most pressing problems.

Remark. In the scenarios we have just described, encryption of tag data is presumed to be performed by some external device, *e.g.*, a point-of-sale device. More sophisticated approaches to privacy protection are possible if tags themselves can perform standard cryptographic operations like encryption. As noted above, however, due to the exigencies of cost, basic RFID tags do not contain a sufficient amount of circuitry to do so.[22] Moore's Law – the long established trend toward a halving of circuitry costs every eighteen months or so – is sometimes viewed as an argument that tags will

[22] S.E. Sarma, S.A. Weis, and D.W. Engels, *Radio-Frequency Identification Systems*, *in* CHES 2002, 454-469, (B. Kalinski, Jr., C. Kaya Koc, and C. Paar, eds., 2002)., LNCS no. 2523; S.E. Sarma, Towards the Five-Cent Tag, Technical Report from MIT Auto ID Center, MIT-AUTOID-WH-006, 2001.

eventually be available to perform cryptographic operations. It is important to keep in mind, however, that cost is likely to trump functionality in RFID tags for quite some time. Given the choice between a five-cent cryptographically enabled tag and a rudimentary one-cent tag, a retailer or manufacturer is very likely to choose the latter, particularly given the volumes and slender margins on which their businesses depend. The situation could change with the development of more compact ciphers, but this is pure speculation. More importantly, even when RFID tags can perform cryptographic operations, it is still not immediately clear how to solve the vital privacy problems.

5. A TECHNICAL PROPOSAL: THE PRIVACY BIT

The remainder of this article will briefly explore the notion of an RFID *privacy bit*.[23] The *privacy bit* is a simple, cost-effective technical proposal by this author for mitigating the problems of RFID privacy while preserving the consumer benefits of RFID. The aim is to strike a good balance between privacy and utility – to eat our cake and have it too.

A privacy bit is a single logical bit resident in the memory of an RFID tag. It indicates the privacy properties of the tag. A tag's privacy bit may be *off*, indicating that the tag is freely subject to scanning, as in a supermarket or warehouse; or it may be *on*, indicating that the tag is in the private possession of a consumer. To permit changes in the privacy properties of an RFID tag, its privacy bit should be writable by an RFID scanner. The operation of changing the privacy bit should naturally require authorization via an RFID-tag-specific PIN – just like the kill command described above.

As an accompaniment to the privacy bit, an additional tag feature is required. An RFID reader is able to scan tags in one of two modes, public or private. When a tag's privacy bit is on, the tag responds only to private-mode scanning. If the privacy bit is off, the tag responds to either scanning mode.

To illustrate how the privacy bit works, let us consider its use in a clothing store of the future – which we'll call ABC Fashions. The RFID tags on ABC Fashions garments initially have their privacy bits turned off. They

[23] A. Juels and J. Brainard, *Soft Blocking: Flexible Blocker Tags on the Cheap,* PROC. 2004 ACM WORKSHOP ON PRIVACY IN THE ELECTRONIC SOC'Y 1-7 (V. Atluri, P. F. Syverson, S. De Capitani di Vimercati, eds., 2004); A. Juels, R.L. Rivest, M. Szydlo, *The Blocker Tag: Selective Blocking of RFID Tags for Consumer Privacy,* 8TH ACM CONF. ON COMPUTER AND COMM. SECURITY 103-111 (V. Atluri, ed., 2003).

remain off at the factories where the garments are manufactured, in the warehouses they pass through, and on the racks and shelves of the ABC Fashions shops. In any of these places, garments may be scanned normally and naturally: The presence of the privacy bit has no impact on the RFID operations of ABC Fashions.

The privacy bit comes into play when Alice purchases a garment at ABC Fashions – say a blouse. At this point, the privacy bit in the attached RFID tag is turned on.

Scanners in the ABC Fashions shops perform public-mode scanning. Thus, these scanners can perceive and inventory unpurchased items on racks and shelves. Likewise, theft-detection portals in ABC Fashions shops can detect unpurchased items. (Note in fact that the privacy bit serves not only to enforce privacy, but also supports electronic article surveillance!) To ensure privacy, scanners in ABC Fashions and other shops do not perform private-mode scanning. Thus ABC Fashions scanners cannot scan purchased items, because the privacy bits of the tags on those items are turned on. The same is true of the scanners in other shops and public locations that Alice might enter while carrying or wearing her blouse: They cannot scan her blouse.

In Alice's home, RFID scanners perform private-mode scanning. Therefore, Alice's RFID-enabled clothing closet can tell her when her blouse needs to be cleaned, can search the Web for suggestions about colors of trousers to match her blouse, and so forth. Her RFID-enabled washing machine can warn her if she has placed her blouse in a wash cycle that might harm the fabric, or with colors that might stain it. If Alice gives the blouse as a gift, her friend Carol can equally well benefit from the presence of the RFID tag on the blouse.

5.1 Basic enforcement

This is all very well, but the privacy-bit concept is only effective if ABC Fashions and others are respectful of privacy-enhancing scanning policies. What is to prevent ABC Fashions from setting its RFID readers to perform private-mode scanning and harvesting information from the private tags of its customers? And then even if public entities behave responsibly, what is to prevent a thief or rogue law-enforcement officer from initiating private-mode scanning?

Thankfully, the privacy-bit concept need not depend on goodwill alone. There are some effective technical mechanisms for enforcing responsible privacy policies. The simplest is to place restrictions on the software (or firmware) capabilities of RFID readers according to their arenas of deployment. This is effectively the approach used today for digital-rights management. A piece of software like Apple's iTunes®, for instance, restricts the ways in and extent to which users can download, share, and play pieces of music. In principle, by changing the underlying software, it is possible to bypass these restrictions. The effort and savvy required to do so, however, serve as barriers to abuse. Similarly, RFID readers might be equipped to perform private-mode scanning only when appropriate. The RFID reader in a refrigerator might perform private-mode scanning, while RFID readers of the type deployed on shop shelves might only be equipped to perform public-mode scanning. With sufficient expertise, someone might bypass restrictions on reader capabilities, but basic technical barriers might suppress the most common forms of abuse.

More importantly, it is possible to audit RFID scanners independently to verify that they comply with desired privacy policies. In order to execute private-mode scanning, an RFID reader must emit a private-mode scanning command. The emission of this command is readily detectable. It should be possible in the near future, for example, for a mobile phone to detect and alert its owner to the emission of an unexpected and possibly invasive private-mode RFID query. The possibility of simple public auditing would serve as a strong check on RFID privacy abuses.

5.2 Blocking

It is possible to achieve even stronger protection against inappropriate scanning by means of a device known as a "blocker."[24] A blocker obstructs inappropriate private-mode scanning. It does not perform true signal jamming, which violates the regulations of most governments. Rather, a blocker disrupts the RFID scanning process by simulating the presence of many billions of RFID tags and thereby causing a reader to stall. (We gloss over the technical details here.) By carrying a blocker, a consumer can actively prevent the scanning of her private RFID tags.

A blocker can itself take the form of a cheap, passive RFID tag. Thus, it would be possible, for instance, for ABC Fashions to embed RFID blocker tags in its shopping bags. When carrying her blouse home, Alice would then be protected against unwanted scanning. When Alice places her blouse in her closet or washes it, however, its RFID tag would remain operative. (When she wears the blouse, she might rely on the water in her body to prevent unwanted scanning – or she might carry a blocker with her.)

For greater range and reliability, a blocker could alternatively be implemented in a portable device such as a mobile phone. In this case, many

[24] *Id.*

nuanced technical mechanisms for policy enforcement are possible. For example, a mobile phone might block private-mode scanning by default, but refrain from blocking if a scanner presents a valid digital certificate authorizing it to perform private-mode scanning. Many other variant ideas are possible.

Remark. Blockers are sometimes objected to on the grounds that they can be crafted to interfere maliciously with public-mode scanning and mount denial-of-service attacks. This is true. But malicious blockers can readily be created whether or not privacy-preserving blockers exist. Malicious blockers are not a good reason for avoiding the use of privacy-preserving blockers.

5.3 Standards support

For the privacy bit concept to reach fruition, it would require support in technical standards, such as those of EPCglobal. Once the problems of consumer privacy become sufficiently apparent to the developers and deployers of RFID systems, this author hopes that EPCglobal and other standards bodies will support the idea.

6. CONCLUSION

We have presented an overview of some of the technical facets of RFID privacy. The most striking lesson here is that while RFID is a conceptually simple technology, it engenders technological questions and problems of formidable complexity.

For this reason, it is unwise to view RFID privacy as a technological issue alone. Policymaking and legislation will also have a vital role to play in the realm of RFID privacy. They must not only supplement the protections that technology affords, but must prove sensitive to its novelties and nuances. Regulating RFID is not like regulating the Internet or the transmission of credit-card information or the use of mobile phones; each technology has its own distinctive characteristics. Moreover, RFID is simultaneously an embryonic and rapidly changing technology, resistant to prognostication. RFID will bring to policymakers the opportunity to enjoy a camaraderie with technologists in grappling with a difficult and stimulating set of problems. Let us hope that they can together achieve the delicate balance between privacy and utility needed to bring RFID to its high pitch of promise.

Chapter 5

GEOLOCATION AND LOCATIONAL PRIVACY

The "Inside" Story on Geospatial Tracking

Mark Monmonier

Department of Geography, Maxwell School of Citizenship and Public Affairs, Syracuse University, Syracuse, NY 13244-1020

Abstract: Radio frequency identification (RFID) and global positioning system (GPS) technologies are complementary strategies for determining a subject's instantaneous location. Whereas RFID tracking requires readers positioned at appropriate choke points in a circulation network, GPS allows continuous tracking, especially if linked in real time to the wireless telephone system. But because of signal attenuation in buildings and multipath-corrupted signals in urban canyons, GPS does not guarantee reliable, uninterrupted tracking. Privacy issues raised by GPS tracking and its amalgamation with RFID include the retention period, the ownership of an individual's locational history, and a "locate-me" button that would extend "opt-in" protection to cellular-telephone users. Potential for abuse heightens concern about locational privacy as a basic right.

Key words: geolocation, Global Positioning System (GPS), indoor GPS, locational privacy, RFID, satellite surveillance, tracking technology

1. INTRODUCTION

Radio frequency identification (RFID) and global positioning system (GPS) technologies are complementary strategies for determining instantaneous location and tracking people, vehicles, or merchandise. An RFID system can record location whenever an RFID tag passes within range of a compatible reader – the scanner's location is the location recorded – and an RFID unit with read-write memory and the ability to record location-specific signals from low-power transmitters along its path could be debriefed periodically by a system designed to reconstruct the tag's route. Whereas RFID tracking requires readers positioned at appropriate choke points in a circulation network, GPS allows continuous tracking, especially if linked in real time to the wireless telephone system. But because of signal attenuation in buildings and multipath-corrupted signals in urban canyons, GPS does not guarantee reliable, uninterrupted tracking. These technical difficulties suggest that an amalgamation of RFID and GPS solutions could improve geospatial tracking as well as pose a significantly greater threat to locational privacy than either technology implemented independently.

This chapter explores the reliability and privacy threats of RFID, GPS, and hybrid approaches to location tracking. In addition to examining technical limitations that make hybrid systems attractive, it looks at geospatial tracking technology as a two-pronged threat to locational privacy, which can be compromised either by revealing or suggesting a person's current, instantaneous location or by recording places visited and routes taken. However valuable for fleet management, inventory control, and emergency response, location tracking raises privacy issues that include retention period, the ownership and control of an individual's locational history, and a "locate-me" button that would extend "opt-in" protection to mobile-telephone users. The potential for abuse heightens concern about locational privacy as a basic right.

2. RFID, GPS, AND GEOSPATIAL TRACKING

RFID's role in geospatial tracking depends on how efficiently and precisely the movement of merchandise or of an ambulatory subject can be predicted and controlled. Tracking pallets in a warehouse is far simpler than monitoring the movement of people or vehicles throughout a city. In a warehouse setting, interior layout and minimal electronic interference facilitate efficient inventory control with inexpensive RFID tags.[1] A reader range of several meters is adequate for recording the entrance or exit of pallets through a small number of loading docks and doorways, each covered by a single fixed monitor. And well-defined aisles with predetermined stacking levels not only limit possible storage locations within the facility but also allow an efficient audit of current inventory.

By contrast, RFID is a wholly inappropriate technology for tracking swiftly moving vehicles across cities, where a more complex street network covers a much larger area. And unless the road network is reconfigured to

[1] Barnaby J. Feder, *Keeping Better Track from Factory to Checkout,* NEW YORK TIMES (November 11, 2004), Circuits section, 7. For a concise introduction to RFID technology, *see* Farid U. Dowla, HANDBOOK OF RF AND WIRELESS TECHNOLOGIES (2004), esp. 417-36.

channel traffic past carefully chosen *choke points*, RFID has little value for monitoring vehicle movement in suburban and rural neighborhoods.[2] RFID-based fee collection of user fees works well for bridges, toll roads, and airport parking garages, but without an extensive network of street-corner and mid-block readers, tags and scanners cannot match the spatial detail of satellite-based vehicle tracking systems like OnStar, which can pinpoint a vehicle's location to within 10 meters.[3] And while a similar RFID technology can support market-based pricing schemes that charge motorists for using downtown streets during rush hour, the cost of installing and maintaining scanners could distort the size and shape of the fee-for-use area unless artificial barriers are added to reduce the number of choke points.[4]

For similar reasons these "passive" RFID systems, so-called because the tag draws its power from the reader, are generally not an efficient strategy for tracking people, pets, or wildlife. And while "active" RFID tags, with their own power supplies and more powerful transmitters detectable by scanners as far as 100 meters away, could support broader areal coverage with fewer receivers, this strategy seems unlikely to provide either the extensive geographic coverage or the refined spatial resolution required for detailed area-wide tracking.

In principle, of course, a tracking unit with its own power supply could support radio triangulation. For example, networked receivers could estimate the position of a mobile transmitter from the difference in time of arrival of its signals – a process more accurately labeled *trilateration* because location is calculated using relative distances rather than estimated angles. (Time difference is a distance surrogate.) Similarly, a mobile tracking unit could fix its location using synchronized signals from multiple transmitters at known locations. Land-based trilateration is one option for cellular-telephone companies, which must be able to provide emergency

[2] Transportation studies that rely on geographically coarse origin-destination data to limn regional commuter patterns are perhaps a notable exception. RFID tags used for toll-collection might also support real-time monitoring of travel time between pairs of RFID reader points. For a concise overview of intelligent transportation system (ITS) applications of RFID and related technologies, see Louis E. Frenzel, *An Evolving ITS Paves the Way for Intelligent Highways*, 49 ELECTRONIC DESIGN 102 (January 8, 2001).

[3] Highway navigation systems improve upon the accuracy of the GPS estimate by snapping the receiver's location onto an electronic highway map. For an overview of Onstar, see Adrian Slywotzky and Richard Wise, *Demand Innovation: GM's OnStar Case*, 31 STRATEGY AND LEADERSHIP 17 (July 16, 2003).

[4] For an overview of market-based road pricing, see Patrick DeCorla-Souza *et al.*, *Paying the Value Price: Managing Congestion with Market-Based Pricing Might Be the Wave of the Future*, 67 PUBLIC ROADS 43 (September-October 2003); and Richard H. M. Emmerink, INFORMATION AND PRICING IN ROAD TRANSPORTATION (1998).

dispatchers with a reliable location for most 911 calls by 2006.[5] Even so, trilateration using terrestrial infrastructure is less geographically precise and thus less desirable than GPS, the other principal approach to mobile tracking. Although tolerant federal regulations for handset geolocation require units that can fix their location within 50 meters 67 percent of the time and within 150 meters 95 percent of the time, GPS can reduce positioning error to less than 20 meters.

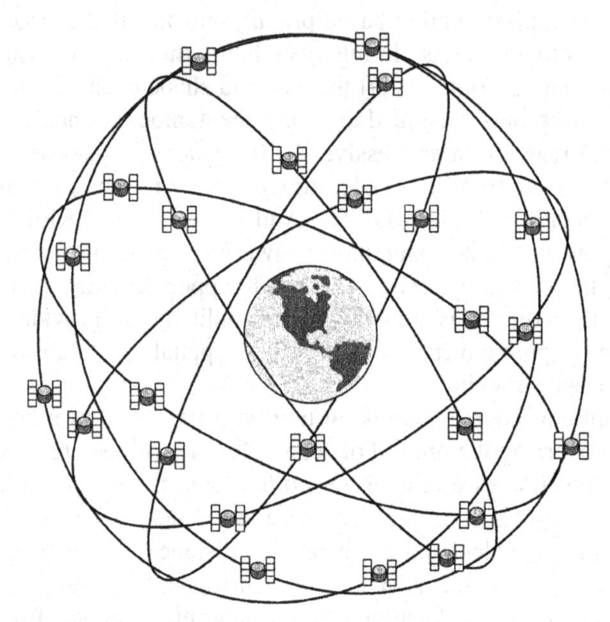

Figure 5-1. The six orbital planes of the Navstar GPS network are inclined 55 degrees to the plane of the equator and positioned 60 degrees apart. Each orbital plane has four satellites.

GPS is a trilateration process based on a constellation of at least 24 Navstar satellites, with a minimum of four in each of six orbits (fig. 5-1) configured so that at any one time a receiver should be able to receive a direct, line-of-sight signal from at least four satellites, all well above the horizon.[6] (Because Navstar satellites circle the planet at an altitude of

[5] For an overview of E-911 technical requirements overseen by the Federal Communications Commission, see Joshua Israelsohn, *On the Edge of Geolocation*, 47 EDN 35 (March 07, 2002).

[6] For an overview of GPS and the Navstar constellation, see Michael Kennedy, THE GLOBAL POSITIONING SYSTEM AND GIS: AN INTRODUCTION (2nd ed., 2002), esp. 1-48; *or* Kevin Monahan and Don Douglas, GPS INSTANT NAVIGATION: FROM BASIC TECHNIQUES TO ELECTRONIC CHARTING (2nd ed., 2000), esp. 1-16.

11,000 nautical miles, a GPS receiver can often "see" as many as twelve satellites, all at least 5 degrees above the horizon.) Each satellite broadcasts an identification code and a time signal from which the receiver estimates its distance from the satellite. Information from four satellites allows the GPS unit to compute its latitude, longitude, and elevation with a high degree of precision – positioning error is largely the result of receiver quality, electromagnetic interference in the ionosphere and troposphere, and trees and other obstructions.[7]

For non-military users, GPS was not always as precise as it is now. Prior to May 1, 2000, the spatial reliability of civilian GPS receivers was constrained by Selective Availability (SA), a Defense Department strategy for denying the enemy the high precision accorded American troops and guided weapons. SA involved two sets of time signals, one blurred by injecting small amounts of error. Under Selective Availability, only GPS receivers specially designed for military users could decode the unadulterated time signal. Civilian users received a randomly distorted time signal that caused the estimated location to oscillate unpredictably, but mostly within a 100-meter radius of the true location. According to official reports, 95 percent of estimated horizontal positions would lie within 100 meters, and half of all readings would fall within a 40-meter circle called the Circular Error Probable (CEP).[8] Averaging a string of instantaneous readings improved the result immensely. Thus, surveyors prepared to occupy a station for several hours and average thousands of readings could estimate position to within a few centimeters. (A military opponent was assumed to have far less patience.) After the Defense Department stopped deliberately blurring the time signal, the CEP shriveled to roughly 12 meters, while the 95-percent circle of error tightened to 30 meters.[9]

SA was abandoned largely because the nascent Location Based Services (LBS) industry convinced the Clinton Administration that the military benefits of SA were appreciably less significant than the economic advantages of a new technology-and-service industry pioneered largely by American firms.[10] The military, eager to tout serendipitous civilian benefits

[7] Even though horizontal location can be fixed using three satellites, estimates based on four satellites are markedly more accurate. And four satellites allow estimation of elevation as well as latitude and longitude.

[8] Monahan and Douglas, *supra* note 6, at 35-38.

[9] Kennedy, *supra* note 6, at 129. For some receivers, the shrinkage was even greater; see Mark Monmonier, *Targets, Tolerances, and Testing: Taking Aim at Cartographic Accuracy*, 7 MERCATOR'S WORLD 52 (July-August 2002).

[10] Steve Poizer and Karissa Todd, *Extending GPS Capabilities*, 16 WIRELESS REVIEW 26 (May 1, 1999).

of defense spending and fully able to deny a foreign enemy the refined GPS signal on a regional basis, conceded the point. Sales of vehicle-mounted and handheld GPS units increased dramatically as prices dropped, partly because of a drastic reduction in the cost of memory.[11] Further evidence of the commercial value of satellite positioning is the Galileo system, a European plan to launch a constellation of 30 GPS satellites and license an encrypted signal providing purely non-military users with more reliable and spatially precise service than Navstar.[12]

Although removal of SA vastly enhanced the spatial precision of commercial GPS units, it did not address a problem largely overlooked by privacy advocates leery of satellite tracking. The delicate signal from a GPS satellite is vulnerable to significant attenuation when concrete, steel, and other building materials interrupt line-of-sight transmission. What's more, reception and reliability are especially vulnerable to "urban canyons," where a GPS signal bouncing off tall, closely spaced buildings suffers multi-path corruption.[13] And because a typical receiver might take a minute or more to detect and lock onto multiple GPS signals – "acquiring the satellites," in GPS parlance – a unit that loses too many of its satellites to obstructions might not instantaneously recover when conditions improve.

Various strategies have been proposed for coping with the "indoor GPS" problem. Local transmitters situated at appropriate locations within a building and broadcasting their own, stronger time signals might suffice, in principle at least, but distances are too short for these Navstar imitators to function like "pseudolite" drones flying well above a battlefield and simulating strong GPS-like signals to thwart a devious opponent intent on jamming satellite transmissions.[14] Promising strategies include improving receiver sensitivity, thereby simplifying the task of screening irrelevant signals.[15] A more practicable approach perhaps is a hybrid GPS-RFID

[11] *Global Positioning Systems: Location-Based Technologies Track Construction Operations*, 252 ENGINEERING NEWS-RECORD 32 (June 21, 2004).

[12] Heinz Hilbrecht, *Galileo: A Satellite Navigation System for the World*, 44 SEA TECHNOLOGY 10 (March 2003).

[13] Youjing Cui and Shuzhi Sam Ge, *Autonomous Vehicle Positioning with GPS in Urban Canyon Environments*, 19 IEEE TRANSACTIONS ON ROBOTICS AND AUTOMATION 15 (February 2003).

[14] Changdon Kee, Doohee Yun, and Haeyoung Jun, *Precise Calibration Method of Pseudolite Positions in Indoor Navigation Systems*, 46 COMPUTERS AND MATHEMATICS WITH APPLICATIONS 1711 (November-December 2003); and Bruce D. Nordwall, *Using Pseudo-Satellites to Foil GPS Jamming*, 155 AVIATION WEEK AND SPACE TECHNOLOGY 54 (September 10, 2001).

[15] Frank van Diggeleen and Charles Abraham, *Indoor GPS: The No-Chip Challenge*, 12 GPS WORLD 50 (September 2001); and Larry D. Vittorini and Brent Robinson, *Optimizing*

receiver that uses RFID to supplement GPS tracking within selected indoor spaces such as shopping malls and office buildings.

3. HYBRID GPS-RFID TRACKING

A hybrid tracking system would supplement the outdoor precision of GPS with an indoor RFID network laid out to accommodate flow patterns within the structure and sufficiently dense to serve specific needs. The indoor component of this hypothetical system could have one of two distinctly different manifestations. One version would cover the interior space with a network of indoor transmitters, each broadcasting an RFID signal from which the tracking device could compute its position by triangulation or trilateration (fig. 5-2). This approach mimics GPS principles insofar as each signal would have to include its transmitter's location, either directly as spatial coordinates or indirectly through an identification number linked through a look-up table listing coordinates. A variation would invert the transmitter-receiver relationship so that a computer monitoring a network of receiver nodes would triangulate or trilaterate the location of a tracking unit broadcasting a signal with a unique RFID signal. This approach mimics the network-based geolocation strategy of wireless telephony, which might provide the indoor infrastructure as well as the mechanism for relaying the tracking unit's instantaneous location to a subscriber – a child's parents, for example, or the county probation office.

Indoor GPS Performance, 14 GPS WORLD 40 (November 2003). The Galileo system that the European Space Agency plans to make operational by 2008 promises a less fragile signal able to penetrate some buildings. Duncan Graham-Rowe, *Europe's Answer to GPS Irks Pentagon*, 178 NEW SCIENTIST 13 (June 7, 2003).

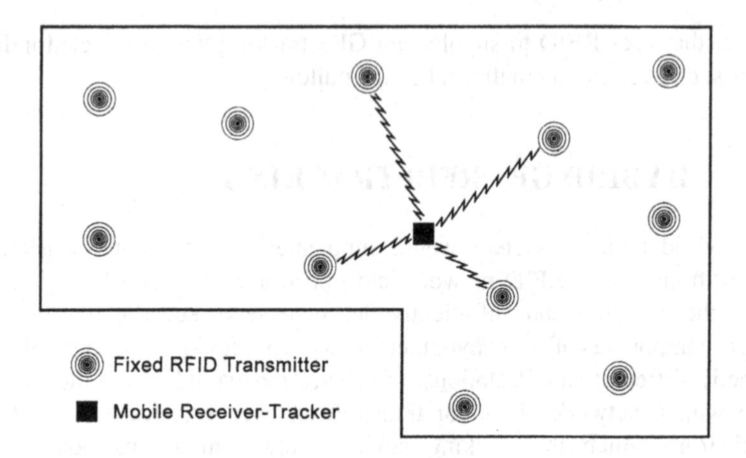

Figure 5-2. Using triangulation or trilateration a tracking device can estimate its current position using signals broadcast from a network of transmitters with known locations. Similarly, a network of receivers linked to a central computer could estimate location by triangulating or trilaterating the signal from the tracking device.

A different approach would have the tracking device collect locational identifiers from a network of fixed RFID units positioned strategically throughout a shopping mall, office building, school, prison, hospital, or indoor sports complex, to name a few likely settings (fig. 5-3). Low-cost stations powered by batteries need not be connected to a telecommunications network. Whenever the tracking unit came within range, it could record the time and identifier of the RFID station as well as the time the unit passed out of range. This information could be transmitted in real time over the wireless network or stored internally, for later electronic debriefing. Comparatively inexpensive RFID debriefing stations could be positioned throughout the building, at choke points and other key locations, and the tracking device interrogated periodically as well as when it left the building. Even if a direct wireless link allowed comparatively continuous reporting, temporary storage of recent RFID encounters could fill in gaps in the locational history caused by interruptions in mobile-telephone reception.

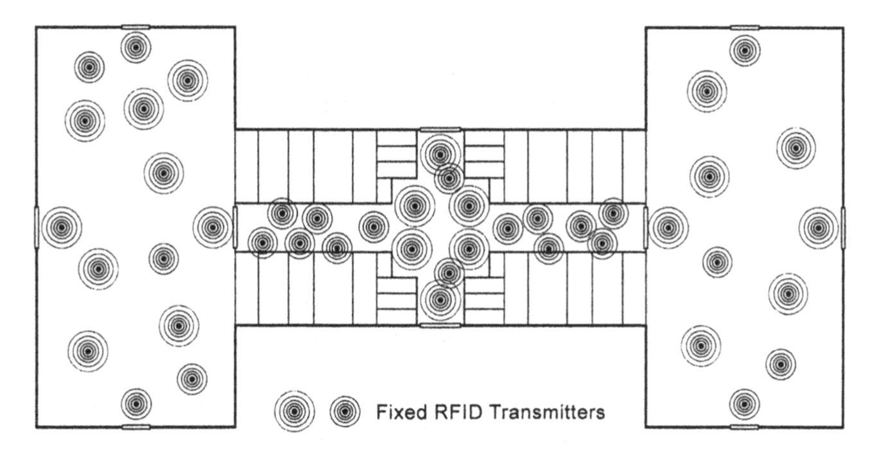

Figure 5-3. A tracking device could compile a history of its movements within an interior space, for example, a shopping mall, by recording locational identifiers from a network of purposefully placed RFID transmitters.

As the foregoing discussion implies, adding an RFID component to a GPS tracking system can significantly diminish the privacy of the subject who hopes to evade the panoptic gaze by ducking indoors. While the added scrutiny might be minimal for small structures – a suburban parent's vigilance or curiosity might be addressed by knowing which neighbor's house a child is visiting – RFID-based indoor tracking could enhance the precision of any tracking system unable to penetrate the interior intricacies of shopping malls and apartment buildings. By making satellite tracking more geographically reliable, hybrid GPS-RFID systems could also promote a wider interest in tracking by parents, employers, parole departments, and family courts, among other likely consumers. Similarly, an RFID extension could also provide a longer, more flexible electronic leash for subjects otherwise prohibited from (or penalized for) entering spaces where poor satellite reception thwarts detailed interior tracking.

4. INTEGRATED MONITORING AND PERVASIVE SURVEILLANCE

RFID poses a further threat if the code embedded in a tiny electronic tag (implanted under the skin or embedded in an ever-present cellphone or credit card) were to announce the subject's presence to an integrated, centrally controlled video surveillance system. In this Orwellian scenario, when John W. Smith (AKA subject #1476-5902-2971) walks into Central City Shoppers' World, portal scanners and RFID readers scattered throughout the mall could direct the facility's video cameras to follow him from store to store. The integrated video-RFID system could then record Smith's pauses, glances, and meetings, and if directional microphones were added, his conversations as well.

If the spymaster's motives are wholly commercial, this information could prompt enterprising sales associates to flatter John W. by remembering his name, recalling his preferences in ties or sports coats, and prompting solicitous inquiries about the health of his aging parents or his child's clarinet lessons. Link one store's data with the shopper's locational history as well as his transaction histories for other retailers, and the potential for manipulation is enormous. Some customers might enjoy this computer-enhanced personal treatment, and sales associates might be more highly motivated if they know, for instance, that Smith prefers blue oxford-cloth cotton dress shirts with button-down collars, which might conveniently have gone on sale just now (a mere second ago) – a strong incentive for embedding a customer-profile RFID chip in J.W.S.'s personal charge card. Electronically primed recognition – merely an option, of course, as described in the merchant's arcane privacy policy – might well evolve as both a shrewd marketing strategy and an incentive for patronizing a particular retailer.

Ostensibly harmless, RFID point-of-sale customer recognition not only promotes growing public acquiescence to all forms of surveillance but also creates a wider commercial market for location tracking. For example, sales associates might appreciate knowing that Smith's normal route through the mall will take him to a competitor's men's department 800 feet ahead. In a world of total shopping-behavior awareness, some small gain might even accrue from knowing where he parked his car.

Linking geospatial tracking to a geographic information system (GIS) affords a more direct and markedly more ominous type of control. Defined as a computerized system for acquiring, storing, retrieving, analyzing, and displaying geospatial information, a GIS can compare the subject's current location with a prescribed route or a set of prohibited locations. If made automatic or algorithmic, geospatial surveillance can more directly constrain a person's movements than a surveillance system designed to influence buying. For instance, the driver of a delivery truck might be given a prescribed route designed by a computer to expedite delivery of a vanload of parcels varying in urgency. An electronic map mounted on the dashboard tracks progress along the route, signals turns and delivery points, detects departures from the prescribed route, and suggests strategies for getting back on course.[16] Instead of merely alerting the driver to wrong turns, the system might report deviations to the local terminal, or even corporate headquarters, where excessive side trips could be grounds for reprimand or dismissal. The driver's knowledge of bottlenecks and reliable places to park becomes secondary to an electronic wizard's take on how the delivery should be orchestrated. Even though allowances might be made for roadwork, traffic tie-ups, and bathroom breaks, the driver might feel more oppressed than empowered when a useful tool becomes an exasperating tattle-tale.

This loss of control troubles geographers Jerome Dobson and Peter Fisher, who coined the term "geoslavery" to describe "a practice in which one entity, the master, coercively or surreptitiously monitors and exerts control over the physical location of another individual, the slave."[17] Willing to concede beneficial Location Based Services applications, they foresee "social hazards unparalleled in human history." Citing the brutal murder of a young Turkish girl who disgraced her family by attending a movie without permission, they embellish the opinion that geoslavery is "perhaps first and foremost, a women's rights issue" with the fable of an enterprising slavemaster-for-hire who invests $2,000 in a computer and GPS-tracking station, amortizes the marginal cost of $100 wrist monitors over ten years, and convinces a hundred families in his village to sign up for affordable, low-cost tracking of wives and female children. One remedy, the authors suggest, is the licensing of encrypted GPS signals under international supervision. Not optimistic that human rights organizations can contain geoslavery, they compare LBS to nuclear energy and conclude with Dobson's comment to a colleague at Oak Ridge National Laboratory: "Invent something dangerous enough, and screw it up badly enough, and you'll have a job forever."[18]

Though geoslavery seems far less menacing than nuclear war and radiological disasters, potential abuses of geospatial tracking clearly exceed the privacy threats posed by most other surveillance technologies. As I argue in *Spying with Maps*, the greatest danger of LBS is its potential for expanding the range of activities that society can suppress through electronic

[16] To me at least, this scenario is highly plausible. I use a small dashboard-mounted GPS navigator on long trips because its maps are useful and I like electronic gadgets. But I often wince at how it registers any departure from the programmed path, whether occasioned by whimsy or road construction.

[17] Jerome Dobson and Peter F. Fisher, *Geoslavery*, 22 IEEE TECHNOLOGY AND SOCIETY MAGAZINE 47 (Spring 2003).

[18] *Id.* at 52.

parole and similar forms of close surveillance.[19] For instance, a despotic government could prohibit face-to-face meetings of dissidents by fitting them with non-removable wrist or ankle bracelets. With continuous geospatial tracking, the system might alert authorities and the subjects themselves whenever two or more of them were within, say, 10 meters – with GPS tracking units reporting everyone's horizontal and vertical coordinates, the system could easily monitor a dynamic table of inter-subject distances. And because GIS is eminently proficient at point-in-polygon comparisons, the system could also prohibit subjects from entering "no-go" zones described by a list of map coordinates or confine them to electronic cellblocks, similarly defined by digital boundaries. Should a warning buzzer or electronic annunciator fail to halt the impending transgression, an electronic warden could automatically ensure compliance by injecting a sleep-inducing drug into the subject's bloodstream.

What is especially daunting is the prospect of acquiescent, perhaps eager acceptance of punitive tracking by a society in mortal fear of religious extremists and sexual predators. And in promising prompt recovery of missing children, wandering Alzheimer's patients, and lost pets, LBS monitoring services like Digital Angel and Wherify Wireless cater to benevolent angst over vulnerable subjects unlikely to object to locational surveillance.[20] While protective tracking systems employing a special wristwatch or pet collar might seem benign, nanotechnology and transducers that convert body heat into voltage could make it practical to geo-chip dogs, cats, young children, and elderly parents. Indeed, the feasibility of a subdermal tracking device has surely not escaped the attention of Applied Digital Solutions, Digital Angel's parent company. In 1999, for instance, the firm bought the patent for a "personal tracking and recovery system" that "utilizes an implantable transceiver incorporating a power supply and

[19] Mark Monmonier, SPYING WITH MAPS: SURVEILLANCE TECHNOLOGIES AND THE FUTURE OF PRIVACY (2002), esp. 174-75.

[20] For concise descriptions of locational services offered by these firms, *see* Charles J. Murray, *GPS Makes Inroads in Personal Security Technology*, ELECTRONIC ENGINEERING NEWS (August 19, 2000) at 2; Will Wade, *Keeping Tabs: A Two-Way Street*, NEW YORK TIMES (January 16, 2003), Circuits Section at 1; May Wong, *Can You Pinpoint Where Your Children Are?* SAN DIEGO UNION-TRIBUNE (October 12, 2002) at E-2; and the Wherify Wireless Location Services Web site, at www.wherifywireless.com. The Digital Angel wristwatch, promoted on the Internet in 2002 as a means of tracking children and Altzheimer's patients, was apparently discontinued after a disappointing marketing effort. See Alina Tugend, *Sensing Opportunity in Mad Cow Worries*, NEW YORK TIMES (February 22, 2004), Late Edition, at C7. As described on the company's Web site, www.DigitalAngelCorp.com, Digital Angel still markets non-GPS systems for tracking pets and livestock as well as tracking systems for "high value mobile assets."

actuation system allowing the unit to remain implanted and functional for years without maintenance."[21] A cumbersome external battery is unnecessary because "power for the remote-activated receiver is generated electromechanically through the movement of body muscle." Linked to a "locating and tracking station" by a network of satellite and ground stations (fig. 5-4), the device "has a transmission range which also makes it suitable for wilderness sporting activities." Although Applied Digital Solutions might have acquired this property right partly to forestall lawsuits alleging patent infringement by its wristwatch tracking system, the company's long-term interest in sub-dermal electronics is apparent in the VeriChip personal identity tags developed by another ADS subsidiary.[22]

However daunting for privacy advocates, geospatial-tracking technology is inherently ambiguous insofar as slavish adherence to prescribed routes and punitive avoidance of "no-go" areas is neither ominously evil nor obviously beneficial. For example, geospatial prowess in thwarting mass meetings of dissidents could prove equally effective in keeping pedophiles out of playgrounds and separating stalkers and abusive spouses from their prey. And for many minor offenders, the constraint of a geospatial leash might seem immeasurably preferable to a prison environment that fosters abusive treatment by guards and other inmates. If it works, the integration of GPS, GIS, and wireless telephony could quite literally keep parolees on straight-and-narrow judicially prescribed paths designed to promote their reentry into society and the workforce. And if it works, geospatial monitoring could not only reduce traffic congestion through time-based road pricing but also provide relentlessly accurate enforcement of traffic regulations that adjust dynamically for weather, time of day, and traffic conditions – a welcome relief from one-size-fits-all speed limits. Moreover, algorithms for detecting erratic, aggressive or intoxicated drivers, could, if reliable, make highways measurably safer.

[21] Quotations are from the abstract of U.S. Pat. No. 5,629,678, awarded May 13, 1997.

[22] Applied Digital Solutions and Digital Angel Corporation are separate publicly traded companies with complex linkages; see Barnaby J. Feder, *Did the F.D.A.'s Approval of Applied Digital's Implantable Chip Help its Long-term Outlook?* NEW YORK TIMES (November 1, 2004), Late Edition, at C2.

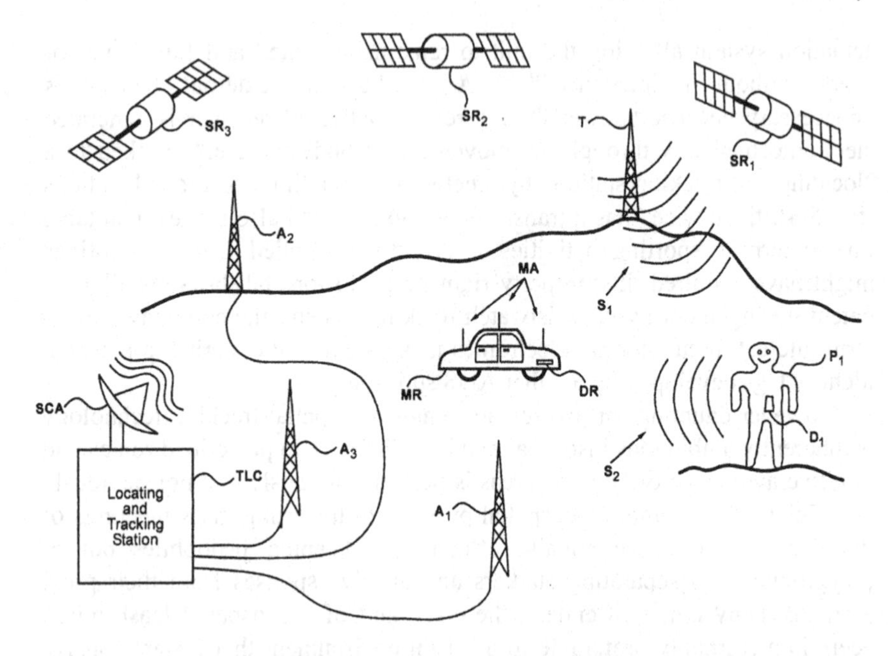

Figure 5-4. Drawing labeled figure 1 accompanying U.S. Patent 5,629,678 shows the "exemplary implantable triggerable transmitting device" linked to "several ground-based receiving antennae and alternate satellite-based receiving antennae, mobile receiving unit, and associated coordinating equipment, which combine to perform the locating and tracking function."

Caveats abound. "If it works" and "if reliable" are worrisome qualifiers that hint of mechanical and electronic breakdowns as well as clever resistance. Privacy concerns are no less important than the questionable dependability of a highly complex technology vulnerable to hackers, power outages, bad chips, anti-satellite weaponry, and electromagnetic pulse (EMP). If geospatial tracking plays out as a twenty-first-century version of the Maginot Line, infamous because of its backers' naïve assumptions about how Germany might invade France, litigation over contractual obligations and product liability could easily eclipse anxiety about privacy and human rights.

5. LOCATION TRACKING AND LOCATIONAL PRIVACY

By facilitating geospatial tracking on an unprecedented scale, satellite technology (perhaps with an assist from RFID) raises the question of whether locational privacy should be a privacy right. Many people think so, and for good reason: if addresses are readily available in printed or online directories, individuals become more vulnerable to stalkers, direct-mail advertisers, bill collectors, and disgruntled clients than if would-be harassers had more difficulty finding them. Sensitivity to these concerns explains why telephone subscribers can choose not to have their addresses listed in phone directories. Geospatial tracking poses a markedly greater threat by revealing not only where we are but also where we've been – a disclosure potentially troubling to cheating spouses, malingering employees, and politicians who frequent strip clubs or X-rated bookstores. As these examples suggest, the threat involves not only current location but also locational history, that is, a list of addresses visited, the duration of each visit, and the routes taken. Much can be inferred (perhaps erroneously) about a person's character or spending habits from his or her locational history, which could be accumulated continuously and archived indefinitely. This information could be useful to retailers curious about the shopping habits of their competitors' customers as well as to anxious parents, divorce attorneys, investigative journalists, paparazzi, political opponents, potential creditors, suspicious employers, and run-of-the-mill locational voyeurs eager to purchase their friends' and neighbors' locational histories. And because the database would be massive, algorithms would no doubt troll for suspicious behavior. If locational histories can be compiled systematically, cost-effectively, and legally, they can be commodified as a location-based service.

Although these futuristic scenarios might seem unduly alarmist, the Congress inadvertently threatened locational privacy by passing the Wireless Communications and Public Safety Act of 1999 (the E-911 act), which requires the wireless telephone industry to geolocate emergency calls – even callers who know where they are occasionally have difficulty giving directions to a 911 dispatcher.[23] A system that can fix the origin of a call to within 10 meters, the current GPS capability, could track a mobile phone, and presumably its owner, day and night, week after week.

The threat of vigilant tracking accompanied by the routine recording of locational histories and their commodification led to the Location Privacy Protection Act, introduced by Senator John Edwards in July 2001 but eclipsed two months later by the hand-wringing that followed the Al Qaeda attack on the World Trade Center and the Pentagon. A key provision of Senate Bill 1164 was the requirement "that all providers of location-based services or applications . . . restrict any collection, use, disclosure of, retention of, and access to customer location information to the specific purpose that is the subject of the express authorization of the customer

[23] 106 P.L. 81; 113 Stat. 1286 (LEXIS 1999).

concerned."[24] Edwards' bill covered wireless E-911 tracking as well as LBS applications not regulated by the Federal Communications Commission.[25] More explicit language in voluntary industry standards proposed the same year by the Wireless Industry Location Association prohibited disclosure of "geographically-tagged personally identifiable data" without the "confirmed opt-in permission" of the customer or user.[26] Implementation might require a "locate-me" button that the customer could depress or release as well as a list of permissible uses that could be checked off interactively.[27] Equally important are whether a locational history can be recorded and, if so, how long records may be retained.

Don't look to federal regulators for guidance. The Cellular Telecommunications and Internet Association, which shares the WILA's concerns but prefers a set of federal regulations binding on all LBS providers, could not convince the FCC to initiate a rulemaking proceeding. In denying the CTIA's request, the commission noted that the E-911 act added location to the list of customer proprietary network information protected by a federal law ambiguous about opt-in protection of locational information.[28] Eager not "to artificially constrain the still-developing market for location-based services," the commission "determine[d] that the better course is to vigorously enforce the law as written, without further clarification of the statutory provisions by rule."[29] But as Commissioner Michael J. Copps noted in a vigorous dissent, existing law does not define "location information" nor does it explicitly rule out "implied consent," an interpretation some of his colleagues apparently favored. With regulators reluctant to promote clarity, privacy advocates are rightly wary that profit will trump privacy. As Marc Rotenberg, executive director of the Electronic

[24] Location Privacy Protection Act, S. 1164, 107th Cong., 1st sess., introduced July 11, 2001; quotation from § 3(b)(1)(C)(i).

[25] Matthew G. Nelson and John Rendleman, *Location-Based Services Can Offer Convenience and Safety, But Customers' Privacy Is a Sensitive Issue – Reaching Too Far?* INFORMATION WEEK (August 20, 2001), at 20.

[26] Wireless Location Industry Association, Draft WLIA Privacy Policy Standards (First Revision), 2001, http://www.wliaonline.org/indstandard/privacy.html (accessed November 24, 2004). The association's Web site seems not to have been updated since 2001.

[27] Mark Monmonier, *The Internet, Cartographic Surveillance, and Locational Privacy, in* Michael P. Peterson, MAPS AND THE INTERNET (ed., 2003), at 97-113.

[28] Customer proprietary network information is defined in 47 U.S.C. § 222 (h)(1).

[29] Federal Communications Commission, Order No. 02-208, in the Matter of Request by Cellular Telecommunications and Internet Association to Commence Rulemaking to Establish Fair Location Information Practices, WT Docket No. 01-72, Adopted July 8, 2002 and released July 24, 2002.

Privacy Information Center, recently noted, "Real-time location data is the Holy Grail in the mobile phone industry."[30]

6. CONCLUSIONS

Because satellite tracking can be ineffective indoors, RFID joins GPS and wireless telephony as a technological threat to locational privacy, a growing concern now that geospatial technology affords real-time tracking and precise locational histories. Reliable, low-cost satellite tracking will not only broaden the market for location-based services but also heighten conflict between privacy advocates and the LBS industry as well as among locational recluses, locational exhibitionists, and locational voyeurs. Latent anxiety over routine, widespread satellite tracking of cellphone users suggests that the public might eventually pressure Congress to confirm locational privacy as a basic right.

[30] Quoted in Will Wade, *Keeping Tabs: A Two-Way Street*, NEW YORK TIMES (January 16, 2003), Circuits Section, at 1.

Chapter 6

PRIVACY INALIENABILITY AND PERSONAL DATA CHIPS

Paul M. Schwartz
Anita and Stuart Subotnick Professor of Law, Brooklyn Law School; Visiting Professor of Law, Boalt Hall, School of Law, U.C.-Berkeley, 2005. This essay is an abridged version of Property, Privacy and Personal Data, 117 Harvard Law Review 2055 (2004).

Abstract: Even as new possibilities for trade in personal information promise new avenues for the creation of wealth, this controversial market raises significant concerns for individual privacy–consumers and citizens are often unaware of, or unable to evaluate, the increasingly sophisticated methods devised to collect information about them. This Essay develops a model of propertized personal information that responds to concerns about privacy and evaluates it in the context of tracking chips. It sets out the five critical elements of such a model, which is intended to fashion a market for data trade that respects individual privacy and helps maintain a democratic order. These five elements are: limitations on an individual's right to alienate personal information; default rules that force disclosure of the terms of trade; a right of exit for participants in the market; the establishment of damages to deter market abuses; and institutions to police the personal information market and punish privacy violations.

Key words: tracking chips, property, inalienability, hybrid inalienability, secondary use, downstream use of personal information, data trade, right to exit, Gramm-Leach-Bliley Act, damages, privacy protecting institutions

1. INTRODUCTION

A privacy-sensitive model for personal data trade should respond to five areas: inalienabilities, defaults, a right of exit, damages, and institutions. A key element of this privacy promoting model is the employment of use-transferability restrictions in conjunction with an opt-in default. This Essay calls this model "hybrid inalienability" because it allows individuals to share, as well as to place limitations on, the future use of their personal information. The proposed hybrid inalienability follows personal information through downstream transfers and limits the negative effects that result from "one-shot" permission to all personal data trade.

In this Essay, I first develop this privacy sensitive model for personal data trade and then apply it to the use of electronic data chips. I then analyze the model in the context of two devices: the VeriChip, an implantable chip, and the wOzNet, a wearable chip. The VeriChip stores six lines of text, which function as a personal ID number, and emits a 125-kilohertz radio signal to a special receiver that can read the text.[1] A physician implants the VeriChip by injecting it under the skin in an outpatient procedure that requires only local anesthesia. A similar device has already been implanted in millions of pets and livestock to help their owners keep track of them. Applied Digital Solutions, the maker of the VeriChip, plans an implantation cost of $200 and an annual service fee of forty dollars for maintaining the user's database.

Whereas the VeriChip involves an implantable identification device, the wOzNet involves a plan to commercialize a wearable identification device.[2] Stephen Wozniak, the famous cofounder of Apple Computer, is the creator of the wOzNet. A product of Wheels of Zeus, the wOzNet tracks a cluster of inexpensive electronic tags from a base station by using Global Positioning Satellite (GPS) information. The broadcast of location information from the chip to the base station is done along the same 900-megahertz radio spectrum used by portable phones. This portion of the spectrum is largely unregulated; the wOzNet will not be obligated to purchase spectrum rights like a cell phone company. A wOzNet product package, including the chip and the base station, is expected to sell for $200

[1] *See* Julia Scheeres, *They Want Their ID Chips Now*, Wired News (Feb. 6, 2002), *available at* http://www.wired.com/news/privacy/0,1848,50187,00.html; Julia Scheeres, *Why, Hello, Mr. Chips*, Wired News (Apr. 4, 2002), *available at* http://www.wired.com/news/technology/0,1282,51575,00.html. The Food and Drug Administration (FDA) has found that the VeriChip is not a "medical device" under the Food and Drug Act, and is therefore not subject to its regulation for security and identification purposes. *See* Julia Scheeres, *ID Chip's Controversial Approval,* Wired News (Oct. 23, 2002), *available at* http:// www.wired.com/news/print/0,1294,55952,00.html.

[2] *See* Wheels of Zeus, Overview, at http://www.woz.com/about.html (last visited Apr. 10, 2004); John Markoff, *Apple Co-Founder Creates Electronic ID Tags*, N.Y. TIMES, July 21, 2003, at C3; Benny Evangelista, *Wireless Networks Could Get Personal*, S.F. CHRON., July 21, 2003, at E1; Associated Press, *Apple Co-Founder To Form Locator Network*, ABCNews.com (July 21, 2003), *available at* http://abcnews.go.com/wire/Business/ap20030721_ 1823.html.

to $250.

2. THE FIVE ELEMENTS OF PROPERTY IN PERSONAL INFORMATION

A dominant property metaphor is the Blackstonian idea of "sole and despotic dominion" over a thing.[3] An equally dominant metaphor is the idea of property as a "bundle of sticks." This idea, as Wesley Hohfeld expressed it, relates to the notion that property is "a complex aggregate" of different interests.[4] There are distinguishable classes of jural relations that relate to a single piece of property; indeed, a person's ability to possess or do something with a single stick in the bundle can be "strikingly independent" of the person's relation to another stick.[5]

2.1 Inalienabilities

Propertized personal information requires the creation of inalienabilities to respond to the problems of market failure and to address the need for a privacy commons. According to Susan Rose-Ackerman's definition, an "inalienability" is "any restriction on the transferability, ownership, or use of an entitlement."[6] As this definition makes clear, inalienabilities may consist of separate kinds of limitations on a single entitlement. In the context of personal data trade, a single combination of these inalienabilities proves to be of greatest significance– namely, a restriction on the use of personal data combined with a limitation on their transferability. This Part first analyzes this combination and then discusses why this hybrid inalienability should

[3] 2 William Blackstone, COMMENTARIES ON THE LAWS OF ENGLAND 2 (facsimile ed. 1979) (1766).

[4] Wesley Newcomb Hohfeld, *Fundamental Legal Conceptions as Applied in Judicial Reasoning*, 26 YALE L.J. 710, 746 (1917).

[5] *Id.* at 733-34, 747. Scholars have expressed views for and against the "bundle of sticks" approach to property. *See* Peter Benson, *Philosophy of Property Law*, *in* THE OXFORD HANDBOOK OF JURISPRUDENCE & PHILOSOPHY OF LAW 752, 771 (Jules Coleman & Scott Shapiro eds., 2002) (arguing that the "incidents" of property are "fully integrated and mutually connected"); Hanoch Dagan, *The Craft of Property*, 91 CAL. L. REV. 1518, 1558-70, (2003) (arguing that the "bundle metaphor" must coexist with the conception of property as forms); A.M. Honoré, *Ownership*, *in* OXFORD ESSAYS IN JURISPRUDENCE 107, 108-34 (A.G. Guest ed., 1961) (discussing the "standard incidents" of ownership).

[6] Susan Rose-Ackerman, *Inalienability and the Theory of Property Rights*, 85 COLUM. L. REV. 931, 931 (1985).

include a recourse to defaults.

Before turning to these two issues, however, it is important to note that propertized personal information, like all property, is necessarily subject to general limitations on account of the public interest. These limitations, in turn, take certain uses of information entirely outside of the realm of property. For example, law enforcement access to personal data should not be structured through recourse to a propertized model in which police are obliged to bid for access to information. Likewise, and more generally, the government's acquisition and use of personal data should not be subject to eminent domain or Takings Clause jurisprudence. Rather, mandatory or immutable rules for data access and privacy are necessary. Other similar limits on propertization may become appropriate when the media obtains personal data; in general, the First Amendment serves as a strong device for removing personal data from the realm of private negotiations and increasing their availability to the public. It is important to note that the focus of this Essay is not on these mandatory legal requirements that remove personal data entirely from the realm of private negotiations. Instead, this Essay focuses on those use and transferability restrictions that allow personal data to remain at least partially propertized.

These restrictions must respond to concerns about private market failure and contribute to the creation of a privacy commons. Regarding privacy market failure, both downstream data use and subsequent transfers of personal information may exacerbate market shortcomings. Thus a variety of devices and systems that commodify information lead to downstream uses and onward transfers. For example, the VeriChip and the wOzNet generate tracking data, and this information is likely to be traded and shared by companies that collect it.

Beyond downstream data use and subsequent transfers, free alienability is problematic because information asymmetries about data collection and current processing practices are likely to resist easy fixes. The ongoing difficulties in providing understandable "privacy notices" in both online and offline contexts illustrate the challenges of supplying individuals with adequate information about privacy practices. As a result, there may be real limits to a data trade model under which consumers have only a single chance to negotiate future uses of their information. To limit the negative results of this one-shot permission for data trade, this Essay proposes a model that combines limitations on use with limitations on transfer. Under this approach, property is an interest that "runs with the asset"; the use-transferability restrictions follow the personal information through downstream transfers and thus limit the potential third-party interest in it.

The model proposed here not only addresses concerns about private market failure, but also supports the maintenance of a privacy commons. A privacy commons is a place created through rules for information exchange. It is a multidimensional privacy territory that should be ordered through legislation that structures anonymous and semi-anonymous information spaces. From this perspective, propertization of personal information should be limited to the extent it undermines the privacy commons.

Problems for the privacy commons can arise regardless of whether a market failure problem exists. Nevertheless, because the coordination necessary to establish a functioning privacy commons may prove difficult to achieve, market failure may have especially pernicious results in this context. As Rose-Ackerman has stated: "The coordination problem arises most clearly in the case of pure public goods . . . consumed in common by a large group."[7] Should market failure continue, the present circumstances are unlikely to yield an optimal privacy commons.

Yet even if market failure ceases to be a problem, a well-functioning privacy market may fail to create public goods. Rose-Ackerman provides an illuminating example of this proposition in her discussion of the problem of settling a new geographic region: "Everyone is better off if other people have settled first, but no one has an incentive to be the first settler."[8] In this context, the market might lead to real estate speculation without any person wanting to move first to the new area. As a further example, a market in private national defense may induce some individuals to purchase protective services, but it may fail to generate an adequate level of nationwide protection. In the privacy context, a market may cause people to sell personal information or to exchange it for additional services or a lower price on products, but it may not necessarily encourage coordination of individual privacy wishes and the creation of a privacy commons.

This Essay proposes that the ideal alienability restriction on personal data is a hybrid one based partially on the Rose-Ackerman taxonomy. This hybrid consists of a use-transferability restriction plus an opt-in default. In practice, it would permit the transfer for an initial category of use of personal data, but only if the customer is granted an opportunity to block further transfer or use by unaffiliated entities. Any further use or transfer would require the customer to opt in–that is, it would be prohibited unless the customer affirmatively agrees to it.

As an initial example concerning compensated telemarketing, a successful pitch for Star Trek memorabilia would justify the use of personal data by the telemarketing company and the transfer of it both to process the order and for other related purposes. Any outside use or unrelated transfers

[7] *Id.* at 939.
[8] *Id.* at 940.

of this information would, however, require obtaining further permission from the individual. Note that this restriction limits the alienability of individuals' personal information by preventing them from granting one-stop permission for all use or transfer of their information. A data processor's desire to carry out further transfers thus obligates the processor to supply additional information and provides another chance for the individual to bargain with the data collector.

This use-transferability restriction also reinforces the relation of this Essay's model to ideas regarding propertization. The use-transferability restriction runs with the asset; it follows the personal information downstream. Or, to suggest another metaphor, property enables certain interests to be "built in"; these interests adhere to the property.

To ensure that the opt-in default leads to meaningful disclosure of additional information, however, two additional elements are needed. First, the government must have a significant role in regulating the way that notice of privacy practices is provided. A critical issue will be the "frame" in which information about data processing is presented. The FTC and other agencies given oversight authority under the Gramm-Leach-Bliley Act of 1999 (GLB Act) are already engaged in working with banks, other financial institutions, and consumer advocacy groups to develop acceptable model annual "privacy notices."[9]

Second, meaningful disclosure requires addressing what Henry Hansmann and Reinier Kraakman term "verification problems."[10] Their scholarship points to the critical condition that third parties must be able to verify that a given piece of personal information has in fact been propertized and then identify the specific rules that apply to it. As they explain, "[a] verification rule sets out the conditions under which a given right in a given asset will run with the asset."[11] In the context of propertized personal information, the requirement for verification creates a role for nonpersonal metadata, a tag or kind of barcode, to provide necessary background information and notice.

[9] *See* Pub. L. No. 106-102, § 501(a), 113 Stat. 1338 (codified at 15 U.S.C. § 6801(a) (2000)). The GLB Act also requires the oversight agencies to establish "appropriate standards" for data security and integrity. *See* § 501(b) (codified at 15 U.S.C. § 6801(b)); *see also* Fed. Trade Comm'n, *Getting Noticed: Writing Effective Financial Privacy Notices* 1-2 (October 2002),
available at http://www.ftc.gov/bcp/conline/pubs/buspubs/getnoticed.pdf.

[10] Henry Hansmann & Reinier Kraakman, *Property, Contract, and Verification: The Numerus Clausus Problem and the Divisibility of Rights*, 31 J. LEGAL STUD. S373, S384 (2002).

[11] *Id.*

A survey of existing statutes finds that the law already employs at least some of the restrictions and safeguards proposed in the model. In particular, certain transferability and use restrictions already exist in information privacy statutes. The Video Privacy Protection Act of 1988 (Video Act) contains one such limitation: it imposes different authorization requirements depending on the planned use or transfer of the data.[12] Moreover, this statute's transferability restriction requires a "video tape service provider" to obtain in advance a consumer's permission each time the provider shares the consumer's video sale or rental data with any third party.[13] This rule restricts data trade by preventing consumers from granting permanent authorization to all transfers of their information.

A second statute incorporating use and transferability limitations is the Driver's Privacy Protection Act of 1994 (DPPA), which places numerous restrictions on the ability of state departments of motor vehicles to transfer personal motor vehicle information to third parties.[14] The statute's general rule is to restrict use of these data to purposes relating to regulation of motor vehicles. Both the Video Act and the DPPA respond to the flaws inherent in one-time permanent authorization under conditions of market failure. Moreover, combined with a default rule, this approach could have the additional positive effect of forcing the disclosure of information about data transfer and use to the individuals whose personal information is at stake. This Essay now turns to the default element of its model for information property.

2.2 Defaults

As a further safeguard to promote individual choice, this Essay supports the use of defaults. It prefers an opt-in default because it would be information-forcing – that is, it would place pressure on the better-informed party to disclose material information about how personal data will be used. This default promises to force the disclosure of hidden information about data-processing practices. Furthermore, such a default should generally be mandatory to further encourage disclosure – that is, the law should bar parties from bargaining out of the default rule. The strengths of the proposed model can be illustrated through a consideration of the design and the effects, both positive and negative, of both a long-established German

[12] 18 U.S.C. § 2710(b) (2000).

[13] *Id.*

[14] 18 U.S.C. §§ 2721-2725 (2000). For a discussion of the DPPA, *see* PAUL M. SCHWARTZ & JOEL R. REIDENBERG, DATA PRIVACY LAW 32-34 (1st Ed. 1996 & Supp. 1998).

statute and a recent American statute.

German law recognizes the need for mandatory protection of certain privacy interests. The Federal Data Protection Law (Bundesdatenschutzgesetz, or BDSG) not only assigns wide-ranging personal rights to the "data subject" but also makes certain of them "unalterable."[15] As the leading treatise on the BDSG states, this statute prevents individuals from signing away certain personal interests in any kind of "legal transaction" (Rechtsgeschäft).[16] The BDSG does so to protect an individual's interest in "informational self-determination," a fundamental right that the German Constitutional Court has identified in the Basic Law (Grundgesetz), the German constitution.[17]

In the United States, the GLB Act removed legal barriers blocking certain transactions between different kinds of financial institutions and provided new rules for financial privacy. These privacy rules require financial entities to mail annual privacy notices to their customers.[18] Moreover, consistent with the model that I have proposed, the GLB Act incorporates a transferability restriction.[19] Unlike the proposed default, however, the Act merely compels financial entities to give individuals an opportunity to opt out, or to indicate their refusal, before their personal data can be shared with unaffiliated entities.[20] Thus, the GLB Act does not have a true information-forcing effect because it chooses an opt-out rule over an opt-in rule.

An assessment of the GLB Act supports the proposition that a use-

[15] Gesetz zum Schutz vor Mißbrauch personenbezogener Daten bei der Datenverarbeitung (Bundesdatenschutzgesetz) § 6, v. 27.1.1977 (BGBl. I S.201), reprinted in v. 14.1.2003 (BGBl. I S.66).

[16] Otto Mallman, § 6, in KOMMENTAR ZUM BUNDESDATENSCHUTZGESETZ 545-47 (Spiros Simitis ed., 5th ed. 2003) [hereinafter BDSG Treatise].

[17] For the critical case in which the Constitutional Court recognized this fundamental right, *see* BVerfGE 65, 1 (43-44). This decision has inspired an outpouring of academic commentary. *See, e.g.*, Paul Schwartz, *The Computer in German and American Constitutional Law: Towards an American Right of Informational Self-Determination*, 37 AM. J. COMP. L. 675, 686-92 (1989); Hans-Heinrich Trute, *Verfassungsrechtliche Grundlagen*, in HANDBUCH DATENSCHUTZ: DIE NEUEN GRUNDLAGEN FÜR WIRTSCHAFT UND VERWALTUNG 156, 162-71 (Alexander Rossnagel ed., 2003); Spiros Simitis, *Das Volkszählungsurteil oder der Lange Weg zur Informationsaskese*, 83 KRITISCHE VIERTELJAHRESSCHRIFT FÜR GESETZGEBUNG UND RECHTSWISSENSCHAFT 359, 368 (2000); Spiros Simitis, *Einleitung*, in BDSG Treatise, supra note 17, at 1, 14-24.

[18] These protections are found in Title V of the GLB Act. *See* Gramm-Leach-Bliley Act, Pub. L. No. 106-102, §§ 501-527, 113 Stat. 1338, 1436-50 (1999) (codified at 15 U.S.C. §§ 6821-6827 (2000)).

[19] *See id.* § 502 (codified at 15 U.S.C. § 6802 (2000)).

[20] *See id.* § 502(a) (codified at 15 U.S.C. § 6802(a) (2000)).

transferability restriction, combined with a default regime, can lead to optimal information-sharing. Consistent with the privacy model proposed by this Essay, the GLB Act obligates the relatively better-informed parties – financial institutions – to share information with other parties. Also, it sets this obligation to inform as a mandatory default: the GLB requires financial institutions to supply annual privacy notices to their customers. A client cannot trade the notice away for more products and services or even opt not to receive the notices because she does not want to receive more paper. Even if many individuals do not read privacy notices, a mandatory disclosure rule is crucial to the goal of creating a critical mass of informed consumers.

Unfortunately, the GLB Act's promise of informed participation in privacy protection has yet to be realized, due in large part to the relative weakness of its default rule, which allows information-sharing if consumers do not opt out. The opt-out rule fails to impose any penalty on the party with superior knowledge – the financial entity – should negotiations over further use and transfer of data fail to occur. Under the Act, information can be shared with unaffiliated parties unless individuals take the affirmative step of informing the financial entity that they refuse to allow the disclosure of their personal data. In other words, the GLB Act places the burden of bargaining on the less-informed party, the individual consumer. Examination of the often confusing or misleading nature of GLB Act privacy notices confirms this Essay's doubts about the efficacy of an opt-out rule: an opt-out rule creates incentives for financial entities to draft privacy notices that lead to consumer inaction.

On a more positive note, the agencies given oversight authority by the GLB Act have engaged in a major effort to find superior ways of providing information through privacy notices.[21] These agencies, the most prominent of which is the FTC, have engaged both privacy advocacy organizations and the financial services industry in a discussion of the design of short forms that will attract greater public attention and convey information in a clearer fashion.[22]

An opt-in rule is therefore an improvement over an opt-out rule. More specifically, an opt-in regime improves the functioning of the privacy market by reducing information asymmetry problems. An opt-in rule forces the data processor to obtain consent to acquire, use, and transfer personal

[21] *Id.* § 501(a) (codified at 15 U.S.C. § 6801(a) (2000)). The GLB Act also requires the oversight agencies to establish "appropriate standards" for data security and integrity. *Id.* § 501(b) (codified at 15 U.S.C. § 6801(b)).

[22] *See* Fed. Trade Comm'n, *Getting Noticed: Writing Effective Financial Privacy Notices* 1-2 (October 2002), *available at* http://www.ftc.gov/bcp/conline/pubs/buspubs/getnoticed.pdf.

information. It creates an entitlement in personal information and places pressure on the data collector to induce the individual to surrender it. In addition to having a positive impact on the privacy market, the opt-in regime also promotes social investment in privacy.

However promising the opt-in default regime may be, it still has some weaknesses and thus should only be one of several elements in any privacy-sensitive propertization scheme for personal data. The opt-in regime's first weakness is that many data-processing institutions are likely to be good at obtaining consent on their terms regardless of whether the default requires consumers to authorize or preclude information-sharing. Consider financial institutions, the subject of Congress's regulation in the GLB Act. These entities provide services that most people greatly desire. As a result, a customer will likely agree to a financial institution's proposed terms, if refusing permission to share information means not getting a checking account or a credit card. More generally, consumers are likely to be far more sensitive to price terms, such as the cost of a checking account, than to nonprice terms like the financial institution's privacy policies and practices. Because better information may not cure market failure, the effect of information-forcing defaults should be bolstered through use-transfer restrictions and other protection mechanisms, such as a right to exit.

2.3 Right of Exit

Consent to data trade should imply not only an initial opportunity to refuse trade, but also a later chance to exit from an agreement to trade. According to Hanoch Dagan and Michael Heller, "[e]xit stands for the right to withdraw or refuse to engage: the ability to dissociate, to cut oneself out of a relationship with other persons."[23] Providing a chance to withdraw is important because current standards afford little protection to privacy. Once companies are able to establish a low level of privacy as a dominant practice, individuals may face intractable collective action problems in making their wishes heard. As a consequence, an information privacy entitlement should include a right of exit from data trades. This right of exit, for example, would allow people to turn off the tracking devices that follow them through real space, to disable spyware and adware on the Internet, and to cancel their obligations to hear compensated telemarketing pitches.

For the privacy market, a right of exit prevents initial bad bargains from

[23] Hanoch Dagan & Michael A. Heller, *The Liberal Commons*, 110 YALE L.J. 549, 568 (2001) (*citing* Laurence H. Tribe, AMERICAN CONSTITUTIONAL LAW §§ 15-17, at 1400-09 (2d ed. 1988)).

having long-term consequences. For the privacy commons, a right of exit preserves mobility so people can make use of privacy-enhancing opportunities and otherwise reconsider initial bad bargains. Dagan and Heller have proposed that exit is a necessary element of a "liberal commons" because "well-functioning commons regimes give paramount concern to nurturing shared values and excluding bad cooperators."[24] A right of exit allows customers to discipline deceptive information collectors. Existing customers will leave as a result of the bad practices, and potential customers will be scared off. In this fashion, a privacy market disciplines deceptive information collectors by shrinking their customer base.

The right to exit also brings with it a related interest: the ability to re-enter data trades. Individuals may wish to alternate between privacy preferences more than once. As an illustration of the implications of the right to re-enter, a wearable chip appears relatively attractive in comparison to the implantable chip because of the lower costs involved should one have a change of heart after an exit. An implantable chip makes it not only more difficult to exit, but also more costly to re-enter and make one's personal data available again to vendors and third parties.

The possible danger of a right of exit, however, is that it might actually encourage, rather than discourage, deceptive claims from data collectors. The risk is that deceptive information collectors will encourage defections from existing arrangements that are privacy-friendly. Something analogous to this phenomenon is already occurring in telephony with "cramming" and "slamming."

Cramming refers to misleading or deceptive charges on telephone bills; it takes place, for example, when a local or long-distance telephone company fails to describe accurately all relevant charges to the consumer when marketing a service.[25] Slamming refers to changes made to a customer's carrier selection without her permission.[26] The response to the risk of such deceptive behavior in the context of information privacy should include legislative regulation of the way that privacy promises are made, including regulation of privacy notices and creation of institutions to police privacy promises. This Essay returns to the issues of notice and privacy-promoting institutions below.

[24] *Id.* at 571.

[25] *See* Fed. Communications Comm'n, *Unauthorized, Misleading, or Deceptive Charges Placed on Your Telephone Bill –"Cramming", available at* http://www.fcc.gov/cgb/consumerfacts/cramming.html (last visited Apr. 10, 2004).

[26] *See* Fed. Communications Comm'n, *Slamming, available at* http://www.fcc.gov/slamming/welcome.html (last visited Apr. 10, 2004).

2.4 Damages

In the classic methodology of Guido Calabresi and Douglas Melamed, "property rules" are enforced by the subjective valuations of a party and injunctions for specific performance.[27] In Property Rules, Liability Rules, and Inalienability: One View of the Cathedral, Calabresi and Melamed argue that in a property regime "the value of the entitlement is agreed upon by the seller."[28] They contrast this approach with a state determination of damages, which they associate with a "liability rule."[29] This Essay's preference when harm occurs to information privacy interests is for state determination of damages, including explicit recourse to liquidated damages. Leaving data sellers and buyers free to set the prices for privacy violations will produce inadequate obedience to these obligations.

First, actual damages are frequently difficult to show in the context of privacy. Already, in two notable instances, litigation for privacy violations under a tort theory has foundered because courts determined that the actual harm that the plaintiffs suffered was *de minimis*. Second, an individual's personal data may not have a high enough market value to justify the costs of litigation. Finally, due to the difficulty of detection, many violations of privacy promises will themselves remain private. Often, identity theft victims do not realize that their identities have been stolen. Spyware provides another example of a privacy invasion that is difficult to notice. If damages are to reflect an implicit price payable for violation of a legal right, this price should be set higher or lower depending on the probability of detection of the violation. Since many privacy violations have a low probability of detection, damages should be higher.

A state determination of damages through privacy legislation is preferable to the Calabresi-Melamed approach of enforcing the subjective valuations of private parties with injunctions. Schemes providing for liquidated damages will assist the operation of the privacy market and the construction and maintenance of a privacy commons. It will encourage companies to keep privacy promises by setting damages high enough to deter potential violators and encourage litigation to defend privacy entitlements. In addition, damages support a privacy commons by promoting social investment in privacy protection. Such damages may also reduce the adverse impact of collective action problems in the privacy

[27] *See* Guido Calabresi & A. Douglas Melamed, *Property Rules, Liability Rules, and Inalienability: One View of the Cathedral*, 85 HARV. L. REV. 1089, 1092 (1972).

[28] *Id.*

[29] *See id.*

market by allowing consumers who do not litigate to benefit from the improved privacy practices that follow from successful litigation.

Existing privacy law sometimes adheres to this path by either collectively setting damages or relying on liquidated damages. Thus, the Video Privacy Protection Act allows a court to "award . . . actual damages but not less than liquidated damages in an amount of $2,500."[30] The Driver's Privacy Protection Act contains for similar language regarding damage awards against a "person who knowingly obtains, discloses or uses personal information, from a motor vehicle record, for a purpose not permitted under this chapter."[31] Finally, the Cable Communications Policy Act, which safeguards cable subscriber information, allows a court to award "liquidated damages computed at the rate of $100 a day for each day of violation or $1,000, whichever is higher."[32]

2.5 Institutions

Institutions shape the legal and social structure in which property is necessarily embedded. Just as Carol Rose speaks of property as "the most profoundly sociable of human institutions," it is also an institution that depends on other entities for its shape and maintenance.[33] For example, intellectual property has been fostered by the performing rights societies such as the American Society of Composers, Authors, and Publishers (ASCAP) and Broadcast Music, Inc. (BMI). These organizations license performance rights in nondramatic musical compositions and distribute royalties to artists. Automobiles are another form of property that is structured by legal obligations; they require title recordings, annual safety inspections, and, depending on the state, different mandatory insurance policies.

These requirements in turn create a dynamic of institution-building. What role should institutions play as part of a system of propertized personal data? Institutions are needed for three general purposes: to provide trading mechanisms (a "market-making" function), to verify claims to propertized personal data (a verification function), and to police compliance with agreed-upon terms and legislatively mandated safeguards (an oversight function). Institutions filling these roles will assist the privacy market by

[30] 18 U.S.C. § 2710(c)(2) (2000).

[31] *Id.* § 2724.

[32] 47 U.S.C. § 551(f) (2000).

[33] Carol M. Rose, *Canons of Property Talk, or, Blackstone's Anxiety*, 108 YALE L.J. 601, 632 (1998).

ensuring that processes exist for the exchange of data and for the detection of violations of privacy promises.

Such entities can also help construct and maintain the privacy commons – the literature on commons, in fact, notes the need for such institutions. Consider how different entities police overfishing of the ocean and seek to detect pollution that degrades the environment. Consider also a fascinating recent examination of an everyday public good – parking rights at a curb – in which Richard Epstein discusses how a move away from "bottom-up rules of first possession" requires construction of parking meters or assignment of stickers to neighborhood residents.[34] Although not the focus of Epstein's analysis, these approaches also require institutions to ticket parking violations and assign parking stickers.

Two additional introductory points can be made regarding institutions. First, this Essay's preferred model involves decentralization of both the market-making and the oversight functions whenever possible. Such decentralization should also include private rights of action so that citizens can participate in protecting their own rights. Second, the Federal Trade Commission (FTC) already plays an important role in privacy protection, and its activities indicate the importance both of the policing of privacy promises and of decentralized institutional infrastructures.

As to the first role of institutions, the "market-making" function is best handled through different centers for information exchange. In contrast to this view, Kenneth Laudon has proposed the establishment of a National Information Market (NIM). In the NIM, "[i]ndividuals would establish information accounts and deposit their information assets and informational rights in a local information bank, which could be any local financial institution interested in moving into the information business."[35] These institutions would pool information assets and sell them in "baskets" on a National Information Exchange. They would also allocate the resulting compensation, minus a charge for their services, to the individuals whose information comprises a given basket.

The NIM would be a centralized market for propertized personal data. This vision necessitates a single institutional infrastructure that would permit "personal information to be bought and sold, conferring on the seller the right to determine how much information is divulged." Unfortunately, this single market might also encourage privacy violations because its

[34] Richard A. Epstein, *The Allocation of the Commons: Parking on Public Roads*, 31 J. LEGAL STUD. S515, S523 (2002).

[35] Kenneth C. Laudon, *Markets and Privacy*, COMMUNICATIONS OF THE ACM, Sept. 1996, at 100.

centralized nature makes it an easy target for attacks. In response to the possibility of cheating and other abuse, Laudon calls for development of "National Information Accounts (NIAs) for suppliers (individuals and institutions) and buyers (information brokers, individuals, and institutions)."[36] He writes: "Every participating citizen would be assigned an NIA with a unique identifier number and barcode symbol."[37]

In contrast, this Essay calls for verification of propertized personal information through an association with nonpersonal metadata. This metadata might contain information such as the database from which the personal information originated, whether any privacy legislation covered that information, and the existence of any restrictions on further data exchange without permission from the individual to whom the data referred. Such a decentralized approach would avoid the possibility of a major privacy meltdown due to the unique identifiers associated with a single NIA. Decentralized data markets also have the potential to develop privacy-friendly innovations in discrete submarkets. Given the novelty of an institutionalized data trade, it makes sense to start with multiple small markets that can draw on local knowledge rather than with Laudon's single NIM.

Data trading laws should also allow private rights of action, including class actions, when privacy rights are violated. Such rights of action can be highly effective in increasing compliance with statutory standards. For example, current rules against telemarketing allow lawsuits against companies that continue to make calls after a consumer has requested that they cease.[38] Such suits have resulted in millions of dollars in fines, and have made the words "place me on your do not call list" a potent request.

All of which is not to say, however, that the FTC and other governmental agencies do not have an important role to play in privacy protection. Here, the FTC's existing activities illustrate the contribution to policing possible from both public sector institutions and decentralized institutional infrastructures. The FTC has acted in a number of instances to enforce the privacy promises of companies that collect personal data, particularly those who do so on the Internet. Its jurisdiction in these cases is predicated, however, on a company making false representations regarding its privacy practices.[39] These false promises must constitute "unfair or

[36] *Id.*

[37] *Id.*

[38] 47 U.S.C. § 227(b)(1) (2000).

[39] For information on the FTC's enforcement role, *see* Federal Trade Commission, *Privacy Initiatives: Introduction*, *available at* http://www.ftc.gov/privacy/index.html (last visited Apr. 10, 2004).

deceptive trade practices" under the Federal Trade Commission Act for the FTC to have jurisdiction.[40] This requirement of deception means that the agency is powerless – absent a specific statutory grant of authority – to regulate the collection of personal data by companies that either make no promises about their privacy practices or tell individuals that they will engage in unrestricted use and transfer of their personal data.

Even with a specific grant of authority, the FTC would likely be overwhelmed if it were the sole institution responsible for policing the personal information market. Innovative approaches involving multiple institutions are necessary. Thus, as noted, this Essay favors a decentralized institutional model. The GLB Act offers an interesting example of this model because it divides enforcement authority between the FTC and other administrative agencies depending on the nature of the regulated financial institution.[41] The Children's Online Privacy Protection Act further decentralizes this institutional model.[42] It permits a state attorney general to bring civil actions "on behalf of residents of the State."[43] Similarly, the Telephone Consumer Protection Act (TCPA), which places restrictions on junk faxes and telemarketing, allows suits by state attorneys general.[44]

In addition, some privacy laws have added a private right of action to this mixture. Such laws include the TCPA, Video Act, Fair Credit Reporting Act, Cable Privacy Act, and Electronic Communication Privacy Act.[45] The private right of action allows individuals to enforce statutory interests. In addition, it overcomes the weaknesses of the privacy tort, which generally has not proved useful in responding to violations of information privacy.

Finally, as part of this decentralized model, the federal government should create a Data Protection Commission. In contrast to existing agencies that carry out enforcement actions, such as the FTC, a United States Data Protection Commission is needed to fill a more general oversight function. This governmental entity would assist the general public, privacy

[40] Paul M. Schwartz, *Privacy and Democracy in Cyberspace*, 52 VAND. L. REV. 1609, 1680 (1999).

[41] Under the GLB Act, regulatory agencies can assess monetary fines for violations of the Act's privacy requirements and even seek criminal penalties. *See* Gramm-Leach-Bliley Act, Pub. L. No. 106-102, § 523(b), 113 Stat. 1338, 1448 (1999) (codified at 15 U.S.C. § 6823 (2000)).

[42] 15 U.S.C. § 6504 (2000).

[43] *Id.*

[44] 47 U.S.C. § 227(f) (2000). For examples of such litigation, *see* Missouri v. American Blastfax, Inc., 323 F.3d 649 (8th Cir. 2003); and Texas v. American Blastfax, Inc., 121 F. Supp. 2d 1085 (W.D. Tex. 2000).

[45] *See* 47 U.S.C. § 227(b)(3) (2000); 18 U.S.C. § 2710(c) (2000); 15 U.S.C. §§ 1681n-1681o (2000); 47 U.S.C. § 551(f) (2000); 18 U.S.C. § 2707 (2000).

advocacy groups, and the legislature in understanding the boundaries of existing information territories. With the exception of the United States, all large Western nations have created such independent privacy commissions.[46]

3. TRACKING CHIPS: WEARABLE VERSUS IMPLANTABLE CHIPS

This Essay now turns to the VeriChip, an implantable tracking device, and the wOzNet, a wearable device. It assesses the proper application of this Essay's five-part model involving inalienabilities, defaults, a right of exit, damages, and institutions.

The VeriChip and the wOzNet share certain characteristics. Both devices are ID chips that allow the tracking either of a bearer (in the case of the implantable VeriChip) or a wearer (in the case of the clip-on wOzNet). These devices raise two issues for consideration: first, whether a distinction should be drawn between the implantable and wearable tracking devices; and second, the extent to which this Essay's five elements for information property can respond to attendant privacy risks from ID chips.

Implantable chips in particular raise such a significant threat to privacy that commercial data trades should not be allowed with them. As a general matter, implantable chips in humans, with or without tracking capacity, represent a wave of the future. Like the VeriChip, chip-based "micro-electromechanical systems" (MEMS) are a clear indication of this trend.[47] Yet the use of implantable chips as part of a scheme for commercial data trade is likely to impact the privacy commons in a highly negative fashion.

An implantable chip for data trade creates the ultimate barcode – one located inside the human body. Once people are fitted with these barcodes, the implantable chip tracks them constantly and collects their data. This tracking, in turn, has the capacity to destroy socially necessary kinds of multidimensional information privacy spaces. For example, implantable chips undercut the protection of any information privacy statutes that restrict personal data use and transfer. These biochips also permit observation of the same individual in all sectors of physical space, and facilitate multifunctional use of their personal information. There is also a threat that companies or individuals may poach the signals from such chips. This kind

[46] David H. Flaherty, PROTECTING PRIVACY IN SURVEILLANCE SOCIETIES: THE FEDERAL REPUBLIC OF GERMANY, SWEDEN, FRANCE, CANADA, AND THE UNITED STATES 394-97 (1989).

[47] *See* Robert Langer, *Where a Pill Won't Reach*, SCI. AM., Apr. 2003, at 50, 57.

of unauthorized behavior could take the form of continuous data collection by unauthorized entities, or even a new form of harassment, which I will term "frequency stalking."

Implantable chips may also resist legislative attempts to preserve a right of exit from data trade. Even if one company promises to turn off its data collection from a given implantable chip, others may continue to collect information by poaching on the first company's system and chips. Moreover, other chip-based systems, such as MEMS, might be detectable by outside companies. As a statutory safeguard, a law might require that all implantable chips be removed as soon as contracts expire or a customer wishes to discontinue data trading. This legal requirement would likely be difficult to enforce, however, and the proliferation of leftover or legacy chips would raise difficult problems. Consequently, it would be advantageous to ban commercial data trade with implantable chips while other uses of these devices, such as delivering medicine through MEMS, should be permissible.

An important distinction can be drawn with wearable chips, however, which can be removed from one's clothing and even thrown out. That a wearable chip may be easily removed means that a right of exit can more easily be maintained for wearable chips. Additionally, a distinction can be drawn between the problems posed by implantable chips and Margaret Radin's concept of the double bind.[48] Concerned with the harm that market valuation may do to nonmarket conceptions of personhood, Radin proposes that "[w]hen attributes that are (or were) intrinsically part of the person come to be detached and thought of as objects of exchange, the conception of the person is problematized."[49] Drawing on examples involving trade in sex, children, and body parts, Radin contends that commodification of certain things will harm us. Radin fears that people will be subject to a so-called "double bind" as a consequence of "a gap between the ideals we can formulate and the progress we can realize."[50] More concretely, the double bind is the "practical dilemma of nonideal justice" – if we compromise our ideals too much, we reinforce the status quo, but if we are too utopian in our ideals, we make no progress.[51] As an example, a ban on surrogate motherhood, which Radin ultimately supports, harms poor women who will miss out on the possible economic benefit from selling their reproductive capacity.

In contrast to surrogacy, a ban on implantable chips will not disadvantage

[48] Margaret Jane Radin, CONTESTED COMMODITIES 123-130 (1996).
[49] *Id.* at 156.
[50] *Id.* at 123.
[51] *Id.* at 124.

poor persons in any meaningful fashion. An opportunity to engage in data trade with wearable chips, for example, will still be available. Additionally, because data trade companies will most likely seek affluent and middle-class individuals as customers, such a ban is unlikely to deprive the poor of a significant income stream. Consequently, a ban on data trade from implantable chips will not create a Radinian double bind.

A model privacy law should also regulate the collection and use of personal data with nonimplantable chips. Such legislation should incorporate the five elements for propertization of personal data, as set forth in this Essay. Such a statute should legislate inalienabilities that place use-transfer restrictions on the personal information generated through wearable GPS devices. In addition, it should set opt-in default rules for transfers of tracking data. A model GPS privacy law should only permit a company collecting personal information with such devices to transfer such information to third party companies following an opt-in. This law should also contain a proscription against further transfers of personal data; this restriction might be modeled on the one found in the GLB Act and might prohibit a receiving third party from disclosing such information "to any other person that is a nonaffiliated third party." These use-transfer restrictions plus the default rule would serve a significant information-forcing function with great benefit to consumers.

As for the right of exit, a model GPS privacy statute should allow individuals who do business with wearable chip companies to turn off or stop wearing their tracking devices at any time. These consumers should also have a statutory right to terminate their contracts, perhaps after thirty days or after one year. In the context of automobiles, lemon laws provide owners with a right to return under certain circumstances. The lemon laws protect consumers who may not be able to assess possible defects in an automobile prior to the purchase. In a similar manner, buyers of data chips may be unable to assess questions relating to privacy before buying the devices. At different stages of their lives, buyers of the chips may also value their privacy differently. A college student might not care that she was being tracked; this same person, who later seeks an abortion or substance abuse counseling, might object to the tracking. Moreover, the threat of "frequency stalking" exists not only for implantable chips but also for wearable ones. As a consequence, legislation should protect the right to turn tracking devices off at any time and to terminate underlying contracts at certain times.

To be sure, the danger exists that deceptive information collectors will encourage consumers to switch away from privacy-friendly arrangements. Regulation of the form of notice given to GPS consumers as well as an institutional role in policing privacy promises can help on this score.

Additionally, legislation should set damages for privacy violations following collection of data by wearable chip companies. A statute should track language found in statutes such as the Video Privacy Protection Act, Driver's Privacy Protection Act, and Cable Communications Act, and should permit liquidated damages.

Finally, institutions must police the privacy promises and practices of wearable chip companies. Institutions are necessary to provide trading mechanisms to help with verification of interests in propertized personal data, and to enforce compliance with agreed-upon terms and legislatively mandated safeguards. As the development of the wOzNet has shown, private entities for collecting and trading information generated from wearable chips are likely to develop. In other words, the private sector can handle a market-making function, but a privacy statute in this area is needed to provide for government enforcement of privacy standards and promises.

4. CONCLUSIONS

A strong conception of personal data as a commodity is emerging in the United States, and individual Americans are already participating in the commodification of their personal data. This Essay's goal has been to develop a model for the propertization of personal information that also exhibits sufficient sensitivity to attendant threats to personal privacy. It developed the five critical elements of its model of propertized personal information. This model views information property as a bundle of interests to be shaped through attention to five areas: inalienabilities, defaults, a right of exit, damages, and institutions.

This Essay has called for an outright ban on data trade through the use of implantable chips. Its concern is that implantable chips will permit tracking that would destroy attempts to build a privacy commons. In contrast, it has argued that data trade through wearable chips should be permissible pursuant to the restrictions found in this Essay's model for propertized personal information.

As Dagan has argued, property is a human creation, the form of which can be altered to meet human needs and reflect human values.[52] In this light, this Essay has sought to develop an ideal conception of personal information as property that responds to privacy market failure and to the need for a privacy commons. A critical challenge remains to persuade policymakers and individual data traders of both the benefits of information markets and

[52] Dagan, supra note 5, at 1532.

the need to set appropriate restrictions on them. Moreover, future technological twists and turns are inevitable. A final cautionary point is therefore in order: in propertizing personal information and opening the door to data trade, the legal system must be willing to revisit its judgments and regulations.

III

Privacy Implications of Biometric Technologies

Chapter 7

BIOMETRICS

Overview and Applications

Ishwar K Sethi
Intelligent Information Engineering Lab, Department of Computer Science and Engineering, Oakland University, MI 48309

Abstract: This chapter provides an overview of the biometric technologies and applications. It discusses different modes of biometric application deployment. A discussion on societal issues pertaining to biometrics is provided.

Key words: authentication, biometrics, identification

1. WHAT IS BIOMETRICS?

Traditionally, the term biometrics has stood for "the statistical study of biological data."[1] However, with the growing usage of automated systems for person identification by government and industry, biometrics is now commonly used to denote "automated methods for recognition or verification to determine the identity of a person based on physiological or behavioral characteristics."[2] Examples of physiological attributes used by current biometric systems include fingerprints, hand geometry, face, retina, and iris. Behavioral characteristics are action or dynamic attributes and have an element of time associated with them. Examples of such attributes used by biometric systems include signature, voice, and keystroke pattern.

[1] THE AMERICAN HERITAGE DICTIONARY (2004).

[2] *Glossary of Biometric Terms*, Association for Biometrics, *available at* http://www.ncsa.com/services/consortia/cdbc/glossus1.htm.

Biometric attributes for recognition or verification in biometric systems are captured live and recognition or verification is done in real or near real time. This distinction thus makes biometrics different from other methods of identification, such as latent prints, dental records and DNA, which are often used in criminal investigation[3].

1.1 Recognition versus Verification

Biometric systems make a clear distinction between recognition and verification.[4] Recognition is also known as identification. Recognition implies that the person to be recognized is known to the recognizer. In terms of a biometric system, it means that a person is in the database of the biometric system and the system is able to pull out the identity of the person by matching his biometric attributes against stored attributes of all persons in the database. Thus, recognition involves *one-to-many search.* Verification implies checking the identity claim presented by a user to the system. Thus, a verification system doesn't need to match the presented biometric attributes against all stored attributes; it simply needs to match a pair of attributes, one representing the claimed identity and the other being measured at the time of the claim. Thus, verification involves *one-to-one search.* Verification is also known as authentication, and such systems are often called biometric authentication systems.

1.2 Why Biometrics?

A number of factors have led to the unprecedented growth of biometrics. Chief among them are decreasing hardware costs, growth in networking and e-commerce, and greater emphasis on security and access control. The terrorist act of September 11 has been another major factor spurring innovation in biometric applications. Biometric industry revenues were $600 million in 2002 and are expected to reach $4.6 billion by 2008 according to International Biometric Group,[5] a leading consulting firm.

The majority of biometric applications involve authentication.[6] Biometric authentication offers tremendous advantages over competing methods for

[3] http://www.biometrics.org.

[4] BIOMETRICS-PERSONAL IDENTIFICATION IN NETWORKED SOCIETY, (A. Jain, R. Bolle, and S. Pankati, eds., 1999).

[5] International Biometric Group, *available at* http://www.biometricgroup.com.

[6] A. Jain, S. Pankanti, S. Prabahakar, L. Hong, A. Ross and J.L. Wayman, *Biometrics: A Grand Challenge*, Proc. before International Conference on Pattern Recognition, 935-942, 2004.

authentication in the networked world in which we live. Imagine being able to access different resources and assets that we currently access through passwords without remembering a single password. Biometric authentication systems make this possible. Not only do we not have to remember passwords, with biometrics there is no need even to worry about the password being stolen. A biometric system also offers more security since a biometric attribute cannot be shared - unlike a password, which can be intentionally divulged to others to provide unauthorized access. The use of a smart card is another popular method for authentication. However, a smart card can be stolen or misplaced - problems that are not present with a biometric based verification system. These advantages, coupled with low costs for capturing and processing biometric information, are leading to new applications for biometrics every day.[7]

2. HOW DOES A BIOMETRIC SYSTEM WORK?

A biometric system is a *pattern recognition* system.[8] Such systems have found numerous applications in the last fifty years. The OCR (optical character recognition) software that comes bundled with document scanners is an example of a pattern recognition system. Another example is the speech recognition software that transcribes spoken words to text. Yet another example is the software that some investors use to recognize trends in stock market data. Thus, the term "pattern" in "pattern recognition" is used to denote both concrete and abstract concepts depending upon the application.

A pattern recognition system typically consists of four stages. The first stage is the sensor or data capture stage. During this stage, the pattern to be recognized is captured in a form suitable for computer or electronic processing. The next stage is the preprocessing stage, wherein the captured data is processed to remove any noise or disturbance that might have made its way through the sensing stage. Often, the data capture and the preprocessing stages are shown merged as a single stage for convenience. The next stage is the feature extraction stage. Here, the captured data is analyzed to measure the extent of important properties present in the data. These properties are called *features*, and the list of properties is fixed *a*

[7] A.K. Jain, A. Ross and S. Prabhakar, *An Introduction to Biometric Recognition*, 14 IEEE TRANSACTIONS ON CIRCUITS AND SYSTEMS FOR VIDEO TECHNOLOGY, SPECIAL ISSUE ON IMAGE AND VIDEO BASED BIOMETRICS 4-20 (2004).

[8] R. Duda and P. Hart, PATTERN CLASSIFICATION AND SCENE ANALYSIS (1973).

priori by the designer of the pattern recognition system. The final stage is the recognition stage. Here, a decision is made regarding the identity of the pattern. This is accomplished by one of two possible approaches: (i) *recognition by classification*, and (ii) *recognition by matching*. Recognition by classification simply uses one or more mathematical functions with the properties measured during feature extraction serving as function arguments. Recognition by matching is performed by comparing the extracted properties of the input pattern against the set of stored properties representing different patterns that the system is capable of recognizing. The set of stored properties of different patterns are called *templates* and the process of matching is often termed *template matching.*[9]

2.1 Biometric System Architecture

The architecture of a typical biometric system[10] is shown in Figure 7-1. The system operates in two modes: the *enrollment phase* and the *identification/verification phase*. The enrollment phase is also known as the *registration phase*. During the enrollment phase, the biometric attributes and other personal information - for example, name, address, and so forth - for each person expected to be recognized or verified are collected. It is common to capture multiple readings of the biometric attribute used by the system, such as a fingerprint for each person, at the time of enrollment to minimize variability in the captured data. The system extracts features from the captured data to create a template. These templates are stored in a database along with other personal information about each individual registered with the system. Enrollment conditions often affect the performance of the biometric system. It is often the first time that most of the users come in contact with a biometric device. If the enrollment is done under one set of environmental conditions and the system is deployed under another set of conditions, the difference between the two sets of conditions can cause performance degradation in many cases.

During the identification/verification phase, the system goes through processing identical to the enrollment phase to form a template of the biometric attributes of the person seeking verification or being identified. The template thus formed is then fed to a template matcher, which performs

[9] J.L. Wayman, *Fundamentals of Biometric Authentication Technologies*, 1 INT. J. IMAGE 93-113 (2001).

[10] D. Maltoni, D. Maio, A.K. Jain, and S. Prabhakar, HANDBOOK OF FINGERPRINT RECOGNITION (2003).

either a one-to-one or one-to-many search to authenticate or identify the person being present at input.

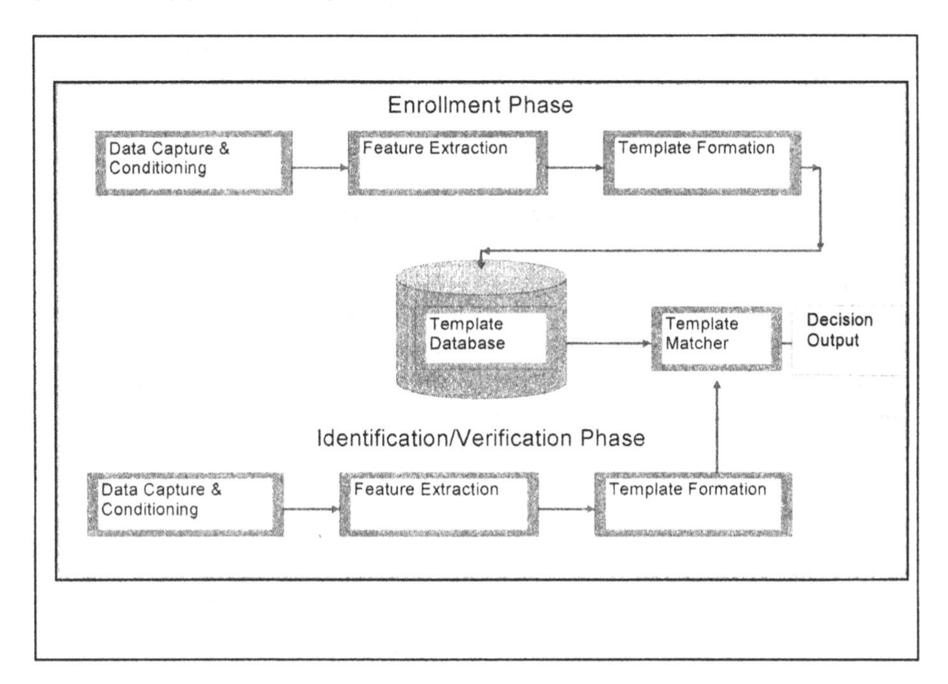

Figure 7-1. Biometric system architecture

Many biometric authentication systems do not make use of a template database.[11] Instead, the user is issued a *biometric identification card*, which stores the user's biometric template. During authentication, the template from the biometric card is read and compared against the template being derived from biometric measurements made at that time. This approach is particularly popular with those biometric attributes, such as hand geometry, for which the template size is small.

A biometric system can be made to operate for *positive* or *negative identification*. While operating for positive identification, the decision output of the system establishes the identity of the subject by matching the captured biometric data with a template in the system's database; that is, the system knows the subject by virtue of having "seen" the subject before. The purpose of operating in the positive identification mode is to limit access to

[11] *supra* note 6 at 935-942.

resources or facilities to specific, known people. In the negative identification mode of operation, the output of the system establishes the identity of the subject as someone who is not enrolled in the system, that is, not known to the system. The purpose of negative identification is to prevent users from having multiple identities to gain advantages or benefits. For example, it can be used to prevent people from submitting multiple claims for social security benefits under false identities.

2.2 Desirable Characteristics of Biometric Attributes

Fingerprints, hand geometry, face, retina, iris, signature, voice, blood vessels, and keystroke pattern are the main biometric attributes used by current commercial systems. Systems using many other attributes such as ear shape, skin reflectance, body odor, and gait are under development. Thus, with a range of available attribute choices, a natural question to ask is "what are the desirable characteristics of a biometric attribute for automatic identification and verification?" There are three characteristics that are deemed most important.[12] They are: *distinctiveness*, *robustness*, and *measurability*.

Distinctiveness implies that the attribute must show great variability over the general population so as to yield a unique signature for each individual. Robustness of a biometric attribute determines how the attribute varies over time. Ideally, it should not change over time. The measurability of an attribute refers to the ease with which it can be sensed. In addition to these three characteristics, it is desirable that an attribute be *acceptable* and *universal.* The acceptability of an attribute implies that no negative connotation is associated with it. For example, fingerprinting often is associated with criminal records. The universality characteristic of an attribute determines the fraction of the general population that possesses it in multiples.

In the following we provide a brief description of commonly used biometric attributes.

2.2.1 Face

Faces are our most visible attribute. Thus, it is not surprising that face is one of the most acceptable biometric attributes. Another factor in favor of face as a biometric attribute is that acquiring face images is non-intrusive.

[12] J.L. Wayman, FUNDAMENTALS OF BIOMETRIC TECHNOLOGIES, *available at* http://www.engr.sjsu.edu/biometrics/publications_tech.html.

Facial features, such as the upper outlines of the eye sockets, the areas surrounding the cheekbones, the lips, and the eyes and nose location, are often used to build face templates. Another approach to building face templates is the use of *eigenfaces*.[13] An eigenface is some sort of a standard face image; that is, a kind of a standard building block. A biometric system will typically use a set of eigenfaces combined via a linear mix to construct face templates for persons enrolled with the system. The different face templates will be realized by changing the proportions of different eigenfaces in the mix. The eigenfaces used by a biometric system are function of the enrollee faces determined by some well-known algorithms.

The performance of face verification and identification systems often is dependent upon changes in facial expressions, changes in lighting conditions, face pose variations, and changes over time - for example, growth of facial hair and changes in hair style. The size of the face template varies typically between 100 to 1,000 bytes depending upon the methodology used to extract and represent facial characteristics for matching.

2.2.2 Fingerprints

The use of fingerprints for person identification dates back to the late eighteenth century with the advent of the Henry Classification System[14] for fingerprint classification, which classifies ridge patterns found on fingerprints into five categories: left loop, right loop, arch, whorl, and tented arch. While the Henry system of classification is useful for large-scale forensic applications, fingerprint-based biometric systems mainly rely on the spatial distribution patterns of discontinuities, known as *minutiae*, in the flow of ridgelines in fingerprints. Many types of minutiae exist, as shown in Figure 7-2.

In current systems, fingerprint sensing is performed either by optical means or through a capacitance sensor embedded in a silicon chip. The optical technology is a mature technology and generally performs well. It, however, can exhibit performance degradation due to prints left on the platen, where a user places her finger for optical imaging, from previous users. The use of silicon chips for fingerprint capturing is growing because these chips provide better images and are smaller in size.

[13] M. Turk and A. Pentland, *Eigenfaces for Recognition*, 3 J. COGNITIVE NEUROSCIENCE 71-86.

[14] *supra* note 10.

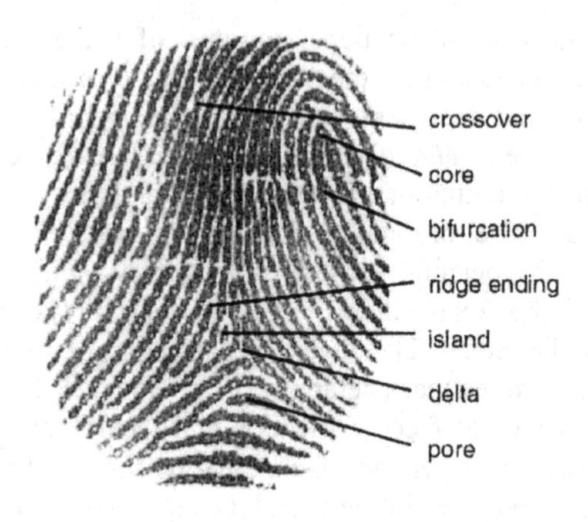

Figure 7-2. Different minutiae types in a fingerprint

2.2.3 Iris

The iris, located between the cornea and lens of the eye, controls the amount of light that enters the eye through the pupil. It has a complex visual texture pattern, which reaches stability by the 8th month after birth and stays the same through a person's entire life. The iris pattern for an individual is unique due to a very high degree of randomness and no two irises are alike. A camera operating in the near-infrared range best captures the iris pattern. An example of an iris pattern is shown in Figure 7-3. Both active and passive methods are available to capture an iris image to perform further analysis. In the active iris image capture method, the camera is located close to the user, about a foot away, and the user is required to move back and forth to get the user's iris in focus. In passive sensing, the user can be up to three feet away from the camera.

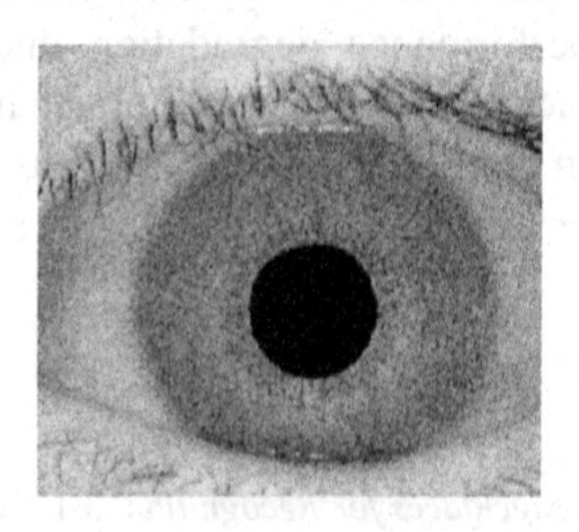

Figure 7-3. Example of an iris pattern

The iris pattern is converted to a unique template for each individual through visual texture analysis. The template size is about 200 bytes. Iris identification is extremely accurate and fast.[15] In its largest deployment, an iris recognition system in the United Arab Emirates does 3.8 billion comparisons per day. The technology is being used for both verification and identification applications.[16] Its main disadvantage is that it is perceived as intrusive and invasive.

2.2.4 Hand/Finger geometry

Hand geometry relies on characteristics such as length of fingers, which are relatively invariant for a person but not very distinctive. The hand geometry data is captured easily and non-intrusively by having the user place his palm flatly on a board with guiding pegs. Hand-geometry-based systems are primarily used only for verification applications because of the limited distinctiveness of hand geometry features. Storage of a hand geometry template requires very few bytes. Finger geometry systems work by looking at the geometry of only one or two fingers. Recently, systems based on finger blood vessel patterns have started to appear in the market.

2.2.5 Retina scan

Retina scan technology is claimed to be most secure biometric measure, as it is almost impossible to change or replicate the blood vessel structure of the eye. Retinal scanning is performed by shining a bright light source into the eye of a user who is standing very close to the imaging device. Despite its extremely high degree of reliability, retina scan technology has not found much acceptance because of its intrusive nature. The retina scan template size is typically between 50 to 100 bytes. An important issue with retina scan technology is that certain medical conditions, such as high blood pressure and AIDS, can be revealed involuntarily through its application.

2.2.6 Voice

Voice recognition is different from speech recognition; here the objective is to identify the speaker rather than to recognize words. Voice as a

[15] John G. Daugman, *High Confidence Visual Recognition of Persons by a Test of Statistical Independence*, 15 IEEE TRANS. ON PATTERN ANALYSIS AND MACHINE INTELLIGENCE 1148-1161 (1993).

[16] Richard P. Wildes, *Iris Recognition: An Emerging Biometric Technology*, 85 PROC. IEEE 1348-1363 (1997).

biometric has several advantages; it is inexpensive, non-intrusive, and widely accepted. Furthermore, voice-based systems can be used from a distance, for example, through telephone lines. However, a voice-based system is not suitable for a large-scale identification application because voice patterns are not that distinctive. Other factors to consider about the use of voice recognition are possible performance degradations due to background noise, changes in microphone characteristics, and an ability on the part of certain people to mimic the voices of others.

2.2.7 Signatures

The use of handwritten signatures for authentication is a long established tradition. In the context of biometrics, an individual's signature is captured as it is being drawn. This allows dynamic information, for example pressure variation, stroke direction and speed, to be captured for feature extraction and template generation. The dynamic signature template size varies between 50 to 300 bytes. A person's signature can vary and is influenced by a number of factors including emotional and physical well-being. Although dynamic signatures are less prone to forgery than static signatures, forgery is still possible.

2.2.8 Keystroke dynamics

Keystroke dynamics basically measure the speed, pressure, and time gap between successive key-hits to identify/verify a person. The main attraction of this behavioral biometric attribute is that it can be measured unobtrusively. The technology is mainly suitable for verification as it is not discriminatory enough for use in one to many searches.

2.3 Performance Measures

The performance of a biometric system is specified through three performance measures.[17] These are: *false match rate* (FMR), *false non-match rate* (FNMR), and *failure to enroll rate* (FER). The false match rate, also known as Type II error, is an indicator of how often the system will match a subject with someone else's template. The FMR is primarily dependent on the distinctiveness quality of the biometric attribute used by the system. The

[17] W. Chen, M. Surette, and R. Khanna, *Evaluation of Automated Biometrics Based on Identification and Verification Systems*, 85 PROC. IEEE 1464-1478 (1997)

false non-match rate, also known as Type I error, of a system is indicative of how often the system will fail to verify a match when it does really exist. The FNMR depends upon the robustness of the biometric attribute used by the system. The actual FMR and FNMR exhibited by a system depend upon a threshold setting that determines the tradeoff between FMR and FNMR. The failure to enroll rate tells how often the system will fail to enroll a person because of unacceptable quality of the biometric sample. The FER is directly related to the universality and accessibility properties of the biometric attributes used by the system.

Getting a realistic indication of performance measures for existing biometric technologies and systems is difficult; vendor estimates are always optimistic and early adopters of the technology generally tend to under-report the real error rates. Furthermore, different vendors use different data sets to assess their systems; thus, making it difficult to perform a meaningful comparison. Sensing a need for a fair comparison, a number of efforts have taken place. For example, a national biometric test center[18] has been established at San Jose State University through federal funds for evaluating different authentication technologies. Another example is the testing done by IBG, which regularly conducts comparative biometric testing using protocols that correspond to realistic usage settings and cooperative users with little or no biometric experience. According IBG, many systems now offer accuracies well above 95%, although performance varies significantly across technologies. Fingerprint systems, for example, provide near zero error rates, while systems using face recognition technology tend to show error rates in the range of 30-40%. The FER for some systems have been found to be as high as 12%. The testing also shows performance deterioration over time.

2.4 Ways of Deploying Biometric Systems

There are several different ways in which a biometric system can be deployed subject to the constraints of the underlying technology. Dr. Jim Wayman, Director of the National Biometric Test Center at San Jose State University, lists seven different modes of applying or characterizing a biometric application.[19] These are: (i) *overt or covert*, (ii) *cooperative or non-cooperative*, (iii) *habituated or non-habituated*, (iv) *supervised or*

[18] J.L. Wayman, *Fundamentals of Biometric Technologies*, *in* BIOMETRICS: PERSONAL IDENTIFICATION IN A NETWORKED SOCIETY, (A. Jain, et al, eds., 1999).

[19] J.L. Wayman, *Fundamentals of Biometric Technologies*, *available at* http://www.engr.sjsu.edu/biometrics/publications_tech.html.

unsupervised, (v) *closed or open*, (vi) *public or private*, and (vii) *standard or nonstandard* environment.

The overt or covert mode of an application refers to whether the biometric data is being captured with the consent and awareness of the person subject to authentication or identification. The most common mode of biometric applications is the overt mode. Authentication applications are almost always in overt mode. A biometric system based on face recognition offers a convenient way of operating in covert mode because a person's face can be easily captured without the person ever becoming aware of being photographed.

The distinction between cooperative and non-cooperative deployment of a biometric system refers to how an imposter will try to defeat the system. Will the imposter gain access by cooperating with the system to get a positive match so as to fraudulently access a secured asset such as a bank account? Or, will the imposter try to avoid a match by not cooperating with the system?

Habituated versus non-habituated deployment of a biometric system refers to the comfort of the intended users. A habituated deployment implies that users come in contact with the system frequently while a non-habituated deployment means users occasionally access the system. The frequency of contact with the system obviously plays a role in the choice of the biometric attribute that should be selected, as well as in determining the performance of the system.

The supervised versus unsupervised distinction concerns whether an operator is present to ensure the proper capturing of biometric data. This distinction is also referred to as attended versus unattended.

The closed versus open nature of a deployment refers to how the biometric templates are going to be used. A closed deployment means that the template database is used only for a single application for which it was created. An open deployment, on the other hand, means that the template database is available for multiple applications, for example for access to buildings as well as to a computer network.

The public or non-public distinction refers to the relationship between the users and the management of the biometric system. If the system is being operated for employees of an organization, it is a private deployment. On the other hand, it becomes a public deployment if the users are the clients or customers of the organization. Public versus non-public deployment basically determines the attitudes of the users towards the biometric system and consequently its performance.

The standard versus non-standard distinction tells us about the deployment environment. If the system is deployed in a controlled environment, for example indoor with proper lighting and temperature control, it is considered a standard application; otherwise it is a non-standard application.

3. EXAMPLES OF BIOMETRIC APPLICATIONS

Biometric applications can be grouped into the following broad categories:[20]

- Access control and attendance. For example, controlling access to physical locations or replacing time punch cards to monitor employees time-in and time-out.
- Computer and enterprise network access control.
- Financial and health services. For example, controlling access to ATMs, or limiting access to health records to comply with privacy laws.
- Government. For example, immigration and border control to verify travel documents, or fraud prevention in social services.
- Law enforcement. For example, surveillance, criminal identification, or drivers' licenses.
- Telecommunications.

The market share for different biometric technologies for the year 2004 for which the data is available from International Biometric Group (IBG) is shown in Figure 7-4.[21] About half of the market share is dominated by fingerprint technology. Some specific examples of applications for selected biometric technologies are provided below.

[20] Biometrics Consortium, *available at* http://biometrics.org.
[21] *See supra* note 5.

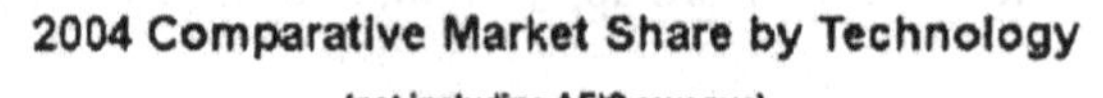

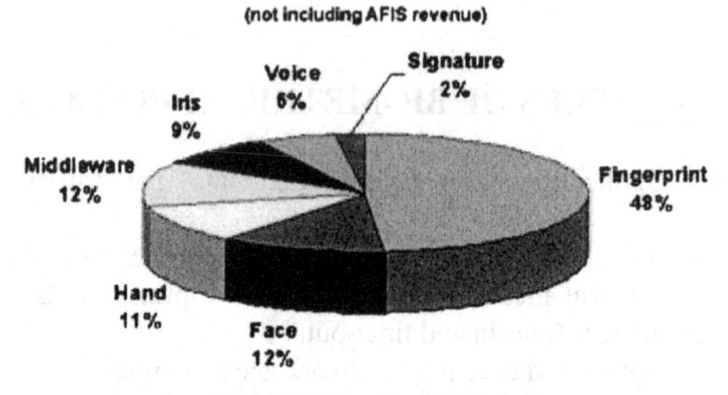

Figure 7-4. Comparative market share of different biometric technologies

Iris recognition technology is being used for passenger authentication programs at selected airports in the U.S., U.K., Amsterdam, and Iceland. At border crossings in Canada, the Netherlands, and the United Arab Emirates, it is being used to help registered travelers cross borders without passing through lengthy immigration checks and lines. Some correctional facilities in the U.S are using it to identify inmates. Iris recognition technology is also being used in Pakistan to identify and track Afghan refugees in the camps run by the Office of the United Nations High Commissioner for Refugees (UNHCR).

Fingerprint technology applications abound in access control, be it physical access to restricted buildings or logical access to enterprise networks and PCs, and to authorize transactions. Fingerprint technology is also prevalent in tracking employee attendance. The US Department of Homeland Security is using it in the US-VISIT system to register and verify visitors from abroad. The US-VISIT entry procedures are currently in place at 115 airports and 14 seaports. By December 31, 2005, US-VISIT is expected to be operational at all 165 land ports of entry. Almost all fingerprint system deployments are for authentication. One of the reasons for fingerprint dominance is the small size of the sensors which allows them to be embedded in devices such as cell phones and keyboards. In some applications, fingerprint technology is also being used for person identification using one-to-many search techniques. For example, the state of Connecticut is using fingerprint-based identification in real time to prevent persons from receiving welfare benefits under multiple identities.

Face recognition is being deployed in screening, surveillance, and law enforcement applications. For example, immigration officials in Massachusetts are using it to prevent duplicate green card applications. In Pinellas County in Florida, face identification is being used to identify criminals upon arrest by matching their faces with face images stored in the system.

A well-known application of hand geometry technology is the INSPASS (Immigration and Naturalization Service Passenger Accelerated Service System) project. The project is a multi-airport project to allow frequent travelers to skip long lines for immigration at international airports in Los Angeles, Miami, Newark, N.J., New York City, Washington, San Francisco, Toronto, and Vancouver. Those enrolled in the program carry cards encoded with their hand geometry information, which they swipe at designated terminals, where their hand geometry is captured and matched with the swiped information. INSPASS has about 50,000 flyers enrolled, and approximately 20,000 verifications take place every month. Hand geometry readers are also in use at San Francisco International Airport to secure 180 doors and verify over 18,000 employees everyday.

4. HOW SECURE ARE BIOMETRIC SYSTEMS?

With so much emphasis on security, it is important to know if biometric systems can be fooled. Since a biometric attribute is measured from a live individual, it might seem that faking a live biometric attribute would be difficult. Unfortunately, many recent studies and demonstrations have shown the vulnerability of many biometric technologies to modest efforts on the part of a person trying to defeat the system. The term *spoofing* is used to demonstrate the fallibility of biometric systems through fake biometric samples. Some of the documented spoofing methods include molding fingerprints from gelatin, activating leftover fingerprints by breathing on the sensor, using a face image, and superimposing iris images atop human eyes. One way to minimize spoofing is to incorporate robust methods for "live-ness" detection, *i.e.,* ensuring that the biometric sample is coming off a live person. Live-ness detection can be accomplished through several means including the comparison of multiple recordings of the biometric attribute separated by very short time intervals, detecting the temperature of the user, and penetrating surface layers of skin.

Another way to compromise the security of a biometric system is to reverse engineer the templates to generate input images, for example, face images. While the industry would have us believe that reverse engineering of templates is not possible, there are reported cases of reverse engineering of templates. In a widely publicized case from Australia, an undergraduate computer science student at Australian National University reverse engineered a fingerprint system as part of his honors thesis in 2002.[22]

One way to improve security is to have systems that use multiple biometric attributes.[23] Another possibility is to combine biometrics with smart cards that store biometric information from the enrollment time along with other information, such as a password or a PIN. Since there is no central storage of templates, smart cards are less susceptible to template reverse engineering.

5. SOCIETAL ISSUES

Biometric technology must be implemented with full awareness of various issues and concerns from a privacy and societal perspective. People want to go about their lives anonymously in the eyes of the government and other organizations. We as a society highly value individual privacy, both physical and informational. Technologies such as biometrics are perceived to impinge upon privacy and thus are often viewed with suspicion.[24]

A report from Rand Corporation on the use of biometrics identifies three categories of societal concerns.[25] These are: *religious*, *informational privacy*, and *physical privacy*.

Religious concerns stem from objections that some potential enrollees in a biometric program may have. Such objections might arise from one or more religious groups. For example, certain Christian groups consider biometrics a "Mark of the Beast" based on language in the New Testament's Book of Revelation. Religious objection led Alabama to abandon a program to include biometric identifiers in drivers' licenses. Although religious objections to biometrics may not be widespread, such objections need to be taken seriously because of respect for sincerely held religious beliefs.

Informational privacy concerns relate to three issues: function creep, tracking, and identity theft. *Function creep* is the term used to denote the process by which the use of information is slowly enlarged beyond the original purpose for which it was obtained. Function creep may be harmful

[22] *Id.*

[23] L. Hong and A. Jain, *Integrating Faces and Fingerprints for Personal Identification*, 20 IEEE TRANS. ON PATTERN ANALYSIS AND MACHINE INTELLIGENCE 1295-1307 (1998).

[24] S. Garfinkel, DATABASE NATION: THE DEATH OF PRIVACY IN THE 21ST CENTURY, (2000).

[25] John D. Woodward, Jr., Katharine W. Webb, Elaine M. Newton, M. Bradley, and D. Rubenson, *Army Biometric Applications: Identifying and Addressing Sociocultural Concerns*, Rpt. MR-1237-A (Rand Publications, 2001.)..

or harmless and could occur with or without the knowledge of the persons providing the information. The use of social security numbers in all kinds of applications is the best-known example of function creep.

Tracking refers to the possibility of linking different transactions to build profiles. With the growth of biometric technologies, tracking places an enormous amount of power in the hands of government and others with access to records to build detailed profiles for individuals. Since some biometric technologies can be used covertly, it is also possible that individuals could be tracked clandestinely.[26] One such widely publicized case occurred during the Super Bowl[27] a few years ago in Tampa Bay, Florida. Law enforcement authorities scanned the crowd using face recognition technology in an attempt to identify criminals and terrorists.

Although the use of biometrics is expected to minimize identity theft, its benefits are still unclear because many biometric systems allow remote authentication. Furthermore, spoofing can also lead to identity theft.

Physical privacy concerns center around three issues. First, there is the issue of cleanliness and hygiene. Users may not feel comfortable getting close to sensors or coming in contact with them knowing that many other users have done so, and may worry that such contact may cause infection. The second physical privacy issue is that of a potential harm ensuing from being scanned for a biometric measurement. Although biometric technologies are known to be fully safe, it is not difficult to perceive a potential harm. For example, a user might perceive a danger to her eyes while they are scanned. The third physical privacy issue is cultural; there might be a stigma attached to the capture of biometric information of certain kinds, such as fingerprints. Overall, physical privacy concerns are easier to deal with than some other issues because they may be addressed through proper education and training.

While many privacy concerns related to biometrics are no different from concerns related to the use of other kinds of identifiers, such as PINs and SSNs, there are some important differences. For example a bank can issue a new PIN in case of an identity theft. Such a simple recourse is not possible in the case of a biometric identity theft. There must be a provision in place for using an alternative biometric attribute. The other important difference is that some of the biometric information can be captured covertly; this makes it feasible to track people without their consent or knowledge. Furthermore,

[26] John D. Woodward, Jr., *Biometrics – Facing up to Terrorism*, Rpt. IP-218, (Rand Publications, 2001).

[27] John D. Woodward, Jr., *Super Bowl Surveillance: Facing up to Biometrics*, Rpt. IP-209 (Rand Publications, 2001).

biometric identifiers are permanent; thus making it impossible to avoid being tracked.

6. SUMMARY

Biometric technology is rapidly maturing and is poised for significant growth in coming years. While the government sector has been the early driver of biometrics, most growth in the next few years is expected to come in the financial, health care, and travel and transportation sectors. Efforts aimed at development and adoption of industry standards are expected to provide a further impetus to the growth of biometrics. To gain a wider public acceptance, there must be efforts aimed at educating people. At the same time, current legal safeguards need to be reviewed to ensure the privacy rights of individuals.

Chapter 8

BIOMETRICS: APPLICATIONS, CHALLENGES AND THE FUTURE

Gang Wei[1] and Dongge Li[2]
[1]Accenture Technology Labs, 161 N Clark, Chicago, IL, 60614; [2]Motorola Labs, 1301 E. Algonquin Rd., Schaumburg, IL 60196

Abstract: Biometrics refers to the science of identifying or verifying individuals based on biological or behavioral characteristics. It is one of the fastest growing technology areas and has the potential to impact the way we live and do business, and to make the world safer and more convenient. However, the technology also raises technical and social issues that must be addressed, such as concerns about insufficient accuracy and privacy. In this chapter, we describe how different biometrics methods work and discuss their respective strengths and weaknesses. A major challenge facing biometrics systems today is the threat of circumvention by criminals. This chapter points out techniques being used to spoof biometrics systems and the countermeasures being taken to prevent them. As facial recognition is considered to be the most non-intrusive and promising biometrics method, this chapter uses it as a case study for a detailed discussion. Despite the difficulties and challenges today, we believe that biometrics technology will gain acceptance and be pervasive in applications in the near future.

Key words: biometrics, person identification, security, access control, fingerprint, iris, facial recognition

1. INTRODUCTION: A SAFER AND EASIER LIFE WITH BIOMETRICS

David arrives at the office building in the morning, and realizes that he left his access badge at home. No problem. At the entrance, he puts his head in front of an iris scanner, and is immediately cleared to enter the building. After lunch, David goes to the coffee shop. While he waits in line, the surveillance system in the store identifies him as a loyal customer and automatically orders his favorite latte for him. In the meantime, his colleague Sarah walks into his office to have a word with him. The camera on top of his desk takes a picture of her face, matches it to David's, and identifies her as a visitor. Instantaneously, the picture is sent to David's cell phone. David runs back to his office to meet Sarah, dropping his credit card on the floor in the rush. Some crook finds it, and decides to try it first at a gas station. Fortunately, each pump in this station is equipped with a fingerprint reader. As the thief hesitates, the pump starts to beep, and he runs away. The credit card company is notified immediately about possible fraud.

Sounds too good to be true? It probably is, as the core enabling technology, namely biometrics, is not yet mature. However, with rapid technological progress, combined with the proliferation of sensors that are getting cheaper and smarter, all of the above scenarios may turn into reality soon. Even today, some pioneers in this field have built prototypes and products that illustrate how biometrics will shape our future.

Biometrics is the science of measuring and statistically analyzing biological data. In information technology terms, it refers to automated methods of identifying or verifying individuals based on biological or behavioral characteristics. The most commonly used characteristics include facial features, fingerprints, the iris, and the retina, while additional characteristics being proposed and tested by industrial and academic researchers include hand geometry, gait, and vein patterns. The biometrics industry is growing fast with the potential to have an impact on applications in security, retailing, and access control. According to the International Biometrics Group, revenues worldwide from biometric technology could reach $4.6 billion in 2008, up from $1.2 billion in 2004.[1] Advances in biometrics technology will have a profound and extensive impact on how individuals live and work, as well as on how businesses and governments operate.

Despite its brilliant future, biometric technology faces several challenges. In Section 2, we will introduce how different biometrics technologies work, and describe their respective strengths and weaknesses. Section 3 will discuss criteria by which to evaluate biometrics methods. Section 4 presents techniques that are being used to spoof biometrics systems and counter-measures used to prevent spoofing. Since person identification through

[1] International Biometrics Group, BIOMETRICS MARKET AND INDUSTRY REPORT 2004-2008 (2004), www.biometricgroup.com/reports/public/market_report.html.

visual analysis (e.g., facial recognition) is considered to be the most non-intrusive and convenient biometric technology, Section 5 is devoted to a case study of facial recognition technologies.

2. AN OVERVIEW OF BIOMETRICS TECHNOLOGIES

A typical biometric system consists of two steps, Enrollment, and Matching. In Enrollment, a biometric sample of a person is captured by sensors. Then the system extracts certain features of the sample, which are stored in a database. Matching is used to determine the identity of a person, or verify that someone is indeed who he or she claims to be. To perform matching, the system takes samples of the same biometrics from the person and extracts the features. The remaining processing depends on the task. For the purpose of identification, the features will be compared with all the records in the database to see if there is a match. In contrast, for verification, the system will retrieve the features of a single person and perform a one-to-one comparison.

Many biometric technologies have emerged in recent years, varying in the theories underlying them, and in aspects such as reliability, acceptability by the public, convenience of use, and application scope. Biometrics can be categorized into two general classes, namely 1) physical characteristics, such as fingerprints, faces or iris patterns; and 2) behavioral features, such as voice, signature, or walking patterns (gait). This section gives a brief introduction to the commonly-used biometrics and how they can be used to identify people.

Fingerprint. Fingerprinting is the most mature and commonly used biometrics technology. A person's fingerprints are genetically decided and remain the same for his or her entire life unless severe injuries occur. Fingerprint identification functions by detecting and matching either the distribution of minutiae or the ridges/valley patterns of fingerprints.[2] It is commonly used as strong evidence in criminal identification and immigration management. As the hardware and software used for fingerprint reading and recognition are getting cheaper, and because of the convenience of use, fingerprinting is gaining increasing popularity in access control, especially for personal property protection. For example, in Asia

[2] A. Ross and A.K. Jain, *Biometric Sensor Interoperability: A Case Study in Fingerprints*, 3087 PROC. OF INT'L ECCV WORKSHOP ON BIOMETRIC AUTHENTICATION 134-45 (2004).

and Europe, many high-end cell phones are equipped with fingerprint readers. According to AuthenTec, a biometrics service vendor, fingerprint scanners are one of the fastest growing segments in the cell phone industry. In the U.S., most Pocket PCs and some new laptop models also have fingerprint readers. In 2004, fingerprinting accounted for 48% of the revenue of the biometrics market and its share will keep growing in the coming years.

While considered to be one of the most authoritative types of biometric evidence, fingerprint matching is not error-proof. As revealed by the Fingerprint Vendor Technology Evaluation by the National Institute of Standards and Technology (NITS) in 2003,[3] although the accuracy of leading fingerprint matching technologies is close to 100% for high-quality images, it drops dramatically as image quality degrades. This degradation could cause serious problems for criminal identification. For example, in 2004 the false matching of fingerprints led to the mistaken arrest of a lawyer in connection with the investigation of train bombings in Madrid, Spain. Moreover, the public is often concerned about privacy and liberty and is reluctant to use fingerprints for identification and verification..

Other technologies that have similar features, advantages, and drawbacks include hand geometry, vein patterns, nail-bed recognition, and the sound transmission characteristics of finger bones.

Iris pattern. The iris pattern is the biometric characteristic with the highest discriminating power. In 2001 it was used to identify an Afghan girl whose picture was shown on the cover of National Geographic 16 years earlier. An iris-based person identification system works by encoding the irises as sequences of 0's and 1's and then comparing these sequences. Authenticating a person involves photographing his eye with a camera that is within 3 feet of the person. Then the system detects the boundary of the iris and applies a 2-D Garbor wavelet operation to perform the encoding. Finally the sequence is matched against the records in the database. Several organizations conducted evaluations of iris-based person identification systems from 2001 through 2003, and no false match has ever been found among the hundreds of millions of tests.

Retina pattern. The blood vessel patterns on retinas can also be used to distinguish people. This technology has comparable accuracy to that of iris pattern matching. It requires the camera to be within 2 inches of the eye, making collecting the sample more intrusive, but the technology is less susceptible to theft due to surreptitious capture of the biometric sample.

[3] National Institute of Standards and Technology, FINGERPRINT VENDOR TECHNOLOGY EVALUATION 2003: SUMMARY OF RESULTS AND ANALYSIS REPORT.

Skin spectrum. Different persons' skins show different absorption spectrums.[4] Therefore, we can identify individuals by sending a low power beam of light to a certain part of the skin and analyzing the reflected light. The advantage of using skin spectrum is the convenience of use. However, the skin characteristics of a person may change with age, injury or sun tans, creating a constant need to update the stored profiles.

Ear pattern. Recognizing a person by the ear may sound funny, however it is useful for some occasions when other biometrics are not available.[5] A straightforward use is to compare the geometric features of the ear by image processing approaches. Alternatively, the acoustic response of the ear canal could be used. This technology faces fewer privacy concerns and is easier for the public to accept because people usually don't consider the ear as a critical biometric feature. Applications are limited because the ears are not always visible due to hair styles or hats.

Gait. In a crowded street, we can often recognize friends from a distance by how they walk. Today, researchers are developing automatic gait-based person identification methods.[6] Although not extremely accurate, they show great long range potential because of their non-intrusiveness and long range applicability.

Face. Identifying people by face has been an active research topic for the past decade. Although progress has been slow due to the complexity of the problem, it is still one of the most promising biometric technologies. Faces are the most visible and available biometric identifiers and usually people show little resistance to face recognition systems in public. With the increasing ubiquity of image capture devices such as digital cameras, camcorders and surveillance cameras, person identification by face will be the fastest growing area and most widely used technology in practical applications. In Section 5 we will describe facial recognition technology in more detail.

Voice. With the development of audio processing algorithms, automatically identifying people by their voices is becoming possible. Known as speaker identification or voice recognition, such systems usually

[4] R.K. Rowe *et al.*, BIOMETRIC IDENTITY DETERMINATION USING SKIN SPECTROSCOPY, http://www.lumidigm.com/PDFs/Biometric%20Identity%20Determination%20using%20Skin%20Spectroscopy.pdf.

[5] M. Burge and W. Burger, *Ear Biometrics, in* BIOMETRICS: Personal Identification in a Networked Society 273-86 (A. Jain et al. eds., 1998).

[6] J. Boyd and J.J. Little, *Shape of Motion and the Perception of Human Gaits*, 98 WORKSHOP ON EMPIRICAL EVALUATION IN COMPUTER VISION, (CVPR 1998).

capture the voice of a person through a microphone, then extract certain features from the signal and compare them to those of known persons in the database. The most commonly used feature is the voice spectrum, while more sophisticated analyses can be used to improve robustness.[7] This method is convenient for access control; however the accuracy could suffer from background noise. People also are sensitive about their talk being recorded.

3. A COMPARISON OF DIFFERENT BIOMETRICS TECHNOLOGIES

Given the numerous person identification methods using biometrics, the reader may wonder which one is the best. Each biometric has its strengths and weaknesses and the choice depends on the application. As mentioned in Section 2, different biometrics vary in terms of accuracy, range of operation, and convenience. The choice of technology will depend on the requirements of a particular application. For example, in access control systems, fingerprints or iris scans should be used if security is very critical; otherwise the more convenient but less reliable facial or voice recognition could be a suitable alternative. To identify criminals at airports, the most feasible approach would probably be facial or gait recognition. There are seven criteria commonly used to evaluate the appropriateness of a biometric technique:

• Universality: Do most people have this trait? For example, recognition by fingerprints may not apply to some handicapped people.

• Uniqueness: Is everyone different in this biometric feature? A person's height and weight are examples of biometrics that lack uniqueness. The other extreme is the iris characteristic, which can be used to distinguish anyone in the world.

• Performance: Accuracy and robustness. Although some biometrics methods, such as facial recognition, have high power to distinguish people theoretically, they are still struggling with bad accuracy in practice.

• Permanence: Many biometrics can change over time or be affected by injuries or other factors. For example, the voice of the same person could be different at different ages, or when he or she catches a cold. In contrast, iris and retinal blood vessel patterns remain the same for life.

[7] B. Raj *et al.*, *Vector Polynomial Approximations for Robust Speech Recognition*, PROC. ETRW (1997).

• Collectability: Can the feature be measured and extracted consistently and easily?

• Acceptability: Are people willing to accept it? Some biometrics capturing devices are considered to be more intrusive than others. For example, in public places such as offices and waiting rooms, people are usually more sensitive to microphones for voice recognition than surveillance cameras.

• Circumvention: Is it easy to fool the system? Almost every biometrics system can be breached, however the cost and difficulty varies, and therefore this is a factor to be considered in applications in which the desired security level is high.

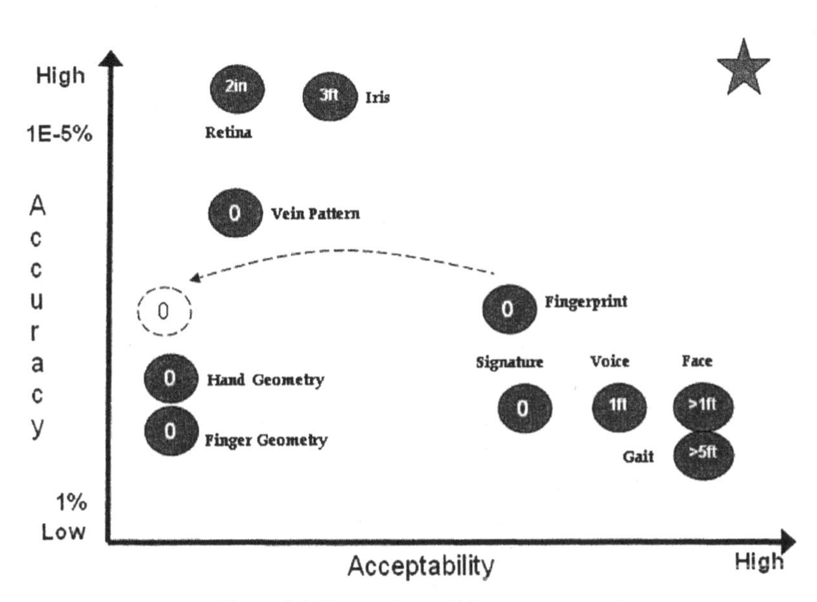

Figure 8-1. Comparison of biometrics methods

Among the above 7 dimensions, the two most important are usually performance and acceptability. Figure 8-1 shows how different biometrics methods are positioned with respect to these two dimensions.

In Figure 8-1, the horizontal axis is the acceptability and the vertical axis is the accuracy of the method that can be achieved by today's technology. The numbers in the circles correspond to the working distances of these methods. An ideal biometrics system should be located at the upper-right corner. Fingerprint-based systems feature well-balanced accuracy and acceptability and are the most widely-used biometrics methods today.

Sometimes the public is not willing to accept fingerprinting because of "privacy concerns." This fact is shown by the dotted arrow in the figure. However, such psychological distaste is likely to fade away as fingerprint systems gain more popularity. It is also worth noting that biometrics methods at the two extremes, *i.e.*, those with high accuracy versus those with high acceptability, are both promising in the future. Iris or retina methods will play essential roles in applications with high security levels, such as military facility access. Biometrics such as facial and gait recognition will be very useful for public area safety because of their long operational range and non-intrusive style. In addition, the face and gait features have very high uniqueness to human perceptual systems? This means that there is still a lot of room to improve the accuracy of existing algorithms, at least in principle. Therefore, such biometrics methods are attracting a lot research interest. We will single out facial recognition for discussion in Section 5.

4. CIRCUMVENTIONS AND COUNTER-MEASURES

While biometric technology has the potential to make the world safer, and life more convenient, it also faces constant circumvention attempts from intruders. The providers of biometric systems have to consistently patch loopholes to foil these spoofing attempts. In this section we will talk about techniques being used to attack biometrics systems and the counter-measures being taken by technology providers to prevent them.

Fingerprints. Two types of devices are commonly used to acquire fingerprints, namely the touch scanner and the swiping scanner. In the former, the user simply puts his finger on the sensor pad. Although convenient, the touch scanner is easily attacked. Such sensors usually work by measuring the capacitance changes between the skin and surface of the sensor caused by the ridges and valleys on the fingerprints. As the user presses the sensor, his fingerprint is printed on the pad, and can be easily "lifted" from the pad by ID thieves. To make things worse, if a criminal just wants to gain access to such a system, he doesn't even have to steal anyone's fingerprint. Instead, there are several handy methods of "Latent Image Reactivation," which use the fingerprint trace left on the sensor by the previous use. To reactivate the sensor and gain access, a potential intruder can 1) breathe on the sensor; 2) put a thin-walled plastic bag of water onto the sensor surface; or 3) dust the senor pad with graphite powder, then attach a clear tape to the dust and press down the sensor. The third approach guarantees almost 100% success rate in many systems.

Swiping scanners require the user to sweep his finger across the scanner bar. This is not a very natural way to capture the fingerprint, however the sensing area is smaller compared to touch pads and the cost is lower. In addition, when sweeping is used, there is no fingerprint left on the scanner and thus Latent Image Reactivation methods don't work with swiping scanners. That's why many electronic devices such as the latest IBM Thinkpad T42 employ swiping scanners for user authentication. Nevertheless, it is still possible to break into such systems if a user's fingerprint model is available. The intruder can map such a fingerprint onto a 3-D mould and create an artificial fingertip. As described in,[8] making a "gummy" finger with plastic and silicone materials usually costs just a few dollars and takes less than 20 minutes. Such a "gummy finger" can fool virtually all fingerprint identification systems, except those combined with thermal sensors.

Iris patterns. Although the iris pattern is considered to be the most reliable biometric, iris-based identification systems are not immune to spoofing. An easy way is to take a good snapshot of a user's eye, then print a high-quality picture at the real size of the eye on inkjet paper. The tricky part is that iris systems detect the iris area by locating the bright spot at the center of the pupil, and that spot printed on the paper is not bright enough. To deal with this, an intruder can cut a miniature hole in the paper at the appropriate spot, and hide behind the paper to let his own pupil shine through.

As even the most accurate biometrics like fingerprint and iris systems can be fooled, other less stringent technologies are even more vulnerable. For example, an attacker can record the voice of a user and play it to fool a speaker identification system. Emulating one's signature or walking is not a hard task for a skilled thief. Facial recognition? Given the robustness of today's technology, a little disguise could fool the most advanced facial recognizers.

As countermeasures, several approaches have been proposed to protect biometrics systems from spoofing efforts. A popular technique known as liveness testing determines if the captured biometric feature is really from a live source. For example, artificial fingers don't moisturize due to perspiration the way that real human fingers do. Based on this observation,

[8] T. Matsumoto *et al.*, *Impact of Artificial "Gummy" Fingers on Fingerprint Systems*, 4677 PROC. OF SPIE (2002).

Schuckers *et al*[9] proposed the analysis of moisture patterns to determine the liveness of a fingerprint. Other examples include the detection of blood pulses or motion. The above methods are called passive liveness testing because they passively collect liveness evidence in addition to the biometric features. An alternative way is to give the person a certain stimulus and check if he has the normal response. For instance, a voice recognition system can randomly pick a combination of words for the user to read. This can prevent the use of recorded voices to fool the system. And in an iris authentication system, we can flash a light at the eye of the user and check if his pupil size changes if he blinks. Such methods are called challenge-response processes.

However, the liveness testing technology is not mature yet. It also introduces costs by requiring additional expensive sensors and stimulus devices. More importantly, liveness testing is an additional verification on top of the biometrics authentication instead of an integral part of the entire system. Such an ad-hoc combination leaves it possible for intruders to find smart ways breach them separately. A more promising anti-counterfeit strategy is multi-modal, fusing more than one piece of biometric evidence collected by heterogeneous sensors in a systematic fashion to make the most accurate and reasonable decision. For example, fingerprint recognition could be used at the same time as skin spectrum recognition, or facial recognition together with an iris scan. In the face recognition method proposed by Li *et al*,[10] a calibrated camera and a weight sensor were used to estimate the height and weight of the person presenting biometric identification, which dramatically improved the reliability of the system. A more loosely-coupled multi-modal architecture is to enable communication between different biometrics systems based on certain domain knowledge to spot possible frauds. For example, if someone uses his fingerprint to withdraw cash at an ATM in Los Angeles, and an iris recognition system at a nuclear plant in up-state New York detects the same person half an hour later, we know something is wrong. As more biometric features and other information sources are taken into account, it becomes increasingly difficult for the intruder to break in.

The multi-modal architecture is not only effective as countermeasures for intrusions. By analyzing and integrating the perceptions of multiple sensors

[9] S. Parthasaradhi, *Time-Series Detection of Perspiration as a Liveness Test in Fingerprint Devices*, IEEE TRANSACTIONS ON SYSTEMS, MAN, AND CYBERNETICS, PART C: APPLICATIONS AND REVIEWS (forthcoming).

[10] D. Li and B. Gandhi, *Identification Method and Apparatus*, U.S. Patent Application No. 20040234108 A1, Motorola Inc.

based on logic and domain knowledge, we can create more accurate and robust systems for access control and surveillance applications. As mentioned in the previous section, visual analysis-based biometrics systems, such as facial and gait recognition, are superior in acceptability and work range, and can be easily and consistently captured with the pervasive availability of cameras. However, such biometrics methods are struggling with bad accuracy today due to the limitations of computer vision technology. The multi-modal approach is a promising solution to this problem. For example, Li et al created a person identification system for TV programs using both facial and voice recognition.[11] With the rapid proliferation of sensors in public places, the ability to identify and track people and other objects creates tremendous opportunities for business and security applications.

Researchers at Accenture Technology Labs proposed a Bayesian framework to integrate the results from multiple cameras and other sensors for surveillance in complex environments.[12] This system includes 30 network cameras, a number of infrared sensors, and a fingerprint reader to cover about 18,000 square feet of office areas. The visual analysis agents monitor the video data from the cameras continuously and calculate the probability that each person will be at different places. The accuracy is low due to various factors, with recall and precision at 68% and 59%, respectively, due to low image quality from the surveillance cameras. The system fuses the results from all the cameras based on domain knowledge, such as the *a priori* probability of each person at different locations (*e.g.*, Jack is more likely to be seen near his own office than the others), and the transitional probability (*e.g.*, the same person cannot be seen at two different places at the same time, and if he is detected at one location, he is likely to stay in the same place or at nearby locations). By integration, the accuracy of the system was improved to 87% and 73% for recall and precision, respectively. The fingerprint reader and the infrared sensors give exact locations of a person; however, they only work occasionally when someone puts his or her finger on the sensor or wears the infrared badge properly. We are currently introducing such accurate but inconsistent information sources into our framework to further boost the performance.

[11] D. Li *et al.*, *Person Identification in TV Programs*, 10 J. OF ELECTRONIC IMAGING 930-38 (2001).

[12] V.A. Petrushin *et al.*, *Multiple-Camera People Localization in an Indoor Environment*, KNOWLEDGE AND INFO. SYSTEMS: AN INT'L J. (forthcoming 2005).

5. FACIAL RECOGNITION: A CASE STUDY

As the most acceptable biometrics method, facial recognition is the second most widely-used biometric technology after fingerprinting, with a projected revenue of $429 million in 2007. With the explosive growth of surveillance cameras, digital cameras and camcorders, facial recognition technology is becoming ubiquitously available. This technology received tremendous attention after the 9/11 tragedy as a possible measure to identify terrorists to improve public security. Such pioneering efforts started as early as January 2001, when a facial recognition system was installed at the Super Bowl in Tampa, Florida. In 2002, Boston's Logan Airport and Palm Beach International Airport both experimented with using facial recognition to find suspects. Unfortunately, these high-profile endeavors have not been very successful. Flooded with false alarms, all the above facial recognition systems were removed quickly. However, progress is still being made gradually, especially outside the U.S. Since November 2002, crew members have been accessing Sydney airport using a facial recognition system called SmartGate. So far nearly 70% (approximately 3000) of all Sydney-based aircrew members are enrolled. According to a recent survey of 530 crew members, 98% prefer SmartGate for border clearance, and 96% of air crew have no privacy concerns. The International Civil Aviation Organization (ICAO) has identified face recognition as the best-suited biometrics for airport and border control applications.

In addition, facial recognition is also being used for retailing and many other applications. This year, the technology vendor Logica CMG tested its facial recognition system for identifying shoplifters in a shopping mall in Rotterdam, and claimed a 90% success rate. It is commonly believed that with the development of computer vision technology, facial recognition will gain more popularity.

Most facial recognition algorithms can be categorized into two general classes, varying in the features on which they are based. The first class uses facial geometrical features. Sometimes multiple cameras can be used to build 3D models of the face to get better accuracy. The second class uses facial thermograms. The temperature patterns of the faces of different people are distinctive, and an infrared camera can capture such patterns. However, the prices of thermo-cameras are still prohibitive for most applications. There are also emerging approaches, including the use of skin details and of the wrinkle changes that occur when a person smiles.

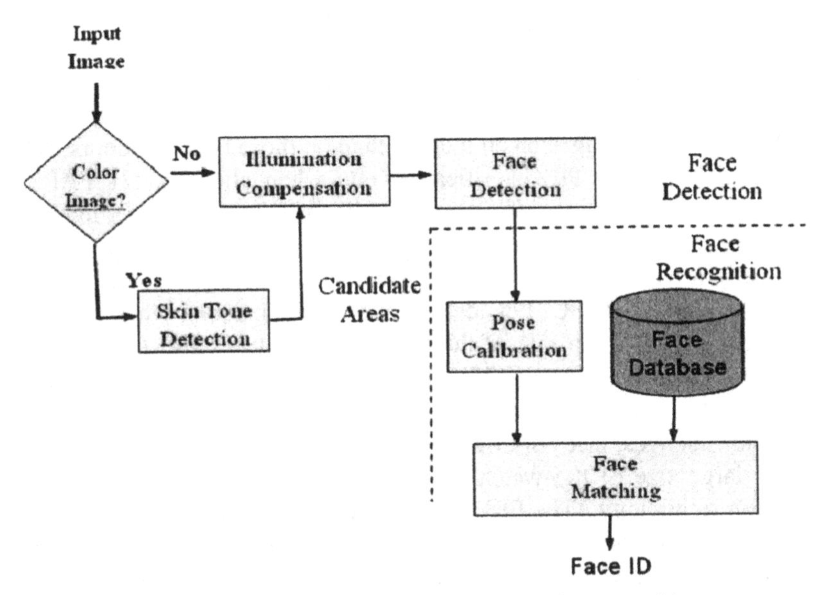

Figure 8-2. High-level block diagram of Motorola face recognition system.

Figure 8-2 shows a high-level block diagram of the system developed at Motorola, which also demonstrates the essential modules within a typical face recognition system. Facial geometry is the most commonly used approach. Due to the complexity of this feature, a promising method is to use machine learning techniques. The Motorola system was trained with 2186 frontal faces, and it uses bootstrap strategy for incremental learning to get 'smarter' after each round. Currently, the system has gone through its third round of learning. Both detection and recognition results are comparable to, if not better than, state of the art techniques. For face detection, the precision and recall are 92 % and 98 %, respectively, and facial recognition has a 94 % recognition rate when tested with 40 individuals.

The biggest obstacle facing massive deployment of facial recognition is its accuracy. To evaluate the state of the art of facial recognition systems, NIST organized the Face Recognition Vender Test (FRVT). The latest test was conducted in 2002.[13] False Rejection Rate (FRR) and False Acceptance

[13] P.J. Phillips *et al.*, *Facial Recognition Vendor Test 2002: Evaluation Report*, NAT'L INST. OF STANDARDS AND TECH. (2003).

Rate (FAR) were used to evaluate the performance of facial recognition systems from 10 participants, among which Cognitec (provider of the SmartGate system mentioned above), Eyematic, and Identix are the top three performers. The experiment used a large dataset – 121,589 operational facial images of 37,437 individuals. It revealed that most systems work much better on face images taken indoors than on those taken outdoors. The best system has about 90% verification rate when allowing 1% FAR in controlled indoor lighting conditions for verification purposes (*i.e.*, to check if the face belongs to a given person). However, the accuracy plunges dramatically when outdoor pictures are used, where the verification rate falls below 50% with 1% FAR. For identification tasks indoors, when allowing 1% FAR, the recognition rate of the best system reaches above 75% when the person watch list is small (25). However, it degrades to barely above 50% when the watch list grows to 3000 people. FRVT revealed that illumination changes, faces of different poses, and the low accuracy resulting from the large size of the watch list were the major challenges for facial recognition technology. The FRVT also uncovered several interesting facts. Usually male faces were easier to identify than female faces, with about 6% ~ 9% higher accuracy, and faces of elderly people were easier than those of younger people. Today's facial recognition technology works sufficiently well for verification and authentication indoors; however there is still a long way to go for surveillance tasks and person identification outdoors. Based on its historical development in the past few decades, the overall accuracy of facial recognition improves 3%~5% every 10 years.

6. CONCLUSION

Biometrics is one of the fastest growing technology areas. By enabling automatic person identification based on biometric characteristics, it can be used to improve security and make our lives easier and safer. Such systems are especially useful when it is impractical to hire people to perform the task, for example, finding criminals in crowded public places in a 24/7 fashion. As the sensors needed to acquire biometrics features get cheaper and smarter, this technology has a greater potential than ever to change how we live, work, and do business. However, biometrics technologies are still in their early stages and there are many challenges to be addressed, of which the most important two are the accuracy and acceptability of the technology. Non-intrusive biometrics methods, such as facial recognition, suffer from low reliability, while accurate ones like fingerprint or iris recognition face public resistance. Nevertheless, with technology advancements, the accuracy will eventually improve to meet the requirements of most applications. On the other hand, as demonstrated by the popularity of credit cards, history tells us that people are willing to trade a certain degree of privacy for convenience and security. Therefore, we are optimistic about the prospects of biometric technology and envision a safer and more convenient world with the pervasive applications of biometric systems.

Chapter 9

CONSTRUCTING POLICY

The Unsettled Question of Biometric Technology and Privacy

Lisa S. Nelson

Graduate School of Public and International Affairs, University of Pittsburgh, Pittburgh, PA USA

Abstract: The creation of privacy legislation specific to biometric identifiers must answer the threshold question of whether such legislation would simply replicate or duplicate existing legislation. The specific privacy protections afforded under the auspices of the Constitution and current statutory privacy protections are certainly applicable to biometric identifiers. The question, however, is how effective the existing legislation will be in combating consumer fears and protecting privacy in the arena of biometric technology. Existing legislation must be evaluated not according to its own terms but according to the standards of protection that it seeks to uphold. New legislative efforts then must strive to effectuate the principles of protective legislation already in place while overcoming its shortcomings in practice. The question considered in this paper is whether biometric identifier privacy legislation is necessary given protections afforded by existing Constitutional and legislative protections. This paper also considers how proposed privacy protections – including consent and procedural due process guarantees – might apply to the deployment of biometric technology.

Key words: biometrics, Fourth Amendment, First Amendment, due process, Privacy Act of 1974, Computer Matching and Privacy Protection Act, Fair Information Practices

1. INTRODUCTION

Privacy concerns have been paramount in the debate surrounding biometric technology. Whether it is in media coverage of biometric technology, or in discussions of its use by members of the biometrics industry or in the halls of academia, privacy is presented as an obstacle to the successful deployment of biometric technology. Yet, it is unclear how privacy is implicated by the use of biometric technology. Part of the problem associated with defining the nexus between privacy and biometric technology is clearly defining privacy and then discerning when privacy is at risk with the use of biometric identifiers.

The most obvious starting point for defining privacy is, of course, legal doctrine. Immediately, however, it becomes clear that the legal doctrine of privacy, whether it be Constitutional or statutory, does not neatly address the issues raised by biometric technology. In fact, a strict reading of Constitutional law or statutory guidelines offers little insight into how biometric technology might be regulated under current doctrine. Despite the ill fit between the "law of privacy" and biometric technology, however, one thing is clear – privacy concerns continue to impede the successful implementation of biometric technology. Thus, the problem faced in this ongoing policy debate is acknowledging the ongoing concern of a loss of privacy while attempting to define the nexus between privacy and biometric technologies. This process is made problematic by the nature of privacy.[1]

In the promulgation of privacy policies and in the debate surrounding biometric technology, normative expectations of privacy are equally as powerful, if not more so, than Constitutional doctrine, statutory guidelines, and state constitutional principles. The distinction drawn by H.L.A Hart between factual and normative principles of privacy is helpful for teasing out the relationship between privacy and biometric technology.[2] According to Hart, the factual principles of privacy are represented by the legal doctrine of privacy and include physical privacy, decisional privacy and increasingly, information privacy and communications privacy. The normative principles of privacy, on the other hand, are perhaps even more pivotal in the public policy debates regarding biometric technology. Normative perceptions of privacy – the more ephemeral and less defined societal perceptions of privacy – are thus perhaps more central to the public debate surrounding biometric technology than legal doctrines.

The relationship between privacy and biometric technology is therefore contradictory. The deployment of biometric technology is largely unaffected by existing legal doctrine; yet normative concerns about loss of privacy remain obstacles to the successful deployment of biometric technology. Protecting privacy in the march to deploy biometric technology is more a

[1] *See* Alan F. Westin, PRIVACY AND FREEDOM 24 (1967) ("The reasons for protecting privacy tend to be familiar to citizens of liberal democracies; thus the specific functions that privacy performs in their political systems are often left unexpressed.")

[2] H. L. A. Hart, *The Ascription of Responsibility and Rights*, 49 PROC. OF THE ARISTOTELIAN SOC'Y 171–94 (1949); Eric Katsh, LAW IN A DIGITAL WORLD (1995).

matter of discerning societal perceptions as a basis for formulating public policy than it is of applying existing legal doctrine. The difficulty of applying legal doctrine is fitting the nuances of biometric technology to the complicated idea of privacy. As a starting point, it is necessary to clarify the purposes for which biometric identifiers are acquired and used.

The gathering of biometric identifiers in the initial process of either authentication or identification requires the individual to reveal a personal identifier, such as a fingerprint, iris scan, hand geometry or other identifier. Biometric identifiers can be captured and used for the purposes of negative identification or positive identification. In the process of negative identification, the captured biometric identifier of an individual is compared to an existing database to exclude an individual from its system. In a process of positive identification, an individual submits a biometric identifier which is then compared to the database of enrollees or registered samples.

Privacy, in its many different forms, is clearly implicated differently depending upon whether the individual willingly enrolls in a biometric system or whether his or her biometric information is gathered by surveillance. If biometric identifiers are used to catalogue individuals or used to increase the visibility of individual behavior or if the information is shared without an individual's knowledge or consent, then there may be some potential privacy concerns which arise independently of how the biometric identifiers are gathered. Beyond the concerns associated with the acquisition of biometric identifiers, are questions surrounding the handling and storage of biometric identifiers, the vulnerability of the data to theft or abuse, the error factor of false identifications or false rejections and possibility of mission creep once the biometric identifiers are acquired in the public and private sectors.

These are complex questions to answer, particularly because the legal protections afforded data like that of biometric identifiers are not well developed. Legislative efforts to create biometric privacy legislation are only now emerging – in Texas and in New Jersey, for example. The paucity of legal doctrine, however, does not preclude consideration of privacy principles which should be employed in the circumstances in which biometric identifiers may be used.

2. GUIDANCE FROM CURRENT PRIVACY LAW

2.1 Constitutional Privacy Protections and Biometrics

Illustrative principles of privacy protection are numerous in the history of American jurisprudence. Though the application of Constitutional law to the use of biometric identifiers may be putting the cart before the horse, it is important to understand how long standing principles of privacy protection may guide both our concerns regarding privacy and the creation of privacy protections. For this reason, the discussion begins with a consideration of Constitutional law as it has protected privacy.

Consider the situation of an acquisition of a biometric identifier without the individual's consent. This might occur in two different settings. One setting might be a surveillance situation. The use of facial recognition technology, for instance, presents this possibility. Setting aside the technological barriers (the difficulty of assuring a sufficiently good image of a face for use in a one-to-many matching situation) there are two potential privacy concerns under current Constitutional doctrine. Not only does the creation of a biometric record create a privacy concern, but the capture of the biometric identifier *potentially* raises Fourth Amendment concerns. While Fourth Amendment jurisprudence imposes limits on the ability of government to carry out surveillance, there are questions as to how well this doctrine of privacy accommodates new technologies such as biometrics. The problematic application of Fourth Amendment jurisprudence to biometric technology is largely defined by the limited notion of a search which does not easily translate to biometric surveillance.

2.1.1 Fourth Amendment Concerns and Biometrics

Technological surveillance, not only of the biometric sort, has posed problems for the interpretive balance between subjective and objective expectations of privacy which defines the scope of Fourth Amendment prohibitions against search and seizure.[3] This is true for several reasons. First, the privacy protections afforded by the Court's interpretation of the Fourth Amendment are diluted as the technology becomes more prominent. When a particular technology is "in general public use" an individual has a weakened claim to privacy. Under this standard, one might argue that as sophisticated technologies, such as biometrics, become more widely utilized, privacy protection diminishes. Second, biometric identifiers tend to rely on personal information that we regularly expose to the public. Whether it is a finger, an eye, a face or a voice, the Court has been reluctant to extend a reasonable expectation of privacy to those features which we regularly expose to the public. For instance, in *United States v. Dionisio* a grand jury subpoenaed individuals in order to obtain voice samples to compare with a

[3] Katz v. United States, 389 U.S. 347 (1967).

recorded conversation that was in evidence.[4] In ruling such a subpoena constitutional, the Court noted that:

> The physical characteristics of a person's voice, its tone and manner, as opposed to the content of a specific conversation, are constantly exposed to the public. Like a man's facial characteristics, or handwriting, his voice is repeatedly produced for others to hear. No person can have a reasonable expectation that others will not know the sound of his voice, any more than he can reasonably expect that his face will be a mystery to the world.[5]

It seems clear enough that those features which an individual regularly exposes to the public are not deserving of Fourth Amendment protection. The collection of biometric identifiers in a public setting does not implicate a privacy interest because the "search" takes place in an open forum where an individual does not retain a very high expectation of privacy.

What of the use of biometric identifiers in other settings? For instance, if iris scans are used to restrict access to schools or fingerprints are used to verify identity before boarding of an airplane, the interest in privacy is balanced by the public policy objective of secure airline travel or safety in the schools. Courts have allowed searches of passengers and baggage because the liberty interest of the right to travel is offset by the need to enhance security. In *United States v. Davis*, the Ninth Circuit Court of Appeals addressed the Constitutionality of airport security screening.[6] The court stated:

> The essence of these decisions is that searches conducted as part of a general regulatory scheme in furtherance of an administrative purpose, rather than as part of a criminal investigation to secure evidence of crime, may be permissible under the Fourth Amendment though not supported by a showing of probable cause directed to a particular place or person to be searched.[7]

The search is justified not because of probable cause but because of the policy objective of protecting the common good against harm. This logic is amplified in *United States v. Edwards*. There the court held:

> When the risk is the jeopardy to hundreds of human lives and millions of dollars of property inherent in the pirating or blowing up of a large

4 United States v. Dionisio, 410 U.S. 1, 3 (1973).

5 *Id.* at 14.

6 United States v. Davis, 482 F.2d 893, 895 (9th Cir. 1973).

7 *Id.* at 908.

> airplane, the danger alone meets the test of reasonableness, so long as the search is conducted in good faith for the purpose of preventing hijacking or like damage and with reasonable scope and the passenger has been given advance notice of his liability to such a search so that he can avoid it by choosing not to travel by air.[8]

Thus, the risk of a hijacking or a terrorist threat minimizes an individual's claim to privacy in the case of an administrative search. The use of biometric technology in this setting is likely to be interpreted as a mere extension of such a search which is necessary to enhance security. Similarly, courts have established that the mere use of fingerprinting for licensing, verification of identity, or employment purposes does not constitute a physical search. Mirroring the logic of an administrative search, courts have held that fingerprinting "does not create an atmosphere of general surveillance or indicate that they will be used for inadmissible purposes." [9] The courts, have, however, generally treated actual physical intrusions into the body, including the drawing of blood or urine analysis, as potential intrusions under the Fourth Amendment. The question not yet answered is whether a retinal scan might prove to be more analogous to a blood or urine test than to a voice sample. Could a retinal scan, in other words, represent a physical intrusion? Biometric technology may be interpreted as an enhancement of the senses in the same manner as a DNA test or urine analysis if the technology advances to the point that it potentially reveals sensitive information about the individual.

There is yet another potential intersection between the use of biometric technology and Fourth Amendment doctrine. On June 21, 2004, the Supreme Court, in *Hiibel v. Sixth Judicial District Court of Nevada,* upheld a conviction under a Nevada law allowing police to arrest an individual, without probable cause, for refusing to provide a name to a police officer.[10] On Hiibel's refusal, he was arrested, and later convicted, for obstructing police business. The Supreme Court rejected the Fourth Amendment challenge to the Nevada statute arguing that "the request for identity has an immediate relation to the purpose, rationale, and practical demands of a *Terry* stop."[11]

The debate in *Hiibel* centered upon whether the divulgence of personal information was appropriate given that probable cause had not been established. If probable cause is not a prerequisite, routine identification

[8] United States v. Edwards, 498 F.2d 496, 500 (2d Cir. 1974).

[9] Thorne v. New York Stock Exch., 306 F. Supp. 1002, 1011 (S.D.N.Y 1969).

[10] Hiibel v. Sixth Judicial Dist. Court, 124 S. Ct. 2451, 2455-56, 2461 (2004).

[11] *Id.* at 2459.

checks might devolve into a significant loss of privacy. This question, however, was not resolved. The opinion of the divided court was narrowly tailored to the statement of a name. There remains an open question with respect to the use of biometric identifiers as a form of personal information and identification. Clearly, the decision reached did not reach the issue of police access to other information, such as biometric identifiers, nor did it answer the question of under what circumstances possible suspicionless seizures are reasonable.

Despite the uneven fit between biometric technology and the protections afforded by Fourth Amendment doctrine, there are implications for the future of privacy policy which can be gleaned from this line of cases. First, courts are clearly reticent to allow technological surveillance which enhances the powers of observation and intrudes upon an individual expectation of privacy. Unclear, however, is how expectations of privacy are evaluated in a swiftly changing context of technological innovation. As technology becomes more prevalent and common in its general usage, is there a concomitant loss of privacy, as some cases suggest? Also, do individuals sacrifice privacy when they knowingly expose themselves to the public through interactions with technology such as internet communications, cell phones or simply walking down a street where facial recognition technology is being used?[12] How do these strands of Fourth Amendment jurisprudence apply to biometric technology?

The answer to the question of whether biometric technology violates privacy as it is defined under the auspices of Fourth Amendment jurisprudence is dependent on the visibility of the technology, the commonality of its usage, the context within which it is used, and the purpose for which it is used. Beyond the explicit doctrinal guidelines, however, there is something more to Fourth Amendment doctrine that is important to consider when discerning the privacy interests at stake. The concept of privacy in Fourth Amendment doctrine is influenced by a whole host of factors which have been framed by the courts as societal or objective standards of privacy. Because a reasonable expectation of privacy is defined not only by the individual's subjective perception but also by an objective standard of reasonableness, the idea of privacy is gleaned from the social and political mores of the times.

Similarly, not only is privacy dependent upon social mores, it is also juxtaposed against policy objectives. A privacy violation is unreasonable if the governmental objective is not legitimate or is not justified. As the development of Fourth Amendment doctrine demonstrates, the societal

[12] Jessica Litman, *Information Privacy/Information Property*, 52 STAN. L. REV. 1283–1313 (2000).

perceptions of privacy and the political objectives of policy are counterpoints in an ongoing effort to strike a balance between policy objectives and the protection of privacy. Such factors as technological innovations or changing political objectives necessarily have profound effects on the creation of privacy policy, not only in terms of Fourth Amendment doctrine but more generally. When political objectives, like those which came to the fore after the events of September 11, loom large, the political landscape of policy objectives is tilted toward security and claims of individual privacy are potentially diminished.

In a recent case challenging the PATRIOT Act, for instance, the balance between the political imperatives of September 11, 2001 and individual privacy was considered. In *Doe and ACLU v. Ashcroft et al.*, Judge Victor Marrero of the Southern District of New York struck down Section 505 of the Act on the grounds that it violated free speech rights under the First Amendment as well as the right to be free from unreasonable searches under the Fourth Amendment.[13] The application of the Fourth Amendment in that case pertained to the procedural protections against "unreasonable searches and seizures." Section 505 of the PATRIOT Act modified section 2709 of the Electronic Communications Privacy Act and allowed the Federal Bureau of Investigation to issue National Security Letters ("NSLs") to facilitate certain investigations under the auspices of the PATRIOT Act.

Administrative subpoenas, such as NSLs, must meet only a relatively low standard of "reasonableness." The Supreme Court has evaluated "reasonableness" by considering whether the administrative subpoena is within the authority of the agency and not too indefinite, and whether the information sought is relevant to the inquiry.[14] In *Doe and ACLU v. Ashcroft,* the issue was not whether the administrative subpoena was substantively reasonable but rather whether an individual must have access to judicial oversight and review to determine the "reasonableness" of an NSL. Longstanding Fourth Amendment principles associated with administrative subpoenas require "judicial supervision" and that they be "surrounded by every safeguard of judicial restraint."[15] In *Doe,* the court concluded that, because the FBI alone had the power to issue and to enforce its own NSLs, the procedural guarantees associated with the Fourth Amendment had been violated.

The *Doe* case points to the tension between the pursuit of a political objective, or common good, and the pursuit of individual privacy. That tension is important to Fourth Amendment jurisprudence, but equally

[13] Doe v. Ashcroft, 334 F. Supp. 2d 471 (2004)

[14] United States v. Morton Salt, 338 U.S. 632 (1950).

[15] *Id.*

important to understanding privacy writ large. "Privacy appears in consciousness as a condition of voluntary limitation of communication to or from certain others, in a situation, with respect to specified information, for the purpose of conducting an activity in pursuit of a perceived good."[16] The pursuit of the common good of preventing harm, which came to the fore after September 11, must be continually tempered by the Constitutional guarantees of privacy.

2.1.2 Informational Privacy and Biometrics

Even if existing doctrine does not directly apply to biometric technology, it nonetheless points to an ongoing need to protect individual privacy even in light of ideologically compelling political objectives. Part of striking this balance involves not only Fourth Amendment-type privacy concerns associated with the acquisition of biometric identifiers but also concerns associated with the protection of biometric information once it has been acquired. While statutory protection of personal information has been steadily rising in the venues of medical and financial privacy, the Constitutional doctrine has been relatively silent.[17] Yet, a growing public awareness of information gathering by governmental entities and private industry feeds the growing sense that a loss of privacy is at stake. There is, in other words, a growing expectation that informational privacy should be afforded protections under the law. This is nowhere more true than with biometric identifiers. This doctrinal void means that one of the central unresolved questions in the contemplation of widespread use of biometric identifiers is what protections are applicable to biometric information once it is acquired. Though the Constitutional doctrine of informational privacy is not well developed, there are some emergent principles with which to guide the usage of biometric technology.

One perceived threat to privacy is the lack of control that an individual has over personal information once it has been gathered. Privacy in this context is conceived of as control over personal data or control over what others (individuals, companies, or the government) know about an individual. Informational privacy, when it is conceived of as control over personal data, is less established in Constitutional doctrine than the Fourth Amendment right to freedom from unreasonable search and seizure, though

[16] Michael A. Weinstein, *The Uses of Privacy in the Good Life*, *in* 56 NOMOS XIII: PRIVACY at 94 (J. Ronald Pennock & John W. Chapman eds., 1971).

[17] Robert C. Post, *The Unwanted Gaze, by Jeffrey Rosen: Three Concepts of Privacy*, 89 GEO. L.J. 2087, 2087 (2001).

its growing importance is evident.[18] This concern is clearly relevant to government-mandated use of biometric identifiers. Unlike the decisional privacy doctrine under substantive due process or Fourth Amendment jurisprudence, the "doctrine" of informational privacy is built largely upon the holding of one significant case.

Whalen v. Roe is the decision noted for outlining the possible contours of a Constitutional right to informational privacy.[19] In this 1977 case, patients and doctors challenged a New York statute that required copies of prescriptions for certain drugs to be recorded and stored in a centralized government computer, arguing that it violated their constitutional right to privacy. Although the Court rejected this claim, a majority opined that under some circumstances the disclosure of health care information might violate a constitutionally protected right to privacy. Justice John Paul Stevens, writing for a majority of the Court, identified informational privacy as one aspect of the constitutional right to privacy. Stevens argued, "The cases sometimes characterized as protecting 'privacy' have in fact involved at least two different kinds of interests. One is the individual interest in avoiding disclosure of personal matters, and another is the interest in independence in making certain kinds of important decisions."[20]

Privacy acts as a safeguard against government surveillance and information gathering, but only to the extent that a societal harm is prevented. In current policy debates, the use of biometric identifiers must exemplify the balance defined in Fourth Amendment jurisprudence and in the logic of *Whalen.* First, there is a recognizable interest in controlling the disclosure of information that is deemed sensitive. Second, this privacy interest is not without constraints. The interest in controlling information to protect privacy is juxtaposed against the political objectives of protecting the common good and preventing harm. In *Whalen,* the Court was persuaded that the centralized database in that case would foster an understanding of the state's drug control laws and ultimately further the "war on drugs."[21] In recognizing the implications of the information being gathered, the Court considered not only the stated justifications but also whether the database represented a "useful adjunct to the proper identification of culpable professional and unscrupulous drug abusers."[22] The Court also acknowledged the sensitivity of this data by considering how well the

[18] Elizabeth L. Beardsley, *Privacy: Autonomy and Selective Disclosure*, *in* 56 NOMOS XIII: PRIVACY (J. Ronald Pennock & John W. Chapman eds., 1971) (referring to informational privacy as "selective disclosure").

[19] Whalen v. Roe, 429 U.S. 589, 591 (1977).

[20] *Id.* at 598-600.

[21] *Id.* at 597-98.

[22] *Id.* at 592 n.6.

information was being protected by procedural and technical safeguards. The impact of emerging technologies was also contemplated in *Whalen.* As Justice Brennan explained "The central storage and easy accessibility of computerized data vastly increases the potential for abuse of that information, and I am not prepared to say that future developments will not demonstrate the necessity of some curb on such technology." [23]

Each of the considerations outlined in *Whalen* is relevant to the use of biometric identifiers by either the public or private sectors. The danger posed by biometric identifiers is defined by their misuse rather than by their mere acquisition. Here, the evaluation of potential misuse must include consideration of their intended uses and of whether they serve a purpose in mitigating harm or advancing a societal good. In addition, once acquired, procedural and technical safeguards are essential to preserving the sanctity of biometric information. The *Whalen* decision thus highlights two important points applicable to the handling of biometric information. The first is that the facilitation of information gathering by technology might necessitate greater Constitutional protections of informational privacy. Second, despite compelling political objectives, personal information must be afforded protections.

2.1.3 Decisional Privacy and Biometrics

When William O. Douglas penned the Supreme Court's majority decision in *Griswold v. Connecticut* in 1965, he established a right to privacy that marked a turning point both in the Court's concept of the law, and, consequently, in the societal concept of the individual.[24] By allowing freedom to make birth control or reproductive decisions, Constitutional doctrine has limited the government's ability to interfere with these types of intimate decisions. One question which is yet unanswered is whether the handling of biometric identifiers potentially implicates decisional privacy.

It might seem that there is little connection between the use of biometric identifiers and the idea of decisional privacy. To follow the logic of decisional autonomy in the Constitution one must understand that there is a protected realm of privacy within which intimate decisions regarding marriage, procreation and birth control are protected. The handling of biometric identifiers may defy the traditional separation of informational privacy from decisional privacy. The widespread use of biometric identifiers to connect personal information has the potential, if unregulated,

[23] *Id.* at 607.

[24] Griswold v. Connecticut, 381 U.S. 479, 483 (1965) (holding that the right to privacy exists in the penumbras formed by the Bill of Rights).

to transgress traditionally protected zones of decision-making. The ideals of ubiquity and anonymity in our daily lives of the public sphere are at issue. For example, a recent federal court decision upheld the privacy rights of clientele of the San Francisco Planned Parenthood organization, against a Department of Justice subpoena of abortion records. The subpoena was part of an attempt to compile information related to the recently passed Partial Birth Abortion Act.[25]

Aggregating data raises the specter of information being used for purposes that were not originally intended. By tracking decisions that have been made, such potential uses might thwart the privacy protections surrounding an individual's freedom to make personal decisions in the first instance. Though the case of the subpoena to Planned Parenthood involved confidential medical records explicitly tied to individuals' decisions to have abortions, biometric information could be used to identify individuals in a similar manner thereby undermining the intimacy of the decision making process.

Another hypothetical intersection between biometric identifiers and decisional privacy is described by John Woodward:

> In response to growing concerns about missing children, a state legislature decides to require all children attending private day care programs to be biometrically scanned for identification purposes. Parents object on the grounds that they are fully satisfied with the less-intrusive security already offered at the private day care facility and that biometric scanning will unduly traumatize their children.[26]

In sum, it is clear that while there are points at which biometric technology and existing Constitutional protections coincide, there are many questions left unsettled in the various Constitutional approaches to privacy. What is fairly certain is that, even if biometric identifiers become commonplace and their usage is justified by political objectives, the issue of protecting biometric identifiers will be a pressing policy imperative. While Constitutional doctrine provides some guidance for future policy development, other guidance may be found elsewhere.

[25] Planned Parenthood Fed'n of Am. v. Ashcroft, 2004 U.S. Dist. LEXIS 3383.
[26] John Woodward *et al.*, IDENTITY ASSURANCE IN THE INFORMATION AGE (2003).

3. DOMESTIC AND INTERNATIONAL GUIDELINES

There are many examples of privacy safeguards which can be used to guide the development of policies to protect the privacy of biometric identifiers. Many are based on codes of fair information practices that have long been an organizing theme of privacy protection. Fair information practices were first set forth in comprehensive form in a United States Department of Health, Education, and Welfare study in 1973. Since this time, these privacy principles have been incorporated into United States law in the Privacy Act of 1974.[27]

3.1 Statutory Regulation of Government Handling of Personal Information

The Privacy Act of 1974 is one piece of legislation which would have direct bearing on biometric identifiers being used for governmental purposes and, as such, provides another point of guidance for future policy development.

Though the Privacy Act of 1974 is designed to provide protection for personal information, its limits are obvious. First, the focus of the Act is on federal agencies.[28] It does not, in other words, provide guidelines for the handling of biometric identifiers in the private sector. Second, the Act is intended to balance an individual's interest in information privacy with the institutional interest in storing data, but it does not necessarily resolve the potential contradiction between the use of biometric identifiers for security purposes and the privacy interests that must be diminished to serve that security purpose.

The Privacy Act was designed to protect the privacy of citizens by requiring governmental departments and agencies to observe certain rules in the use of personal information concerning individuals. The purpose of the Act was to safeguard the interests of citizens in informational privacy with the creation of a code of fair information practices that delineates the duties owed to individual citizens by federal agencies that collect, store, and disseminate information. It is important to note that the Privacy Act does not preclude governmental collection, maintenance, or use of private information about individuals generally or federal employees in particular. Instead, it imposes restrictions on federal agencies to curb abuses in the

[27] The Privacy Act of 1974, 5 U.S.C. §552a.

[28] *Id.* at §552a(a)(1).

acquisition and use of such information. In principle, the Privacy Act limits the disclosure of personnel records and forbids disclosure to any person by a Federal Governmental agency of any record regarding an individual which is contained in a system of records without that individual's express permission.[29] Yet, as discussed below, the express permission of the individual is subject to various exemptions which may undermine the Act's perceived effectiveness in protecting privacy interests.[30]

At first reading, the Privacy Act of 1974 would seem to offer expansive protection of the biometric identifiers which could be considered "records" under the auspices of the Privacy Act; however, the Act provides for several interpretational limitations that affect the strength of its protection of privacy and thus, it may not provide the level of normative privacy protection expected by United States citizens.

One of the potential weaknesses involves consent. The Act prohibits any federal agency from disclosing any record contained in a system of records without written consent of the individual to whom the record pertains.[31] The seemingly expansive scope of this prohibition on nonconsensual disclosure is subject to numerous exemptions.[32] Specific exemptions exist for disclosure to agency employees, the Bureau of the Census, the National Archives and Records Administration, Congress or its committees, the Comptroller General, and the consumer protection agencies. Also exempted is disclosure required under the Freedom of Information Act, for statistical research or law enforcement purposes, in response to emergency circumstances, or pursuant to court order.[33] Finally, the broadest exemption is for disclosure pursuant to a "routine use."[34] Routine use requires that the agency publish the provisions mandating disclosure in the Federal Register. Additionally, routine uses should be narrowly defined and permit disclosure only when justified by substantial public interest. Lastly, the reasons for disclosure should be consistent with those reasons stated in the applicable regulations.

In practice, however, the requirements of notice and prior consent are often circumvented. For instance, federal agencies have used the notice requirement to bypass the nondisclosure provision through the publication of broadly worded routine use notices. Law enforcement agencies have also employed the routine use exemption to avoid the restrictive language of an applicable disclosure exemption.

[29] *Id.* at §552a(b).
[30] *Id.* at §552a(b)(1)-(12).
[31] *Id.* at §552a(b).
[32] *Id.* at §552a(b)(1)-(12).
[33] *Id.* at §552a(4), (6), (9), (10), (12).
[34] *Id.* at §552a(3).

In addition to the routine use exemption, records may also be disseminated pursuant to a court order where the need for the disclosure outweighs the potential harm of disclosure. In the case of airline travel, one might easily infer that the need for disclosure, to protect against a hijacking or terrorist threat, far outweighs the potential harm of disclosure.[35] In a closely related exception, information can also be disseminated to other parties without prior consent where the records are being used for law enforcement purposes.[36] This exception obviously applies to law enforcement agencies; however, it also applies to other agencies whose functions normally do not include law enforcement. In short, the protections of non-disclosure offered by the Privacy Act of 1974 could easily be outweighed by the need of law enforcement agencies to ferret out potential terrorists. While the exemption is necessary, the public may view its presence as undermining the protections afforded them under the Privacy Act of 1974.

Another procedural safeguard in the Privacy Act of 1974 is intended to constrain federal agencies to gather only that information necessary to accomplish their purposes. In principle, this language imposes a limit on the collection, maintenance and dissemination of information, but, in practice, an expansive interpretation of "purpose" translates into a broad discretion.

Moreover, general and specific exemptions allow the heads of federal agencies to promulgate rules that exempt their agencies' systems of records from provisions of the Privacy Act. The general exemptions provide that systems of records maintained by the Central Intelligence Agency or criminal law enforcement agencies may be excused from compliance with the access, collection, and use provisions of the Privacy Act. In addition, there are seven specific exemptions that excuse systems of records from compliance with these provisions of the Privacy Act. The specific exemptions apply to record systems containing: classified material, investigatory materials for law enforcement agencies, information used to protect the President, statistical records, investigatory materials for civil employment, military service or government contracts, employment testing materials, and armed service evaluative materials for promotion.[37]

The Computer Matching and Privacy Protection Act of 1988 (PL 100-503) amended the Privacy Act of 1974 by adding new provisions regulating the federal government's use of computer matching – the computerized comparison of information about individuals, usually for the purpose of

[35] *Id.* at §552a(b)(11).
[36] *Id.* at §552a(b)(7).
[37] *Id.* at §552a(k).

determining the eligibility of those individuals for benefits.[38] This Act mirrors the logic of the Privacy Act of 1974 by requiring notice, data protection, and data verification to prevent any adverse effects on an individual's benefits.

In sum, the Privacy Act of 1974 and the Computer Matching and Privacy Protection Act of 1988 offer guidelines for the development of a sound privacy policy applicable to biometric technologies. However, they may not, in their current form, afford the requisite protections for privacy.[39] While Constitutional principles of privacy and the Privacy Act of 1974 offer some protections against governmental misuse of biometric identifiers, there still remains the issue of private industry. Here, there is a patchwork of regulation that does not necessarily alleviate public concerns regarding the use of biometric identifiers yet, at the very least, they offer guidance.

3.2 Private Industry and Regulating Biometric Technology

Statutory protections for personal information in the private sector rest on the same core set of "fair information practices," which were developed in response to the move from paper to computerized records. The first "code of fair information practices," developed in 1973 by an advisory committee of the then-Department of Health, Education, and Welfare (HEW), provided a core statement of principles. These principles set out basic rules designed to minimize the collection of information, ensure due-process-like protections where personal information is relied upon, protect against secret data collection, provide security, and ensure accountability. In general, the principles emphasized individual knowledge, consent, and correction, as well as the responsibility of organizations to publicize the existence of a record system, to assure the reliability of data, and to prevent misuse of data.

This code formed the basis for future fair information practices in the public and private sectors, as well as future United States laws (including the Privacy Act passed by Congress in the following year.) The basic principles of the code reflected modern privacy concerns: no personal data record-keeping systems should exist in secret; individuals must be able to discover what information is in their files, and to what use it is being put; individuals must be able to correct inaccurate information; organizations creating or maintaining databases of personal information must assure the reliability of the data for its intended use and prevent misuse; and individuals must have

[38] The Computer Matching and Privacy Protection Act of 1988, 5 U.S.C. §552a note (1988).
[39] *Id.*

the ability to prevent information being used for a purpose other than that for which it was collected.

These principles were also reflected in the Guidelines on the Protection of Privacy and Transborder Flows of Personal Data published by the Organization for Economic Cooperation and Development (OECD) in 1980.[40] The OECD guidelines set up standards that served as the principles for privacy policies, and are the foundation of current international agreements and self-regulation policies of international organizations. The contents of the OECD guidelines are somewhat vague, but provide an indication of the framework for protecting privacy in light of biometric technologies. The OECD code emphasized eight principles: collection limitation, data quality, purpose specification, use limitation, security safeguards, openness, individual participation, and accountability.

Under the OECD guidelines, the collection of personal data is limited; data is obtained by lawful means and with consent where appropriate. Data must be relevant to the purpose intended, and must be accurate and up to date. The purposes of collection must be specified at the time of collection or earlier, and use must be limited to those explicit purposes or other related purposes. Data is not to be disclosed or made available without either consent of the subject or by authority of law. Data must be protected by security safeguards against risk of loss or unauthorized access, modification, or disclosure. A policy of openness should ensure that there exist available means to establish the existence and nature of personal data, their purposes, and the identity of the data controller. Individuals should be able to obtain confirmation as to the existence of their personal data records, have access to that information or be given adequate reasons if access is denied and be able to challenge the denial, and be able to challenge the existence or correctness of the data. Finally, data controllers should be accountable for enforcing the above principles.

The Federal Trade Commission has argued based upon these principles that notice/awareness; choice/consent; access/participation; integrity/security; and enforcement/redress are central to the development of privacy policy.[41] Acknowledgement of information sharing and the threat it poses to privacy is reflected in the most recent formulation of privacy principles articulated at the conferences of the Asia-Pacific Economic

[40] *OECD Guidelines on the Protection of Privacy and Transborder Flows of Personal Data*, Organization for Economic Co-operation and Development (Sept. 23, 1980), *available at* http://www.oecd.org/documentprint/0,2744,en_2649_34255_1815186_1_1_1_1,00.html.

[41] Federal Trade Commission, Privacy Online: A Report to Congress (June 1998), *at* http://www.ftc.gov/reports/privacy3/toc.htm.

Cooperation.[42] At the 2003 Singapore conference, 21 member states began to formulate a set of privacy principles which attempt to achieve a balance between informational privacy and government interests. Reflecting a utilitarian framework, the APEC draft attempts to prevent harm while maximizing benefit in society. With these two counterpoints in mind, privacy protections are to be designed to realize the benefits of both information flow and individual privacy.

The APEC principles mirror those of the OECD. Security of information; safeguards against access or disclosure; and proportionality of harm and sensitivity of the information are considered. Notice of collection, purpose, and disclosure, as well as the identity and contact information of the data controller are standard operating procedures while the principle of choice gives data subjects the right to know what options are available to them in limiting the use or disclosure of their information.[43] Other limits on the actions of data controllers in the APEC framework include limits on collection and use. Collection must be relevant to the purpose, and use must be limited to the purpose for which the data was collected, unless the consent of the data subject is obtained, a service is being provided to the subject, or the use is under authority of law. Accuracy of information is required, as is the ability of data subjects to obtain, correct, complete, or delete information that is held about them.

These privacy guidelines represent an international trend afoot to shift the responsibility for privacy protection to governmental and private entities. Unlike the Constitutional doctrines discussed above, which define privacy according to the expectations of privacy formulated by the individual or defined within the realm of intimate or informational decision making, the guidelines outlined in the APEC principles and OECD guidelines focus on procedural protections for the handling of information.

This difference in protections is significant for understanding the protections which may be necessary for biometric identifiers. The Constitutional treatment of privacy is founded upon the assumption that an individual is aware of how his or her expectation of privacy is affected by a "search or seizure" or is cognizant of how intimate decisions or information gathering might be affected by outside forces. Yet, in light of the rise of the Information Age and the many possible uses of biometric identifiers, it may be necessary to shift the burden of protection to the purveyors of biometric identifiers.

[42] Asia-Pacific Economic Cooperation, APEC Privacy Principles: Consultation Draft (March 31, 2004), *at* http://www.export.gov/apeccommerce/privacy/consultation-draft.pdf.

[43] *Id.*

Primarily this change in responsibility is prompted by the inability of an individual to fully comprehend the reach of biometric technology or to anticipate its possible uses. Thus, the expectation of privacy cannot serve as a bulwark of protection. When an expectation of privacy cannot be adequately formed or is mistaken as to its power of protection, then it does little to afford safeguard privacy. The widespread use of biometric identifiers may require the protections be legislated which do not necessarily require the individual to be the sole guardian over his or her information. Instead, specific guarantees of accuracy, transparency, accountability and security particular to the uniqueness of biometric identifiers may be necessary.

Biometric technology, as a security-enhancing technology, also poses the problem of balancing privacy while enhancing security. The privacy standards of the above-mentioned frameworks are not without exception, and the most high profile exception to the privacy framework is the acknowledgement of law enforcement and national security needs. The need for flexibility in light of national security is a balancing factor when attempting to protect privacy. Any exemptions for national security should be "limited and proportional to meeting the objectives to which the exceptions relate."[44] Obviously, the use of biometric identifiers implicates the interests of privacy and security and any protective legislation will have to reflect these two countervailing principles.

4. NORMATIVE AND FACTUAL DEMARCATIONS OF PRIVACY

Beyond the doctrinal, or factual, principles of privacy, there is also the normative expectation of freedom from government, technological or physical intrusion and the right not to have information used for the purposes of discrimination. These are but a few of the markers for privacy that figure either explicitly or implicitly in the policy debates surrounding privacy. Here it is important to draw a distinction between privacy as a factual condition of life and privacy as a legal right.[45] Privacy as a factual condition of life is demarcated by the perception that it has been altered or lost by the actions of others. The perception that we face a loss of privacy in light of the information age is a factual condition of privacy loss and is attributable

[44] Asia-Pacific Economic Cooperation. Privacy Principles, Version 9 (Consultation Draft), 27 February 2004. (III.9)

[45] H. L. A. Hart, *The Ascription of Responsibility and Rights*, 49 PROC. OF THE ARISTOTELIAN SOC'Y 171–94 (1949).

to our normative expectations of privacy. This is where the call for greater regulation and safeguards occurs, because a factual condition of privacy loss is distinct from a violation of the legal right of privacy in that it may be felt even when there is no trespass of legal privacy protection.

The normative expectation of privacy is significant when developing new trajectories of public policy, and it is arguably more influential in today's debate regarding the protections that should be afforded privacy. The question for the current policy debate is twofold. Prior constitutional doctrine and legal policy can guide our understanding of where the violation of the legal right of privacy occurs and where political authority must be reigned in. Beyond this, however, is the more elusive, but nonetheless important, factual condition of the loss of privacy founded in normative expectations. This is not a new challenge in a democratic polity. As Mill pointed out, "there is a sphere of action in which society, as distinguished from the individual, has, if any, only an indirect interest; comprehending all that portion of a person's life and conduct which affect only himself, or if it also affects others, only with their free, voluntary, and undeceived consent and participation."[46] In this sphere, "the inconvenience is one which society can afford to bear, for the sake of the greater good of human freedom."[47]

Tracing the normative expectations of privacy is a difficult task because our notions of privacy are complicated by the tension between our notion of privacy as an individual right and the pursuit of a common good, each of which can be affected by monumental events like those of September 11, 2001. Humphrey Taylor, working with privacy expert Alan Westin, reported in a 2003 Harris Interactive Poll[48] that respondents fell into three general categories: "Privacy fundamentalists" (about one-quarter of those polled) feel they have already lost a great deal of privacy and are resistant to any further erosion. "Privacy unconcerned" constituted about 10 percent of respondents; they have little anxiety about the collection and use of their personal data. Sixty-five percent of those polled were considered "privacy pragmatists," who are concerned about protecting their information from misuse but are willing to provide access to and use of their information when there are clear reasons and tangible benefits, and when there are safeguards against misuse. Regarding the control of personal data, 69 percent felt that consumers have lost control of companies' use of this information. Over half disagreed that companies use information responsibly, and over half disagreed that existing laws and practices provide adequate protections

[46] John Stuart Mill, ON LIBERTY 17 (Gateway 1955) (1859).

[47] *Id.* at 120.

[48] Humphrey Taylor, *Most People are "Privacy Pragmatists" Who, While Concerned about Privacy, Will Sometimes Trade It Off for Other Benefits*, HARRIS INTERACTIVE, March 19, 2003, at http://www.harrisinteractive.com/harris_poll/index.asp?PID=365.

against misuse. In the context of the war on terrorism, however, 45 percent felt the government should have the ability to track internet activity at all times, and 66 percent agreed the government should make it easier for law enforcement to track online activities without notice or consent.

Similar results were found in a 2002 Harris Poll[49] conducted on behalf of the Privacy and American Business think tank. According to that survey, a majority of consumers (57 percent) did not trust businesses to properly handle personal information, and 84 percent preferred independent examination of privacy policies. Respondents reported concern for the following privacy risks: that companies will sell data to others without permission (75 percent); and that transactions are not secure (70 percent). A majority (63 percent) felt that existing law does not provide enough protection against privacy invasions. A large majority of respondents believed that access within businesses to collected information ought to be limited, and that individual permission (or legal justification) ought to be necessary for an exchange of information outside the company.

However, privacy concerns shifted considerably after the terrorist attacks in 2001. Several polls conducted in the months after the attacks sought insight into public opinion about how the government's antiterrorist measures would affect privacy. In general, the public was supportive of measures that would be limited to tracking and deterring terrorism, but skeptical that technological means would be limited to the pursuit of that end. A December 12, 2001, New York Times report gave survey results on privacy concerns in the political landscape after September 11. Sixty-five percent of respondents reported they did not want the government to monitor the communications of ordinary Americans to reduce the threat of terrorism. Forty-eight percent supported wiretap surveillance by the government to deter terrorism, while 44 percent opposed such surveillance on the grounds that it constitutes a civil rights violation.[50] A similar Harris Poll conducted in October 2001 showed Americans to be generally supportive of new surveillance technologies when applied to deterrence of terrorism, but also

[49] First Major Post-9/11 Privacy Survey Finds Consumers Demanding Companies Do More to Protect Privacy; Public Wants Company Privacy Policies to Be Independently Verified, HARRIS INTERACTIVE, February 20, 2002, at http://www.harrisinteractive.com/news/allnewsbydate.asp?NewsID=429.

[50] Robin Toner & Janet Elder, *A Nation Challenged: Attitudes; Public Is Wary but Supportive on Rights Curbs*, N.Y. TIMES, December 12, 2001, *at* http://query.nytimes.com/gst/abstract.html?res=F20617FF3E5B0C718DDDAB0994D940 4482.

concerned that law enforcement applications of the technologies would be extended beyond that purpose.[51]

In practice, the distinction between normative and factual expectations of privacy requires the public policy debate to account for constitutional and legislative limits on political authority that may impinge upon privacy, but also to take the measure of public perceptions. The context of the action, the prevalence of technology, the war on terror, and societal and individual expectations of privacy all figure into the proper balance between biometric technologies and privacy. The changing nature of privacy is the only constant in the ongoing debate regarding privacy and biometric technology. Although there are protections afforded by constitutional doctrine and federal protections of privacy, we must also be cognizant of the fact that technological innovation, surveillance, information gathering, and the conditions of the post–September 11 environment are also parts of the new landscape of society which will not disappear and which, more importantly, must be understood. In short, protecting privacy in light of biometric technologies requires us to negotiate a swiftly changing landscape of interests, technological innovations, and, most importantly, public policy solutions which reside in understanding both legal doctrine and our normative expectations of privacy and the common good.

[51] *Overwhelming Public Support for Increasing Surveillance Powers and, Despite Concerns about Potential Abuse, Confidence that the Powers Will be Used Properly.* October 3, 2001, at http://www.harrisinteractive.com/news/allnewsbydate.asp?NewsID=370.

Chapter 10

FINDING WALDO

Face recognition software and concerns regarding anonymity and public political speech

John A. Stefani
DePaul University College of Law, Class of 2004

Abstract: Should public surveillance cameras equipped with face-recognition software be used to "search" public political rallies? Using established First and Fourth Amendment law, this paper argues that people participating in public political speech should retain their anonymity with respect to the state.

Key words: First Amendment, Fourth Amendment, face recognition, biometrics, free speech, political expression, anonymity, privacy, search and seizure

Waldo, the title character in the popular children's books "Where's Waldo?" spends his time trying to hide in large crowds of people. Try as he might, no matter how large the crowd, nor how well he hides, Waldo is always found. Children of all ages delight in their ability to find him.

Though the books provide a lighthearted game for children, Waldo's situation can be frightening to imagine because of the constant surveillance under which he must live. His is a world devoid of privacy. And while privacy may not matter in Waldo's fictitious world, privacy certainly does in ours. "Privacy is essential to preserving people's ability to form bonds and associate with others having similar beliefs and views."[1] This ability plays a crucial role in the development of the self. Privacy in the public sphere, as is the case with Waldo, is a slightly different matter.

Public privacy, which can be identified with anonymity, "occurs when the individual is in public places or performing public acts but still seeks, and finds, freedom from identification and surveillance . . . he does not expect to be personally identified and held to the full rules of behavior and

[1] Daniel J. Solove & Marc Rotenberg, INFORMATION PRIVACY LAW 402 (2003).

role that would operate if he were known to those observing him."[2] Anonymity also exists as a means whereby the expression of political beliefs can be made without fear of repercussion.

This paper will attempt to outline the importance of maintaining the right to political anonymity amidst the dawn of technological advances that may make in-person identification unnecessary. Drawing from both First and Fourth Amendment law, a nexus will be established between the right to public anonymity, with respect to political speech, and expectations of privacy in public.

1. LEGAL PROTECTION OF PUBLIC ANONYMITY UNDER THE FIRST AMENDMENT

American courts have a long history of protecting anonymity. This "right" has been frequently protected under the First Amendment. The rights of free speech and peaceful assembly have been interpreted to encompass the right to remain anonymous in certain circumstances. Thus, the right of anonymity in one's actions has been established in three primary areas: anonymous writings, anonymous distribution, and anonymous membership. In each of these areas the individual seeks to use the cloak of anonymity in increasingly public forays. Alan Westin states that "[t]his development of individuality is particularly important in democratic societies, since qualities of independent thought, diversity of views, and non-conformity are considered desirable traits for individuals. Such independence requires . . . the opportunity to alter opinions before making them public."[3] Because certain activities important for individual development require group participation, anonymity must sometimes substitute for complete secrecy. Certain political activities, such as rallies, protests, and pamphlet distribution, take place primarily in the public arena. Anonymous writing, distribution, and membership have all been addressed and endorsed by the courts; several such cases are discussed below.

Talley v. California [4] determined the constitutionality of a Los Angeles ordinance requiring any publicly distributed handbill to identify the person who wrote or printed the handbill as well as the distributor. The handbill in question, urging the boycotting of certain merchants and businessmen, listed only the distributing organization's name and address and did not identify

[2] *Id.* at 28-29, *citing* Alan Westin, PRIVACY AND FREEDOM (1967).
[3] Solove, *supra* note 1 at 29-30, *citing* Westin, *supra* note 2.
[4] *Talley v. California*, 362 U.S. 60 (1960).

the person distributing it. The municipal court found Talley guilty of violating the ordinance and fined him ten dollars.

In deciding that the ordinance did not pass constitutional muster, the Court clearly stated that any requirement of identification ". . . would tend to restrict freedom to distribute information and thereby freedom of expression."[5] The Court examined the historical importance of anonymous pamphleteering in the development not only of the United States, but also of humankind in general. "Anonymous pamphlets, leaflets, brochures and even books have played an important role in the progress of mankind. Persecuted groups and sects from time to time throughout history have been able to criticize oppressive practices and laws either anonymously or not at all."[6] The Court reasoned that ". . . identification and fear of reprisal might deter perfectly peaceful discussions of public matters of importance."[7]

The Court again addressed the issue of anonymous political literature in *McIntyre v. Ohio Elections Commission*.[8] The Court looked at whether an Ohio statute prohibiting distribution of anonymous campaign literature abridged the freedom of speech guaranteed by the First Amendment. McIntyre distributed a pamphlet she had authored asking people to vote against a school tax levy. Though the statute required her signature, she did not sign many of the pamphlets, and was ultimately fined one hundred dollars.

In rendering its opinion, the Court stated that "[w]hen a law burdens core political speech, we apply 'exacting scrutiny,' and we uphold the restriction only if it is narrowly tailored to serve an overriding state interest."[9] The Court did not find such a state interest. Indeed, it found that ". . . the interest in having anonymous works enter the marketplace of ideas unquestionably outweighs any public interest in requiring disclosure as a condition of entry."[10] The majority discussed several reasons why one might choose to publish anonymously: "The decision in favor of anonymity may be motivated by fear of economic or official retaliation, by concern about social ostracism, or merely by a desire to preserve as much of one's privacy as possible."[11]

McIntyre expounded on the reasoning used in *Talley*, noting that *Talley* "embraced a respected tradition of anonymity in the advocacy of political

[5] *Id.* at 64
[6] *Id.*
[7] *Id.* at 65.
[8] *McIntyre v. Ohio Elections Comm'n*, 514 U.S. 334 (1995).
[9] *Id.* at 347.
[10] *Id.* at 342.
[11] *Id.* at 341-342.

causes."[12] Noting *Talley*'s expansion of the freedom to publish anonymously beyond the literary realm and into the political realm, the Court suggested an additional motivation for anonymity besides the standard "fear of persecution." "[A]n advocate may believe her ideas will be more persuasive if her readers are unaware of her identity. Anonymity thereby provides a way for a writer who may be personally unpopular to ensure that readers will not prejudge her message simply because they do not like its proponent."[13]

Talley and *McIntyre* express the Court's strongly held position that the right to anonymous writings will be protected. Unless the state exhibits an "overriding interest," authors may inject their political writings into the marketplace of ideas without attaching their identities.

Whether a person publicly distributing such anonymous political writings can be required to wear a name badge was discussed in *Buckley v. American Constitutional Law Foundation, Inc.*[14] *Buckley* looked at three issues concerning the distribution of an initiative petition in Colorado. The context for the Court's discussion was an analysis of the constitutionality of a state law requiring handbill distributors to wear identification badges. The Court ultimately agreed with the Tenth Circuit that the badge requirement "discourages participation in the petition circulation process by forcing name identification without sufficient cause."[15]

Witnesses at the *Buckley* trial testified as to the ramifications of the name badge requirement. One such witness commented that the requirement "... very definitely limited the number of people willing to work for us and the degree to which those who were willing to work would go out in public."[16] "Another witness told of harassment he personally experienced as a circulator of a hemp initiative petition."[17] And, regarding the reluctance of potential circulators to face retaliation when discussing volatile issues, a witness observed that "with their name on the badge, it makes them afraid."[18]

The majority was unequivocal in stating, "The injury to speech is heightened for the petition circulator because the badge requirement compels personal name identification at the precise moment when the circulator's interest in anonymity is greatest."[19] Though he dissented, Justice Rehnquist

[12] *Id.* at 343.
[13] *Id.* at 342.
[14] *Buckley v. Am. Constitutional Law Found.*, 525 U.S. 182 (1999).
[15] *Id.* at 200.
[16] *Id.* at 198.
[17] *Id.*
[18] *Id.*
[19] *Id.* at 199.

agreed with the majority that the identification badge requirement was unconstitutional.[20] Concurring in part and dissenting in part, Justice O'Connor elaborated: "I agree . . . that requiring petition circulators to wear . . . name badges . . . should be subject to, and fails, strict scrutiny. . . . Requiring petition circulators to reveal their names while circulating a petition directly regulates the core political speech of political circulation. The identification badge introduces into the one-on-one dialogue of petition circulation a message the circulator might otherwise refrain from delivering, and the evidence shows that it deters some initiative petition circulators from disseminating their messages."[21]

NAACP v. Alabama[22] considered whether Alabama could compel the National Association for the Advancement of Colored People to disclose to the state's attorney general the names and addresses of its Alabama members. The NAACP argued that such disclosure ". . . would violate rights to freedom of speech and assembly guaranteed under the Fourteenth Amendment to the Constitution of the United States."[23] The Court agreed, viewing the right to associate expansively, so as to include the right to do so anonymously. "Effective advocacy of both public and private points of view, particularly controversial ones, is undeniably enhanced by group association . . ."[24] The Court cited its earlier decision in *American Communications Association v. Douds*,[25] commenting that "[a] requirement that adherents of particular religious faiths or political parties wear identifying armbands, for example, is obviously of this nature."[26] Finally, holding that Alabama could not compel the disclosure of the NAACP's membership lists, the Court made the correlation between anonymity and privacy: "Inviolability of privacy in group association may in many circumstances be indispensable to preservation of freedom of association, particularly where a group espouses dissident beliefs."[27]

In these decisions, the Court has shown cognizance of a right to privacy in the public sphere, through anonymity, within the contexts of political speech and association. Taking solace in the knowledge that authorities may not legally demand that they reveal their identities, citizens are free to test their political selves, evolving their citizenship into a more active and participatory endeavor. However, recently developed technologies may

[20] *Id.* at 232.
[21] *Id.* at 217.
[22] *NAACP v. Alabama*, 357 U.S. 449 (1958).
[23] *Id.* at 453.
[24] *Id.* at 460.
[25] *American Communications Assoc. v. Douds*, 339 U.S. 382 (1950).
[26] 357 U.S. at 462.
[27] *Id.* at 462.

permit the government to track and record the identities of people who are exercising their right to anonymous public political participation without requiring name badges or otherwise having to request the information from the individuals. Face-recognition, a biometric technology, may evolve to permit the state to watch and record political rallies and protests, and then to match the surveillance record to a database of identified faces. Such government activity would pierce the veil of public anonymity by circumventing the need to approach citizens and ask them to identify themselves.

2. FACE-RECOGNITION TECHNOLOGY

"Biometrics" refers to techniques and methods used to identify individuals based on their particular physical characteristics.[28] Current biometric technologies include face recognition, voice recognition, retinal scans, hand geometry, and even gait recognition (recognition of a person's unique manner of walking). These technologies typically involve four steps: (1) a physical characteristic or trait is scanned, (2) the trait is translated into digital code, (3) the code is recorded and stored in some type of database, and (4) the database and digital code are then accessed and used at a later time to identify the individual.[29]

Face-recognition technology involves first the capturing and recording of an individual's face. This can be accomplished through video cameras used for surveillance purposes or "pre-existing photos such as those in driver's license databases."[30] Once a snapshot of the individual's face is obtained, computer software is used to analyze the face, establishing "over eighty nodal points . . . such as the distance between eyes, width of the nose, and depth of the eye sockets."[31] This process distills the human face into a faceprint. In situations where an accurate measuring of every nodal point is not possible, the faceprint is created by the software program using "between fourteen and twenty-two of the measured nodal points . . ."[32]

Once the faceprint is properly digitized and stored in a database, it is available for matching with video surveillance systems. The software allows

[28] Robert H. Thornburg, *Face Recognition Technology: The Potential Orwellian Implications and Constitutionality of Current Uses Under the Fourth Amendment*, 20 J. MARSHALL J. COMPUTER & INFO. L. 321, 323 (Winter, 2002).

[29] *Id.* at 323.

[30] *Q&A On Face-Recognition*, September 2, 2003, http://www.aclu.org/news/NewsPrint.cfm?ID=13434&c=130 (last visited Jan. 30, 2005).

[31] Thornburg, *supra* note 28 at 325.

[32] *Id.* at 325-326.

cameras scanning crowds to locate an individual face and match it with the faceprints on file. This matching of a face in a crowd allows authorities, in effect, to locate a needle in a haystack.

Face-recognition technology has many potential applications. It is currently being used to verify identification for access to weapons, biohazards, nuclear materials, money, or criminal evidence.[33] Casinos use it to identify card counters and other "undesirables."[34] "Since 1998, the West Virginia Department of Motor Vehicles has been using the technology to check for duplicate and false driver's license registrations."[35]

Increasing secretive use of the technology in public spaces has prompted various civil liberties groups to challenge its use. These challenges have revealed the shortcomings of face-recognition systems. According to one report by the American Civil Liberties Union (ACLU), government use of the technology on the streets of Tampa, Florida has been an "overhyped failure." "System logs obtained by the ACLU through Florida's open-records law show that the system never identified even a single individual contained in the department's database of photographs."[36] The Electronic Privacy Information Center (EPIC) notes on its web site that Privacy International gave its 2001 Big Brother Award to the City of Tampa for spying on Super Bowl attendees.[37]

Touted as a way to weed out suspected and potential terrorists, face-recognition software has been employed by airports. However, in his article critiquing the technology, Philip Agre offers a statistical reason why this solution may not be effective:

> "Let us assume, with extreme generosity, that a face recognition system is 99.99 percent accurate. In other words, if a high-quality photograph of your face is not in the 'terrorist watch list' database, then it is 99.99 percent likely that the software will not produce a match when it scans your face in real life. Then let us say that one airline

[33] Philip E. Agre, *Your Face is Not a Bar Code: Arguments Against Automatic Face Recognition in Public Places*, May 5, 2003, http://dlis.gseis.ucla.edu/people/pagre/bar-code.html (last visited Jan. 30, 2005).

[34] John D. Woodward, Jr., *Super Bowl Surveillance: Facing Up to Biometrics*, 2001 RAND ARROYO CENTER 4.

[35] *Id.* at 4.

[36] *Drawing a Blank: Tampa Police Records Reveal Poor Performance of Face-Recognition Technology*, January 3, 2002, http://www.aclu.org/news/NewsPrint.cfm?ID=10210&c=39 (last viewed Jan. 30, 2005).

[37] *Face Recognition*, August 21, 2003, http://www.epic.org/privacy/facerecognition (last viewed Jan. 30, 2005).

> passenger in ten million has their face in the database. Now, 99.99 percent probably sounds good. It means one failure in 10,000. In scanning ten million passengers, however, one failure in 10,000 means 1000 failures – and only one correct match of a real terrorist. In other words, 999 matches out of 1000 will be false, and each of those false matches will cost time and effort that could have been spent protecting security in other ways."[38]

Agre agrees that 1000 false alarms might be worth it to prevent one hijacking. However, the enormous number of false matches could condition security personnel to assume all matches are false.[39] If this is the case, the benefits of face recognition are negated.

3. GOVERNMENT MONITORING OF ANONYMOUS PUBLIC POLITICAL ACTIVITY

The public sphere is much more than a search area for fugitives, however. As illustrated by the line of cases above, public space is frequently used for the purpose of political speech. Whether it be attending rallies for a political party or cause, or protesting about any number of issues, politically expressive conduct occurs in the public sphere. A system of surveillance cameras equipped with face-recognition software might enable the government not only to view and store visual records of these gatherings, but also to keep track of the participating faces.

This concern that the government might take on a Big Brother role is more than just wild conspiracy theory. Local authorities as well as federal agencies, in the ostensible pursuit of public safety, are expanding their activities to include monitoring peaceful political gatherings such as rallies and protests.

Such monitoring is evidenced by an FBI Bulletin, obtained by the American Civil Liberties Union, which was labeled "LAW ENFORCEMENT SENSITIVE" and disseminated amongst state and local police departments.[40] The October 15, 2003 bulletin, labeled with a threat level of "YELLOW (ELEVATED)," informs police of upcoming "mass

[38] Agre, *supra* note 33.
[39] Agre, *supra* note 33.
[40] FBI Intelligence Bulletin no. 89, October 15, 2003, *available at* http://www.aclu.org/Files/OpenFile.cfm?id=14451 (last viewed Jan. 30, 2005).

marches and rallies against the occupation in Iraq."[41] Even though the FBI admits that "[m]ost protests are peaceful events," and that it "possesses no information indicating that violent or terrorist activities are being planned as part of these protests," the bulletin notes that "[e]ven the more peaceful techniques can create a climate of disorder."[42]

The Bulletin notes that "activists are usually reluctant to cooperate with law enforcement officials."[43] It advises law enforcement agencies that protestors rarely carry identification and "often refuse to divulge any information about themselves or other protestors."[44] Aware of this, the bulletin concludes by requesting local agencies to "report any potentially illegal acts to the nearest FBI Joint Terrorism Task Force."[45]

The bulletin shows the federal government's willingness, through the FBI, to characterize political protests as potentially criminal acts. This portrayal can cause quite a chilling effect on potential attendees. The intended secrecy of the directive only increases public concern that Big Brother is watching. The validity of such public concern was evident in a case brought by the ACLU of Colorado against the Denver Police Department.[46] That case challenged the practice of "monitoring and recording the peaceful protest activities of local residents and keeping criminal intelligence files on the free-speech activities of law-abiding advocacy groups, some of whom were falsely labeled as 'criminal extremist.'"[47]

In discovery, the ACLU found that the Denver Police Department was recording information about specific individuals in "permanent" files.[48] The police recorded information including: membership in various groups (such as the American Friends Service Committee, a Quaker organization); organizing and speaking at Amnesty International events; participating in Washington, D.C., protests against the International Monetary Fund and the World Bank; being "seen" at certain protests; license plate numbers of vehicles used by identified participants in "peaceful protest activities"; and

[41] *Id.*
[42] *Id.*
[43] *Id.*
[44] *Id.*
[45] *Id.*
[46] *ACLU and Denver Officials Agree to Resolve Lawsuit over Notorious Police "Spy Files,"* April 17, 2003, *available at* http://www.aclu.org/news/NewsPrint.cfm?ID=12396&c=25 (last visited Jan. 30, 2005).
[47] *Id.*
[48] *ACLU Calls for Denver Police to Stop Keeping Files on Peaceful Protesters*, March 11, 2002, *available at* http://www.aclu-co.org/news/pressrelease/release_spyfiles.htm (last visited Jan. 30, 2005).

home addresses and personal descriptions of "individuals engaged in lawful expressive activity."[49]

Mark Silverstein, the ACLU's Legal Director, underscored the concern for the chilling effect produced by such spying:

> "Denver residents should feel free to join a peaceful protest without fear that their names will wind up in police files. By monitoring lawful expressive activity in this manner and by falsely branding law-abiding organizations as criminals and extremists, the police will make Denver residents afraid to express their views and afraid to participate fully in our democracy."[50]

The parties ultimately settled the case, with the ACLU receiving a promise that procedural changes and safeguards would be implemented by the Denver police to protect Denver residents from further illicit surveillance.

Taken together, the FBI Bulletin and the Denver Police "permanent files" illustrate that, especially after the 9/11 terrorist attacks, government officials may view certain public displays of political speech as potentially criminal. This type of government response to political activity is certainly a cause for great concern. As Silverstein argued, it may create a chilling effect by frightening away individuals who might otherwise seek political involvement.

While the potential for prosecution by the State may be small, the more common, and greater, fear is that the authorities will have a permanent record of individual political activity. This file can be kept, added to, and used at any time deemed necessary. This Damoclean dagger then hangs over every person who has attended or might attend a protest or political rally.

Moreover, the FBI bulletin reveals the government's concern with rallies and its encouragement of local authorities to assist in gathering information on the participants. In effect, the Bulletin asks local police to "search" the crowds and "seize" the protestors' identities. As established in the above line of Supreme Court cases, citizens in a democracy must be allowed to express their political views freely without fear of state reprisal. Reprisal is made easier when those participants are identified. The question is whether such identification procedures – especially as enhanced by biometric

[49] *Id.*
[50] *Id.*

technology – are consistent with the Constitutional limitations on government actions.

4. FOURTH AMENDMENT LIMITATIONS ON SEARCH TECHNOLOGIES

The general right of law enforcement agencies to conduct criminal investigations is limited by constitutional issues and statutory regulations. In any event, privacy rights are at stake whenever a search is conducted. These privacy rights tend to weaken, however, as the searches are conducted in increasingly public areas. The evolution of various sense-enhancing technologies has made this issue more complicated.

The First Amendment cases cited above hold that an individual's anonymity is sometimes protected even in the public sphere while engaging in political speech. This protection prevents the government from mandating some forms of direct identification, including name badges or a signature on a pamphlet. But, the courts have been silent regarding the question of whether the government may circumvent this prohibition using advanced technological means.

The courts have, however, addressed the government's use of sense-enhancing technology with regard to searches in criminal investigations. An examination of how Fourth Amendment law has dealt with the issue of technological advancement may provide insight into how anonymity law might do so.

The Supreme Court considered the prospect of privacy in public in *Katz v. United States*.[51] Katz was suspected of transmitting wagering information via telephone, in violation of a federal statute. Hoping to obtain evidence, FBI agents attached an electronic listening device to the outside of the public telephone booth Katz was known to use. Based on the recordings, thus obtained, Katz was convicted. On appeal, he raised the issue of whether a public telephone booth is a constitutionally protected area that would protect the user's right to privacy.[52]

The government argued that the public setting of the telephone booth, as well as its transparency, precluded any claim of privacy. The Court disagreed with this argument and made its famous statement that "the Fourth Amendment protects people, not places."[53] The Court reasoned that, even in

[51] *Katz v. United States*, 389 U.S. 347 (1967).
[52] *Id.* at 349.
[53] *Id.* at 352.

the home or office, "[w]hat a person knowingly exposes to the public . . . is not a subject of Fourth Amendment protection. But what he seeks to preserve as private, even in an area accessible to the public, may be constitutionally protected."[54] The Court further explained, "What he sought to exclude when he entered the booth was not the intruding eye – it was the uninvited ear . . . One who occupies [the booth and] shuts the door behind him . . . is surely entitled to assume that the words he utters into the mouthpiece will not be broadcast to the world."[55]

The Court held that the actions of the FBI agents, listening to and recording Katz's conversation, violated the privacy upon which he "justifiably relied" while in the telephone booth.[56] In his concurrence, Justice Harlan explained the critical test as having two prongs: "first, that a person have exhibited an actual (subjective) expectation of privacy, and, second, that the expectation be one that society is prepared to recognize as 'reasonable.'"[57] For Justice Harlan, the key to *Katz* was "not that the booth is 'accessible to the public' at other times, . . . but that it is a temporarily private place whose momentary occupants' expectations of freedom from intrusion are recognized as reasonable."[58]

In *Katz,* the Court began laying the ground for protecting an individual's privacy rights from the technologically advanced prying ear of the government. In *Kyllo v. United States,*[59] the Court confronted law enforcement's prying eye.

Kyllo presented the Court with the issue of whether the use of a thermal imaging device to detect radiating heat from a private home constituted a search for Fourth Amendment purposes.[60] Police suspected Kyllo was growing marijuana in his home. Without a warrant, they situated themselves outside the residence and aimed a thermal-imaging device at Kyllo's home. The device showed varying shades of black, white, and gray, depending on the levels of heat radiating from the home. Based on the results, the authorities obtained a search warrant and subsequently found more than 100 marijuana plants being grown.

Similar to its argument in *Katz,* the government contended, and the District Court agreed, that the device was non-intrusive and revealed no intimate details. The crucial point for the topic of this paper is that the device did not intrude upon Kyllo's home, but merely interpreted thermal

[54] *Id.* at 351.
[55] *Id.* at 352.
[56] *Id.* at 353.
[57] *Id.* at 361.
[58] *Id.* at 361.
[59] *Kyllo v. United States,* 533 U.S. 27 (2001).
[60] *Id.* at 29.

radiation that emanated from it into a public space. The thermal-imaging device merely accomplished using thermal radiation what the naked eye accomplished with visible light shining from a house – creating an image. The Court thus had to deal with a new technology that allowed the government to observe the inside of a person's home without stepping foot inside. The Court had to wrestle with the long-standing view that a "visual observation is no 'search' at all . . ."[61]

Ultimately, the Court reconciled the traditional view with the advancing technology. "It would be foolish to contend that the degree of privacy secured to citizens by the Fourth Amendment has been entirely unaffected by the advance of technology . . . The question we confront today is what limits are placed upon this technology to limit encroachment on the realm of guaranteed privacy."[62] The Court determined that maintaining a strict interpretation of its previous holdings would fly in the face of reason when confronted with new technologies that made a physical intrusion by law enforcement officials unnecessary. Thus, *Kyllo* modified the traditional view of privacy in light of technology in a manner similar to the way in which *Katz* had accommodated the technology of public telephones. The Court concluded that the Fourth Amendment prohibits the warrantless use of search technology, even based on publicly accessible physical phenomena, when that search technology is not in "general public use."[63]

5. A PROPOSED APPROACH TO GOVERNMENT USE OF BIOMETRIC IDENTIFICATION TECHNOLOGY

Combining the logic of *Katz* and *Kyllo* creates a clearer picture of a possible Constitutional basis for prohibiting the use of face-recognition software in conjunction with public political speech. *Katz* sets the test for expectation of privacy with a two-pronged approach. Generally speaking a person is unlikely to expect privacy in public and few would find any expectation reasonable. However, every day people consciously or subconsciously rely on anonymity while in public. In a recent article, Professor Christopher Slobogin defends the right to public anonymity stating that it "provides assurances that, when in public, one will remain nameless –

[61] *Kyllo*, 533 U.S. at 32.
[62] *Id.* at 33-34.
[63] *Id.* at 34.

unremarked, part of the undifferentiated crowd – as far as the government is concerned."[64]

If, however, police officers were to follow people at political rallies, taking photos and names, few could disagree with the resulting chilling effect on political speech. Certainly, there is a strong argument that *Buckley*, *Talley*, and *McIntyre* should be read to protect individuals attending political events from being forced to identify themselves. Face-recognition technology allows authorities to accomplish the same feat without ever appearing in person. Although he was discussing a search of a home, Justice Scalia's comment is equally applicable here: "[w]here . . . the Government uses a device . . . to explore details . . . that would previously have been unknowable without physical intrusion, the surveillance is a 'search' and is presumptively unreasonable without a warrant."[65]

Public anonymity is important even for everyday public activities. As Professor Slobogin puts it, "[c]ontinuous, repeated or recorded government surveillance of our innocent public activities that are not meant for public consumption is neither expected nor to be condoned, for it ignores the fundamental fact that we express private thoughts through conduct as well as through words."[66] Constitutional protection for such everyday public anonymity may be difficult to assert, however. Whatever the result regarding everyday public behavior, a person's right to anonymity regarding political speech is of Constitutional significance. An individual may seek anonymity while engaged in political speech for numerous reasons. *McIntyre* alludes to the reasoning that "[t]he decision in favor of anonymity may be motivated by fear of economic or official retaliation, by concern about social ostracism, or merely by a desire to preserve as much of one's privacy as possible."[67] *Talley* continues, commenting that "identification and fear of reprisal might deter perfectly peaceful discussions of public matters of importance."[68]

For any, or many, of these reasons, people may be reluctant to participate in political rallies or protests if their anonymity cannot be preserved. Thus, protecting that anonymity becomes crucial. As with the petition circulators in *Buckley*, a participant in a political rally would be harmed by a requirement to identify himself to authorities.

Face-recognition technology could allow law enforcement agencies to effectively record the identities of protesters without the need for a formal

[64] Christopher Slobogin, *Public Privacy: Camera Surveillance of Public Places and the Right to Anonymity*, 72 MISS. L.J. 213, 238 (Fall, 2002).

[65] *Kyllo*, 533 U.S. at 40.

[66] Slobogin, *supra* note 64 at 217.

[67] *McIntyre*, 514 U.S. at 341-342.

[68] *Talley*, 362 U.S. at 65.

name badge requirement. The cameras would be able to capture a digital image of a person's face. Once captured and recorded, the image could later be mapped and analyzed. Once this is accomplished, the software could run the image against various databases to find a match. In principle, merely possessing a driver's license could result in a match. Once the match is made, the authorities would then have a name to accompany the face. All this could possibly be accomplished with no acquiescence on the part of the individual.

Just as *Kyllo* prohibits the warrantless use of sense-enhancing technology to "replace" a physical intrusion, it can be argued that face-recognition software must be prohibited when it replaces a name badge requirement. Just as a thermal-imaging device eliminates the need for the police to enter a home, face-recognition eliminates the need for police to ask a person's name.

> Critics will argue that, assuming *arguendo* that anonymity in public is constitutionally protected, certain loopholes should be created in the interests of public safety. Professor Slobogin, for example, agrees that the right to public anonymity is not absolute, mentioning that the "right is surrendered only when one does or says something that merits the government's attention, which most of the time must be something suggestive of criminal activity."[69] Public political protest does not meet this requirement.

Moreover, when regarding free speech and association, the courts have been consistent that "[w]hen a law burdens core political speech, we apply 'exacting scrutiny,' and we uphold the restriction only if it is narrowly tailored to serve an overriding state interest."[70] Because of the burden on the right to anonymous political speech that would result from extensive use of face-recognition technology, this test should be applied to formulate safeguards and provisions protecting against the use of face-recognition software at political gatherings. The strict scrutiny test has two elements. First, any use of the technology must be "narrowly tailored." Second, the software may only be used to serve an "overriding state interest."

The most likely state interest to be considered "overriding" is security. Public safety is certainly worthy of protection and courts have been consistent in declaring it a strong state interest. However, as has been illustrated here, security justifications may also be abused by law enforcement agencies. Denver officials recorded vast amounts of personal

[69] Slobogin, *supra note 64* at 238-239.
[70] *McIntyre*, 514 U.S. at 347.

information on individuals who could hardly have been considered dangerous. Likewise, the FBI Bulletin shows that the federal government may view even peaceful demonstrations as potentially violent events. The critical requirement here is that any use of "public safety" or "security" as a justification for surveillance using face-recognition technology must be subjected to exacting scrutiny because of the important First Amendment values involved.

Face-recognition technology continues to improve and may allow law enforcement officials to scan large crowds for particular individuals while expending minimal resources. However, a problem arises when such technology can be used in public areas, constantly surveying innocent passersby, including those engaged in protected political expression.

The First Amendment cases discussed show a growing acceptance of a right to public anonymity. But how far can this right extend? At what point can technology substitute for police officers in gathering identification? *Kyllo* demonstrates the Court's attempt to reconcile Constitutional protections with rapidly evolving sense-enhancing technology in the Fourth Amendment context. The Court holds that technology cannot be used merely to circumvent Constitutional restrictions.

From these two distinct lines of Constitutional law, one can draw an argument that public anonymity should be protected with the same vigor with which the law protects against improper searches. The potential chilling effect government surveillance has on public political gatherings is far too great to ignore.

The circle returns to Waldo. He is easily recognizable, his striped hat and spectacles simultaneously giving away his location and identity. Children find him precisely because they look for a specific picture. The technology may soon exist – if it does not already – to allow the government to scan large crowds and pick out a single individual – based on a specific picture that offers easily recognizable traits. As a fictitious character, Waldo does not suffer when he is found. In the context of political expression, though, people are most certainly harmed when they are found, their identities recorded, and their records kept.

IV

Privacy Implications of Data Mining and Targeted Marketing

Chapter 11

DATA MINING AND PRIVACY: AN OVERVIEW

Christopher W. Clifton,[1] Deirdre K. Mulligan,[2] and Raghu Ramakrishnan[3]
[1] *Department of Computer Sciences, Purdue University;* [2]*School of Law, University of California, Berkeley;* [3]*Computer Sciences Department, University of Wisconsin, Madison*

Abstract: The availability of powerful tools to analyze the increasing amounts of personal data has raised many privacy concerns. In this article, we provide an overview of data mining, aimed at a non-technical audience primarily interested in the social and legal aspects of data mining applications.

Key words: data mining, privacy, law, terrorism

1. INTRODUCTION

Data mining has emerged as a powerful tool for analyzing information, but its use in some controversial projects has led to an association of data mining with invasion of privacy and personal profiling. This is unfortunate on several counts: (1) Data mining is a technology, not a specific application, and its importance is directly tied to the growing flood of data being generated in virtually every major scientific and commercial enterprise. As data grows in volume and complexity, instead of offering more insight, it can actually *hinder* our understanding of the underlying phenomena unless our ability to condense it and analyze it can keep pace with the growth. (2) Restricting research on data mining does little to address privacy concerns. By driving researchers to rely on private funding, restricting public funding for data mining research may have the ill effect of limiting the attention paid to privacy and the information about data mining technology available to the public. (3) Restricting research on data mining, on the other hand, does limit our ability to understand vital data in almost every field of human activity, including medicine, science, and commerce.

The controversy on whether we should develop more powerful data mining techniques not only undermines an important and promising direction for research, it also obscures a sobering reality – current technology (with no further research or development) can already compromise privacy if irresponsibly applied without appropriate controls on data collection and analysis. Continued efforts in developing data mining as a technology are necessary, matched with parallel efforts by policy makers, privacy experts and data mining researchers in the private and public sector to develop a legal framework and supporting technology that allows us to leverage data while ensuring desired levels of privacy and security. All sides of the debate on appropriate uses of data mining must understand that achieving (many of) their goals require neither unfettered access to data nor a complete lockdown to prevent misuse.

2. WHAT IS DATA MINING?

Data mining is the analysis of one or more large datasets to discover interesting and useful relationships or trends. It is related to the sub-area of statistics called *exploratory data analysis*, and to the sub-area of artificial intelligence called *machine learning*. The distinguishing characteristic of data mining relative to these other areas is that the volume of data is very large. An algorithm is *scalable* if the running time grows (linearly) in proportion to the dataset size, holding the available system resources (e.g., amount of main memory and CPU processing speed) constant. Old algorithms must be adapted or new algorithms developed to ensure scalability when discovering patterns from large datasets.

Finding useful patterns in datasets is a rather loose definition of data mining; in a sense, all database queries can be thought of as doing just this. Indeed, we have a continuum of analysis and exploration tools with SQL queries at one end, spreadsheet-like OLAP tools in the middle, and data mining techniques at the other end. SQL queries are constructed using relational algebra (with some extensions) and are widely used for reports; OLAP provides higher-level idioms for summarizing and exploring hierarchically organized subsets of data; and data mining covers the use of a broad range of statistical and learning techniques to build descriptive or predictive models of the data.

2.1 Steps in Data Mining

It is useful to recognize that data mining involves many steps[1] extending long before and after the actual application of data mining algorithms:

1. **Data collection:** Data mining can be carried out on existing data, but data is often gathered to enable subsequent, specific analysis. It is important to ask whether collecting the data is justified, and whether the subjects involved understand how and why the data is being collected and consent; this has a bearing on what uses of the data are permissible. The *quality* of data and how it was gathered could also influence how we interpret the results of data mining.
2. **Data integration:** Often, data is aggregated from multiple sources to facilitate mining. The relevant questions are whether the organization consolidating previously gathered data has the right to do so, and whether the terms of the original data collection permit its re-use in such integrated forms for the specific objectives of the mining effort. In addition, it is important to ensure that data is appropriately correlated, that integrated data is well secured, and access is appropriately restricted.
3. **Pattern finding:** This is the step of running one or more mining algorithms to discover interesting patterns (see Section 2.2 for examples of patterns). Most "data mining" research addresses this step, which, ironically, is the least controversial. Existing data analysis techniques (e.g., queries in an off-the-shelf database system) can be used to severely breach privacy if care is not taken in the other steps.
4. **Pattern interpretation:** The patterns discovered through data mining, in contrast to those from simpler reporting systems, often require a degree of technical expertise to understand correctly. The algorithms involved sometimes make subtle assumptions, and the mathematics underlying predictive models often requires careful interpretation to understand the limitations and exceptions. This raises concerns about whether a given decision-maker is qualified to interpret data mining results, and whether decisions based on such patterns are sufficiently transparent for outside oversight and legal recourse.
5. **Acting on discoveries:** Ultimately, data mining is a tool in the hands of someone addressing a larger issue. Given a set of patterns, that person must decide what is interesting, how reliable the conclusions are, and

[1] *See e.g.*, Usama M. Fayyad *et al.*, *From Data Mining to Knowledge Discovery: An Overview*, *in* ADVANCES IN KNOWLEDGE DISCOVERY AND DATA MINING 1-34 (1996).

what external evidence is required to support different levels of subsequent follow-up action. The issues here are numerous, and include checks and balances to ensure that the results of data mining are appropriately interpreted; that subsequent actions based on data mining results conform to due process; and indeed, that the legal and social framework is adequate to support such checks.

Up to a point, the concerns raised by each of the above steps are not unique to data mining. The issues raised by the first two steps are fundamental to any effort that gathers sensitive data. The issues raised by Step 4 are similar to those present already whenever controversial decisions are arrived at through a complex human-centered process. The issues raised by Step 5 are present whenever we use evidence of any kind to justify subsequent actions. However, the nature of digital data gathering, the specifics of technical pattern discovery, and the use of resulting patterns as evidence, all lead to social and legal gray areas that require particular attention.

2.2 Data Mining Techniques and Patterns

In this section, we briefly discuss one class of patterns, classification rules, and how they can be learned. The emphasis is on highlighting the underlying assumptions and the issues involved in acting upon such patterns.

2.2.1 Classification and Regression

Consider a table in which each record contains a customer's age, gender, race, income, and the type of car driven, a count of the number of moving violations within the past three years, and a flag that indicates whether the customer was profitable over the past three years. We would like to use this information to identify rules that predict the insurance risk of new applicants for car insurance. We know all fields except for the last one, of course, and the goal is to predict who is likely to be a profitable customer in future; the field to be predicted, profitability in our example, is called the *class label.* An example rule for our insurance data might be: "If the customer is a male aged 25 to 45 with no moving violations, and drives a 4-door sedan, he is likely to be profitable." Such rules are called *classification rules*, and are "learned" by applying machine learning algorithms to the given data.

Classification rules can be organized into a hierarchical structure, represented as a graph. A learning algorithm begins by partitioning the given data according to some condition; there are nodes in the graph corresponding to the original data and to each partition, and an edge from the first node to each partition's node. Each partition is similarly subdivided, repeatedly. The final hierarchical graph is called a "tree" in computer science terminology, leading to the name "decision tree."[2] If we start at the first node and follow a path down the tree to a "leaf" node, *i.e.*, a node without children, we effectively obtain a subset of the original data that satisfies all the conditions used to partition nodes on the path from the original node to the given leaf. By design, almost all the data in the subset associated with a given leaf node has the same value in the class label field. A new data record that satisfies the same condition is "predicted" to have this class label value as well.

While the details of decision tree algorithms are outside the scope of this discussion, the following points are worth noting:

- In order for classification rules to be accurate predictors, a key assumption must be met – the given data must be representative of future customers as well. Further, care must be taken not to over-fit, or over-generalize, the given data.
- Even the best classifiers are ultimately making educated guesses, and could be wrong, just like a human decision-maker. This leads to several questions when classifiers are used in decision making: How much reliance should we place upon a prediction, e.g., that a particular customer is not likely to be profitable? How can we understand why a prediction was made, and look for additional data that supports or contradicts the prediction? For example, by looking at the path taken in the tree for a new data record, we see why a given class label value was predicted; let's say that a new customer was considered unprofitable because of low income. If we know that the customer has a large bank account, or has just inherited a fortune, our estimate of their profitability is likely to be quickly revised upward.
- Decision trees are popular because the tree structure is intuitively understandable; we "understand' why a prediction is being made. While this is a desirable property, the patterns learned by a learning algorithm cannot always be understood as naturally. The *explainability* of a predictive model is important when deciding to act on a prediction, as we saw above. It is also important when reviewing decisions. For example, if a decision tree significantly influenced decisions on mortgage applications, it might be important to understand the logic behind it to see if it reflects racial bias. Even without a "race" field in the data, decisions based on collections of fields that correlate highly with race

[2] J. R. Quinlan, *Induction of Decision Trees*, 1 MACHINE LEARNING 81-106 (1986).

may induce a racial bias (as with the outlawed practice of mortgage "redlining" of minority neighborhoods.)

2.2.2 Other Patterns and Learners

Many other kinds of patterns have been studied and a wide range of learning algorithms is commercially available.[3] For example, if the class label field contains a number denoting how profitable a customer is, we can learn a tree that predicts a profitability number (or range) for a new customer. The principle is similar; each leaf node in the tree now has an associated numeric range for the class label field. Such trees are called *regression trees*. *Bayesian* learners try to learn what events cause other events using a probabilistic approach. *Frequent item set* learners find which items co-occur most often, *e.g.*, which items are most likely to be purchased together by a customer. *Clustering* techniques try to group data into groups so that two observations in a group are similar, while observations from different groups are dissimilar. *Link analysis* also seeks to discover groups of individuals or events, but by analyzing a graph of interactions. *Collaborative filtering* algorithms make suggestions (*e.g.*, books you might want to purchase at Amazon) based on what people with similar tastes have liked.

Extensive research has been carried out on how to make learning algorithms scale to large datasets, and for most classes of patterns, scalable learning algorithms are now available.

3. EXAMPLES OF DATA MINING APPLICATIONS

We now consider several example applications of data mining, many of which are widely acknowledged to be valuable, and free of controversy. Where there is controversy, it often has to do with the nature and origin of the data, or the role played by public sector agencies (see Section 4).

3.1 Applications Using Uncontroversial Data

Data mining is a fundamental tool for using data collected in a variety of domains either to improve system performance or to understand central

[3] *See, e.g.*, J. Han & M. Kamber, DATA MINING: CONCEPTS AND TECHNIQUES (2000), and D. J. Hand, H. Mannila & P. Smyth, PRINCIPLES OF DATA MINING (2001).

phenomena of interest. Examples of such applications are numerous, and increasing rapidly.

A growing emphasis in computer system design is to make systems self-managing and robust. It is desirable that systems be able to improve their performance over time, monitor themselves to detect and repair errors, and to do so with minimal human intervention. Corporations such as IBM and Oracle have made initiatives to develop such systems centerpieces of their technical strategies. The key to building such systems is to instrument them to continuously gather data that can be used to measure all aspects of their performance, and to (automatically) learn predictive models of system behavior that can be used to tune the system and to detect breakdowns. Clearly, data mining techniques play a central role in such applications.

Large enterprises rely on data mining techniques to optimize their supply chains. The goal is to minimize the amount of time goods are stored on shelves, while simultaneously minimizing the possibility of running short of some goods. Traditionally, mathematical programming techniques from operations research have been used for this purpose, but data mining techniques are now being used as well. Again, the key is to constantly gather data about every facet of their organization's supply and demand for various goods, and to learn predictive models that suggest when goods must be re-stocked at various points in the supply chain.

Data mining is emerging as an essential tool in science and medicine. Large astronomy datasets have been successfully analyzed using data mining. In recent years, data mining techniques have been applied to microarray data analysis, protein folding, environmental monitoring, improving searches over medical literature and detecting malignant tumors, and it is clear that such use of data mining will accelerate further in coming years, as more and more data is generated in these fields, making manual analysis all but impossible.

In most such applications, the use of data mining is uncontroversial, and in fact, widely acclaimed as one of the most promising approaches to understanding data and utilizing it effectively. However, even when the ownership of data and the appropriateness of mining it are not at issue, there are some data mining applications that raise social and legal concerns, e.g., a bank that uses recommendations of a predictive model to assist in mortgage decisions must put checks in place to ensure that decisions are not biased, racially or otherwise. Social scientists have expressed concerns that such an approach makes it difficult for an outside observer to detect bias, or to establish that there is bias when it exists, given the need to argue about the filtering criteria latent in a sophisticated predictive model, especially one whose workings are not easy to interpret.[4]

3.2 Controversial Data Acquisition and Reuse

We now turn to applications where the nature of the data used in mining raises privacy concerns.

Companies routinely use data gathered during the normal course of business for marketing campaigns and personalization. For example, users who browse websites leave behind "click streams" of activity that could be used to generate special offers, perhaps in conjunction with their past buying history at the site. Data mining to tailor customers' experiences and support future marketing efforts has generated privacy concerns as well as concerns about preferential treatment for privileged customers. Today, such use, where it is clearly disclosed and consent has been obtained, is, by and large, considered reasonable.

However, suppose that the site now uses this data to learn predictive models that are applied in a way that may limit an individual's options in another area, for example, to make a decision about whether to offer an individual a company credit card. The use of the data for this unrelated purpose, which in some instances will limit the individual's economic opportunities, raises privacy and fairness questions.

Access to broad data sets is essential for medical research, and for many applications of data mining in this domain. There is clearly great value in using such data to advance medicine; yet there is a risk that an individual's medical history could be leaked, even from supposedly "de-identified" data. Under what conditions should such a trade-off be considered reasonable, who makes the determination, and according to what social and legal framework?

4. GOVERNMENT DATA MINING

When the Government uses data mining to identify and prosecute terrorists it raises many of the concerns that we pointed out earlier (*e.g.*, in Sections 2.1 and 3.2), and introduces additional concerns about due process. There is no comprehensive statutory framework to govern data mining, and it is unclear whether data mining could be considered a search under the

[4] A. Danna & O. Gandy, *All that Glitters is not Gold: Digging Beneath the Surface of Data Mining*, J. BUS. ETHICS (2002).

Fourth Amendment. What are the appropriate checks and balances on the use of data mining by the Government that will allow reasonable uses while protecting an individual's right to privacy and due process?

While government efforts to use data mining to identify terrorists and terrorist activity have been the privacy lightning rod, the use of data mining techniques by both the public and private sector in a variety of application areas has triggered concerns about loss of privacy, lack of due process protections, and potential discrimination.[5]

Data mining is a powerful tool for law enforcement, but also holds significant risks for individual privacy and procedural protections. Our goal is to highlight the issues that must be considered when weighing the benefits and risks of a particular application of data mining.

4.1 What Data Can the Government Mine?

The data to be mined by or on behalf of the government in the context of terrorism includes the information left behind in various commercial transactions, communications, and interactions, as well as data compiled by the government for the purpose of detecting and preventing terrorist acts. Suppose that a public sector agency integrates data from a variety of sources, including private sector sources such as credit card transactions. Does the mining of this aggregate data constitute a search of the individual that violates their Constitutional rights? Or is their Fourth Amendment right to privacy limited in this case by their voluntary disclosure of the data to a third party (the business that collected the data)?

The personal information revealed in the majority of marketplace transactions is not subject to statutory privacy rules that limit government or private sector use, including data mining, of personal data. In some contexts certain personal information collected in the private sector is governed by statutes based on the Fair Information Practice Principles (FIPS) developed in the 1970's to address the privacy threats posed by the collection and use of data and its storage in main frame databases.[6] The FIPS require entities

[5] Birny Birnbaum, *Credit Scoring: 21st Century Redlining and the End of Insurance*, *in* Panel, Overseeing the Poor: Tech. Privacy Invasions of Vulnerable Groups, Conf. on Computers, Freedom, and Privacy (2004) *at* http://www.cfp2004.org/program/materials/p1-birnbaum_cfp_talk_040421.pdf.

[6] *See* CODE OF FAIR INFORMATION PRINCIPLES, which was developed by the Department of Health Education and Welfare in 1973. *See generally* U.S. DEP'T OF HEALTH, EDUC. & WELFARE, SECRETARY'S ADVISORY COMM. ON AUTOMATED PERSONAL DATA SYSTEMS, RECORDS, COMPUTERS AND THE RIGHTS OF CITIZENS 41-42 (1973); and the Organisation for Economic Co-operation and Development Guidelines on the

collecting data to notify individuals of the data collection and of its purpose, to limit the use of the data for the stated purpose absent the individual's consent to other uses, to provide access and correction rights to the data, to minimize the collection of data, to ensure that the adequacy of the data matches its intended use, to provide adequate security, and to be accountable for the implementation of these rules. Laws regulating the private sector's handling of personal information contain exceptions providing access to law enforcement and intelligence agencies.[7] While government data mining of private sector data is not governed by statutes based on FIPS, the limitation on reuse and the adequacy and correction principles are particularly relevant to the debate about privacy and data mining.

4.2 Data Aggregation

The Supreme Court has noted that integrating multiple sources of data, or *data aggregation*, raises privacy considerations. Even where the underlying data is publicly available, the Court has found that its compilation alters the privacy consideration. In the leading case on this issue the Court made a "distinction, in terms of personal privacy, between scattered disclosure of the bits of information . . . and revelation of the [information] as a whole."[8] The economics of information retrieval and analysis – core elements of data mining – influenced the Court: "Plainly there is a vast difference between the public records that might be found after a diligent search of courthouse files, county archives, and local police stations throughout the country and a computerized summary located in a single clearinghouse of information."[9]

Mining aggregated data raises due process concerns. Principles of due process require individuals to be able to challenge evidence against them. When an individual comes under suspicion because of data mining over aggregated data, the data searched may be in the hands of multiple third parties, and might include personal data about other individuals. How and under what terms could an individual choosing to challenge a decision based on third party data go about getting access to the appropriate data?

Protection of Privacy and Transborder Flows of Personal Data 1980, available at http://www.oecd.org/document/18/0,2340,en_2649_34255_1815186_1_1_1_1,00.html.

[7] Current Legal Standards for Access to Papers, Records and Communications: What Information Can the Government Get About You, and How Can They Get It, Center for Democracy and Technology, August 2004, http://www.cdt.org/wiretap/govaccess/govaccesschart.html.

[8] U.S. Dep't of Justice v. Reporters Comm., 489 U.S 749, 764 (1989).

[9] *Id.* at 770.

4.3 Monitoring

The notion of "voluntary disclosure,"[10] which arguably places all data generated in typical business transactions outside the sphere of privacy protected by the Fourth Amendment, is at odds with the collection of information through non-obvious means such as sensor networks and surveillance cameras. In such instances the mere presence of an individual in a space results in the collection of information. Individuals have neither knowledge of nor the ability to limit data collection.

When data is gathered through surveillance cameras in a public forum, what are the implications of mining it for patterns suggesting terrorist activity or associations? Is data gathered from an Internet newsgroup similar to data gathered from a surveillance camera or is it more akin to voluntary disclosure? While such data may aid, *e.g.*, in detecting terrorist activity, what is its legal status? Is such information completely unprotected? Or do the privacy implications of the aggregation of such data warrant protection?

4.4 Data Mining and Data Quality

Law enforcement decisions affecting the liberty of individuals have been questioned where the data relied upon is known to be inaccurate.[11] Extrapolating, one could posit that the government has some duty to make sure that the data it is relying upon – whether in the public or private sector – meet some level of accuracy, and that the data mining techniques themselves

[10] In a series of cases in the 1970s, the Supreme Court held that the Fourth Amendment does not apply to personal information contained in records held by third parties. Once an individual voluntarily discloses information to a business, the Court reasoned, the individual no longer has a reasonable expectation of privacy in the data and the government can access the record without raising any constitutional privacy concerns. In *Couch v. United States*, 409 U.S. 322 (1972), the Court held that subpoenaing an accountant for records provided by a client for the purposes of preparing a tax return raised neither Fifth nor Fourth Amendment concerns. In *United States v. Miller,* 425 U.S. 435 (1976), the Court held that records of an individual's financial transactions held by his bank were outside the protection of the Fourth Amendment. Lastly, in *Smith v. Maryland,* 442 U.S. 735 (1979), the Court held that individuals have no legitimate expectation of privacy in the phone numbers they dial, and therefore the installation of a technical device (a pen register) that captured such numbers on the phone company's property did not constitute a search. See generally, Deirdre K. Mulligan, "Reasonable Expectations in Electronic Communications: A Critical Perspective on the Electronic Communications Privacy Act," 72 G.W. L. Rev. 1557 (2004); James X. Dempsey, "Communications Privacy in the Digital Age: Revitalizing the Federal Wiretap Laws to Enhance Privacy," 8 Albany L. J. of Science and Tech. 65 (1997).

[11] Arizona v. Evans, 514 U.S. 1 (1995)

meet some level of "accuracy," perhaps established through limits on false positives or false negatives, depending on the application.

For example, we know that credit reports, one of the most highly regulated forms of personal information, contain errors. Consumers are allowed to challenge and remove data and credit bureaus are required by law to fix mistakes. Most private sector data collectors are subject to no such requirement. The inaccuracies in underlying data are of several sorts. Some data is inaccurately reported or collected. Some data is inaccurate due to an incorrect attribution to the wrong person or file. Finally some data is ambiguous, and while the ambiguity may be tolerable for one application it may raise accuracy problems in another context.

Should there be some check on the accuracy and adequacy of data used for various data mining activities?

4.5 Interpretation of Data Mining Results

Acting upon the results of data mining involves interpretation of technical results, and can require expertise in the technology and the underlying application domain. What are reasonable standards for acting on conclusions drawn from data mining, rather than direct evidence?

To illustrate the potential for error, consider the case in which the USDA (in 2001 and 2002) sent letters to three Somali stores in Seattle, informing them that they were permanently disqualified from participating in the food stamp program because EBT transaction records showed a suspicious pattern of trading. The patterns used to detect fraud were, however, typical of legitimate purchasing habits in this immigrant Somali community. The disqualification precluded 80% of the stores' patrons from making purchases. The USDA eventually reversed the ruling.[12]

Difficulties in interpreting patterns obtained by data mining also raise due process concerns. In other areas where we rely on forensic "sciences" an individual may be able to commission independent studies and reach contradictory conclusions. For this to be possible with data mining results, the underlying data and data mining techniques must be made available. If this is done, *e.g.*, with a suspected terrorist, there is the risk of revealing patterns that others could then work to circumvent, thereby diminishing the value of future mining. How can we support due process while preserving necessary levels of secrecy?

[12] Florangela Davila, *USDA drops case against embattled Somali grocers*, SEATTLE TIMES, July 17, 2002.

As an example, consider "no-fly" lists.[13] Without access to the data or the logic used to generate such a list, individuals' ability to challenge their inclusion is clearly impaired. The ability to access information about oneself in the system, have information corrected or removed, and challenge the method used to assign suspicion are important procedural protections.

Going beyond secrecy of data and mined patterns, if the underlying technology supports an automated system that can engage in learning, and refine and run new queries based on discovered knowledge, no one, including the agency engaging in the data mining, may be able to accurately describe the process by which an individual has been identified. Developing techniques that make the results of data mining "explainable" (see Section 2.2) is therefore important for responsible use of data mining results and supporting due process.

4.6 Lack of Particularity

In some instances of data mining, both the development of patterns and the latter application of those patterns are by nature non-particularized searches of non-particularized data. In order to develop patterns, data about both "good" and "bad" activities must be explored. Once patterns are developed they are applied to ambiguous data which might include data on people who have done nothing to arouse suspicion.

Lack of particularity raises concerns if the government is allowed to use mined data to create the predicate for an arrest or a warrant, i.e., the mined data creates the suspicion, and is the probable cause supporting the warrant. When can data mining results, without additional substantive direct evidence, serve as probable cause for a warrant? If data mining results are used in addition to direct evidence, how do we quantify the added degree of suspicion supported by the mining results? These questions are closely tied to the issues of data quality and interpretation of mining results, which we discussed earlier in Sections 4.4 and 4.5.

In commercial settings, profiling is used to identify credit risks. In such settings there is an enormous number of similar transactions that have been mined for years to separate the "normal" from the "suspicious." Based on mined patterns indicating potential fraud, credit card companies call customers to confirm card usage. This provides continued feedback into this

[13] The "no fly list" contains the names of individuals that commercial airlines may not transport, while the "selectee list" contains the names of individuals who must be subject to additional security procedures. Green *et al.* v. Transportation Security Administration *et al.*, *available at* http://www.aclu.org/Files/getFile.cfm?id=15424.

system, and historical data on the predictive quality of mined patterns. In terrorism cases there is, thankfully, at this time a paucity of data on instances of terrorist activity and a rather broad range of patterns. Given the paucity and variety of data it would appear that patterns arising from it may be broad and inclusive. Without reliable techniques for interpreting and quantifying the evidentiary value of mined patterns, using such patterns to obtain a warrant, especially without additional direct evidence, raises serious particularity questions.

4.7 Function Creep

A technical system created to enable queries over distributed private sector datasets is likely to be under enormous pressure to support additional activities. Given a system aimed at supporting the identification and prevention of terrorism, agencies charged with traditional law enforcement activities and perhaps others are likely to want access to these powerful tools and data sets as well.

4.8 Effect of Data Mining on Legal Activity

If the public becomes aware that certain kinds of activities are likely to trigger scrutiny – purchasing one way tickets, reading certain books, pre-paid calling cards, membership in certain organizations – or even if they wrongly believe this to be the case, government data mining of information collected in the private sector may inhibit the public's engagement in perfectly legal and desirable activities.

5. DATA MINING RESEARCH TO PROTECT PRIVACY AND SECURITY

Ongoing research seeks to develop technology to mitigate a number of the privacy risks associated with data mining. This includes work on detecting correlations between data mining models and protected attributes such as disabilities or age; making data mining results more explainable; reducing the rate of "false positives" (e.g., individuals who fit a pattern that suggests terrorist activity, but are not terrorists); and preventing the creation of data mining models that allow accurate predictions of information about specific individuals when that is not permissible. Work in statistics, particularly dealing with census data,[14] provides a model for showing that data mining results do not violate privacy.

To illustrate how new data mining technology can reduce privacy risks, we briefly discuss some recent results that allow us to discover data mining patterns without creating a centralized database. The process of gathering and integrating data to use as input to the data mining process often produces vast datasets (or profiles) of individuals, and the potential for misuse of these datasets is seen as one of the biggest privacy risks. Thus, the ability to discover specific, relevant (and permissible) patterns while minimizing the amount of sensitive data gathered in a central database or revealed to the data miner is a valuable tool.

We assume that each individual item in an individual's record can be seen by someone without violating privacy, although nobody might be able to see all the data, even that pertaining to a single individual. In practice, the creator of a data value is generally authorized to see it. We also assume that the patterns produced by data mining do not inherently violate privacy. (As discussed in previous sections, applications in law enforcement, homeland security, or even areas such as credit approval may demand careful thought on this score.) The challenge, then, is to obtain valid data mining results when the party doing the data mining is not authorized to see all of the data.

Two approaches to solving this challenge, both introduced in 2000 under the title "Privacy-Preserving Data Mining", addressed the construction of a decision tree. However, the assumptions, and resulting approaches, were fundamentally different. One is based on the cryptographic notion of Secure Multiparty Computation, with the assumption that the parties authorized to see each subset of the data will collaborate to obtain data mining results without disclosing their data.[15] The second approach assumes a single "data miner", but instead of giving this data miner actual values, individuals inject noise into their data to preserve privacy.[16] Both techniques control the information disclosed (thus protecting privacy), while still enabling the generation of valid data mining results.

To give a brief idea of how these techniques work, suppose that we want to compute the average salary of a group of individuals. In the Secure Multiparty Computation approach, all individuals participate in a protocol to obtain the result. As an example protocol, let the first individual choose a

[14] *E.g.*, CONFIDENTIALITY, DISCLOSURE AND DATA ACCESS: THEORY AND PRACTICAL APPLICATIONS FOR STATISTICAL AGENCIES (Pat Doyle *et al.*, eds., 2001).

[15] Yehuda Lindell and Benny Pinkas, *Privacy Preserving Data Mining*, *in* ADVANCES IN CRYPTOLOGY – CRYPTO 2000 36-54 (2000).

[16] Rakesh Agrawal and Ramakrishnan Srikant, *Privacy-Preserving Data Mining*, *in* PROC. OF THE ACM SIGMOD CONF. ON MGMT. 439-50 (2000).

random number and add their salary to it. They give this sum to the second individual, who adds their salary and passes it to the third individual. At the end, the last individual has:

$$\text{sum}=R_1+salary_1 + salary_2 + salary_3 +\ldots+ salary_n$$

This sum is given to the first individual, who subtracts the random number to get the sum of salaries and divides by n to get the average. Under certain restrictions, it can be proven that what each individual sees gives them no knowledge about the individual salaries. Yao described the general approach;[17] the challenge for privacy-preserving data mining is to develop efficient and practical solutions.

The second approach, data perturbation, can also be demonstrated with the "compute the average" example. Instead of a cooperative protocol, each party provides their information to a central data miner. The key is that they don't provide their *actual* salary; they first add random noise. In this example, assume that each individual i chooses a random R^i, from a distribution centered on 0, and sends $salary_i+R_i$ to the data miner. The data miner now computes:

$$\sum_i (salary_i + R_i) = \sum_i salary_i + \sum_i R_i$$

Since the random values are centered around 0, for a sufficiently large number of individuals we can expect the sum to be close to 0, and we can ignore the last term to obtain the sum of salaries. However, each individual's salary is masked by a random number unknown to the data miner, so individual privacy is protected. The actual techniques for real data mining problems are much more involved than these simple examples.[18]

The ideas described above can be used to reduce the privacy risks posed by data integration, and technical tools may help with other risks as well, although a full solution clearly requires more than just technology.

[17] Andrew Yao, *Protocols for Secure Computations*, PROC. OF THE IEEE SYMP. ON THE FOUNDS. OF COMPUTER SCI., 160-64 (1982).

[18] For a more extensive treatment of both approaches, see Jaideep Vaidya and Chris Clifton, *Privacy-Preserving Data Mining: Why, How, and What For?*, 2(6) IEEE SECURITY & PRIVACY 19-27 (2004).

6. CONCLUSION

The following recommendations were made by the Technology and Privacy Advisory Committee (TAPAC) in its recent report[19] on the use of advanced information technologies to help identify terrorists:

- The Department of Defense (DOD) should safeguard the privacy of U.S. persons when using data mining to fight terrorism.
- The Secretary of Defense should establish a regulatory framework applicable to all data mining conducted by, or under the authority of, DOD, that is known or reasonably likely to involve personally identifiable information concerning U.S. persons.
- DOD should, to the extent permitted by law, support research into means for improving the accuracy and effectiveness of data mining systems and technologies, technological and other tools for enhancing privacy protection, and the broader legal, ethical, social, and practical issues in connection with data mining concerning U.S. persons.

We share the belief that protecting privacy in certain data mining applications requires both a legal framework and privacy enhancing technologies and techniques. Concurrent advances in privacy policy frameworks and technology to support privacy can enable greater rewards from data mining than seem possible today, while at the same time providing stronger privacy guarantees than currently available. To move forward, lawyers must concretely and specifically identify the privacy interests to be protected and technologists must turn their attention to designing data mining architectures and techniques that protect these privacy interests. Law and technology must advance hand-in-hand. All sides of this debate must understand that achieving (many of) their goals require neither unfettered access to data nor a complete lockdown to prevent misuse. We hope this paper contributes to this necessary and urgent conversation.

[19] Report of the Technology and Privacy Advisory Committee, Safeguarding Privacy in the Fight against Terrorism, March 2004.

ACKNOWLEDGMENTS

We thank David Jensen and Michael Pazzani for their contributions to a panel on data mining and privacy at the KDD 2004 conference, which led to this paper.

Chapter 12

ONLINE PRIVACY, TAILORING, AND PERSUASION

Tal Z. Zarsky
Lecturer, University of Haifa – Faculty of Law (Israel); Fellow, Information Society Project, Yale Law School; LL.B. Hebrew University; LL.M., J.S.D. Columbia University Law School

Abstract: This chapter tackles a somewhat neglected realm of the information privacy discourse, by directly examining the specific detriments arising from the systematic uses of personal information collected online. The chapter begins by drawing out the flow of personal information in today's digital environment, while emphasizing the collection, storage, analysis and subsequent uses of such data. The chapter then focuses on a specific use stemming from the information flow – the ability of online content providers to tailor advertisements and marketing materials for every user. The chapter argues that these forms of advertising are more effective than those practiced in other media, and at times might prove to be unfair and manipulative. Therefore, the chapter states that at times regulatory steps must be taken to mitigate these concerns. Finally, the chapter mentions a recent incident in which the tailoring of advertisements on the basis of personal information has caused a somewhat surprising public outcry, and compares these events with the dynamics addressed above.

Key words: information privacy, data mining, online advertising, direct marketing, Internet law and policy

1. INTRODUCTION – OR THE ELUSIVE MEANING OF PRIVACY

Over the last several years, legal scholars have been grappling with the concept of online information privacy – or lack thereof. Online privacy has stirred a great deal of controversy and an endless number of articles in both the professional and popular press have been devoted to this issue.[1] Such increased attention surely stemmed from the ease with which vast amounts of personal information are collected, and subsequently stored and analyzed by the commercial entities operating today's dominant websites. These privacy concerns have also arisen in view of the common contrary belief – that actions carried out online are anonymous, secret, and allow users to separate themselves from their physical/offline personae – a belief that is constantly shattered to the dismay of many netizens.

As the debate regarding information privacy unfolds, regulators and legal scholars are still striving to establish the fundamental building blocks of the discussion: the *reasons* such privacy is essential, the *definition* of information privacy and the direct and precise *detriments* of its overall deterioration. In the process of achieving these theoretical objectives (which nonetheless have practical ramifications), the privacy discourse commonly refers to abstract notions such as liberty or autonomy, and to somewhat anecdotal episodes of privacy breaches and their unfortunate outcomes. To promote the enactment of privacy-enhancing regulation, the literature addressing information privacy also frequently refers to the negative visceral response that arises when the public is confronted with contemporary practices of personal data collection, aggregation and use. Evidence of such responses is gathered from public opinion polls[2] – a process that presents a garden variety of flaws. While these forms of policy analysis provide some advantages and are both interesting and informative,[3] I believe arguments premised on the foundations mentioned above are insufficient;[4] the

[1] For a leading academic example of this issue, see Volume 52 of the STANFORD LAW REVIEW (2000) that is, in part, devoted to a thorough discussion of this matter.

[2] A leading researcher in this area is Dr. Alan Westin; For example, see: *Opinion Surveys: What Consumers Have to Say About Information* Privacy, 107th Congress (2000) (statement of Dr. Alan Westin).

[3] For a discussion as to these trends (and others) in the privacy discourse, see Tal Z. Zarsky, *Desperately Seeking Solutions – Using Implementation-Based Solutions for the Troubles of Information Privacy in the Age of Data Mining and the Internet Society*, 56 ME. L. REV.13, 32 (2004).

[4] I believe that explanations relying on abstract notions and the other elements mentioned in the text are insufficient for mainly two reasons: (1) normatively, it would be unfair to disregard the interests of these large corporations and their stockholders while relying on these elements alone; and, more importantly, (2) practically, these powerful entities will strongly object to privacy-enhancing policies and will do all that is in their power to stall them. Only by presenting powerful arguments that are based on actual and concrete detriments, will privacy advocates overcome this powerful and influential opposition.

conclusions drawn from today's privacy debate might lead to strict data protection rules which, in turn, will cause substantial losses to large commercial entities that benefit directly from the use of personal data. Therefore the foundations of information privacy policy analysis must include an understanding of the actual and systematic (rather than anecdotal) detriments arising from the use of previously collected personal data. In this chapter, I intend to bridge the crucial gap between the commonly used and somewhat abstract rationales for information privacy, the visceral feeling and public uproar associated with breaches of public trust, and the actual and concrete detriments associated with today's personal information market. Only by bridging this gap can the privacy discourse move on to discuss specific forms of privacy-enabling regulation to resolve the issues at hand.[5]

As an overall analysis of all possible detriments is beyond the scope of this chapter, I devote the following pages to a brief discussion concerning certain commercial entities such as online marketers, advertisers, content providers, which constantly interact with the public, and for a variety of reasons have vested interests in influencing it in several ways. At times, their actions amount to an attempt to persuade a specific group of individuals to purchase a product or service, or to adopt a different lifestyle and set of preferences. As I will demonstrate below, today's new digital environment[6] places these commercial entities at a crucial nexus, where they are capable of both collecting vast amounts of personal information and using such data in an efficient manner. In view of these capabilities, I argue that a proper understanding of today's privacy concerns leads to acknowledging the new powers of persuasion vested with these commercial entities, which are at times overwhelming and abusive. These problems amount to actual

[5] It should be noted that this perspective is far from conventional, as the general trend in privacy scholarship is focused on the detriments arising from the mere collection and storage of personal data, and the effects such actions might have on the data subjects' state of mind. At times, concerns stemming from mere collection are indeed sufficient – such as when these concerns focus on the fear that such information will eventually be transferred to the hands of the government. However, it is my opinion that the privacy discourse should pay closer attention to the actual detriments stemming from the personal information flow. This chapter is one attempt to articulate such a specific detriment.

[6] Throughout this chapter, I use the terms "Internet Society" and "digital environment" interchangeably. In general, I refer to elements of our society that are demonstrating traits once solely associated with the Internet – such as connectivity, omnipresent surveillance, cheep and easy storage of readable information and others. Thus, the spreading of other technological innovations, such as digital TV and TiVo, should also be regarded as part of this overall trend.

detriments and concerns associated with data collection and use in the digital age, and therefore mandate regulatory intervention.

It should be noted that the following analysis focuses on the data subject – the individual, or consumer, who is subjected to data collection on the one hand, and is the recipient of content that is tailored on the basis of such data, on the other. An analysis of these issues from the perspective of the opposing commercial entities presents compelling questions as to these entities' property (regarding their data) and free speech (regarding their use of such data) rights. In addition, an overall discussion of this issue must address the various market forces existing within these personal information markets, and whether they tend to mitigate or exacerbate these matters. Such questions, as well as others, will be addressed elsewhere.

2. PERSONAL DATA FLOWS IN THE INTERNET SOCIETY

To correctly identify the detriments stemming from the actions of such commercial entities, I begin by addressing the overall flow of personal information that consists of the collection, storage, analysis and use of personal information – which is again followed by the subsequent collection of feedback regarding the use of personal information, thus generating a new (and yet more advanced and efficient) cycle of data flow. I hereby examine this information flow in its entirety, while taking into account recent business practices and technological advances.[7]

2.1 Collection and Storage

Let us start with *collection* and *storage*. In general, personal data collection carried out by commercial entities has become (and will continue to be), a striking phenomenon for several reasons. First, there are the sophisticated means of actual surveillance – such as cameras - that have shrunk in size, have decreased in price, and are placed almost everywhere by both the state and private individuals. Second, there are the many business and technological innovations that allow private entities to track the locations and actions of their users: the cell phones that transmit their users'

[7] I provide a thorough analysis of such data flows in Tal Z. Zarsky, *Desperately Seeking Solutions – Using Implementation-Based Solutions for the Troubles of Information Privacy in the Age of Data Mining and the Internet Society*, 56 ME. L. REV. 13 (2004), Part I.

locations, the E-Z passes, SmartCards, credit cards, MetroCards and supermarket cards – all means of convenience that in addition to serving their various purposes also facilitate the gathering of personal information by one central entity.[8] Such ongoing collection is also made possible by the decrease in the costs associated with computer memory and storage, which enables firms to save many terabytes of information for which they currently have no apparent use.[9] Data collection also utilizes the constantly improving communication infrastructure that allows personal information to flow quickly and efficiently from its place of collection to other locations, where it is warehoused.

When focusing our attention online, it is apparent that the shift from the "physical mall" to the virtual one provides a huge leap in both the quality and the quantity of collectable personal data. Where in physical space most of the data collected will be confined to a final transaction or purchase, the virtual realm allows for the collection of information pertaining to every act of the "browsing" (pun intended) customer – every click of a mouse or indication of interest.[10] Cookies and other technical means[11] allow Internet content providers to precisely track every user's online conduct, through the various sessions of Internet use and the IP addresses visited. These enhanced abilities to engage in constant, omnipresent surveillance stem from the fact that the Internet provides an overall digital framework, which enables such ongoing tracking and recording.

Online content providers, in their capacities as portal mangers, email providers and e-commerce facilitators gladly partake in this collection frenzy.[12] They gather information regarding their users' online and offline conduct both directly and through the thriving secondary market for personal

[8] For extensive descriptions as to these means of surveillance, see Jeffery Rosen, *A Watchful State*, The New York Times, Oct. 7, 2001 (cameras); Simson Garfinkel, DATABASE NATION: THE DEATH OF PRIVACY IN THE 21ST CENTURY 277 (2000) (MetroCards).

[9] Wal-Mart is the classic example of a corporate entity storing and thereafter using vast amounts of personal information. According to a 2002 report, there are 500 terabytes of data in Walmart's Bentonville data warehouse center (in comparison, the IRS has 40 terabytes stored).

[10] For an excellent description as to the shift between the physical and virtual world (with regard to information privacy), see Jerry Kang, *Information Privacy in Cyberspace Transactions*, 50 STAN. L. REV. 1193 (1998).

[11] For instance, many websites, such as that of the New York Times (www.nytimes.com), require a login name and password prior to reading the newspaper's full articles. In that way, the website operator can easily register all articles accessed by the specific user.

[12] It should be noted that each of these entities might be governed by different forms of privacy rules and regulations. For more on this issue, see Daniel J. Solove & Marc Rotenberg, INFORMATION PRIVACY LAW 572 (2003).

information. By applying the various means of online data collection, content providers not only learn where their users receive their information, but what forms of information they do and do not consume, at what times they consume it, and what other patterns of consumption they display. With such data in hand, content providers can construct an extensive profile for each customer. This profile provides content providers with sufficient data so to effectively map out their users' preferences, interests, habits, and possibly even personal traits, and serves as an adequate foundation to the subsequent stages of data analysis and use I now address.

2.2 Data Analysis

Beyond mere collection and storage, the efficient flow of personal information is made possible through the introduction of advanced means of data *analysis*. These new tools allow data collectors to capitalize on the vast datasets they now control. Clearly, the ability to collect great amounts of personal information is almost worthless if the collectors cannot utilize the data they obtain. Yet this is easier said than done. Constantly gathering data pertaining to every step of every user creates huge databases with billions of entries. Data collectors in general and online content providers in particular, eager to utilize this information gold mine, may not know where to start analyzing this data and may not have the manpower to tackle this feat. These and other difficulties in analyzing today's vast databases, led to the development of a new generation of data analysis applications, generally referred to as Knowledge Discovery in Databases ("KDD") tools, or data mining applications.[13] The development of these new tools has been made possible thanks to significant progress in the fields of computer science and mathematics. Data mining applications are non-hypothetical driven – they do not require that an analyst post an initial query or suggest how the database should be divided into subsections. Instead, such applications employ sophisticated computer algorithms to run through the entire database automatically, searching for patterns and close correlations between various factors.

Data mining tools provide a variety of benefits. The applications require fewer analysts to deal with more data and allow the process of data analysis

[13] For a discussion as to the various traits and benefits of data mining applications, and their impact on privacy concerns, see Tal Z. Zarsky, *"Mine Your Own Business!": Making the Case for the Implications of the Mining of Personal Information in the Forum of Public Opinion*, 5 YALE J.L. & TECH. 4 (2003), Part I. For a technical discussion as to data mining applications and technologies, see David Hand et al., PRINCIPLES OF DATA MINING (2001).

to be carried out automatically. Moreover, when utilizing these applications, database holders are not confined to their traditional concepts and schemes of thought and analysis that might be limited or outdated; data mining can reveal patterns of which both users and collectors were previously unaware. Furthermore, data mining applications engage in both descriptive and predictive tasks. They allow collectors to use partial information about specific users to make educated guesses as to their future actions by combining their data with patterns and clusters derived from previously collected information. Therefore, content providers can make use of data mining applications to predict the tastes and preferences of users without the need to collect an entire array of information that pertains to them. In summation, data mining applications close the sophistication gap and turn vast amounts of collected personal information into manageable sources of knowledge and insight.

2.3 Uses of Personal Information

Returning to our overall review of personal information flow, it is imperative to consider and understand the *uses* of personal information by the content providers mentioned above, while emphasizing the Internet's ability to substantially alter the ways in which these firms interact with individuals. To do so, we must first establish the new role of media in the Internet society. Until recently, providing knowledge to the masses required the use of "mass media" which distributed the same form of content to a large group of people. As media markets became more specialized and shifted toward public segmentation, content providers aimed to provide every segment of the population with its own tailored content. This was achieved using local radio, as well as newspapers, magazines, and personal mailings reaching out to specific crowds (young women, middle-aged men, people that enjoy fishing). Yet these initiatives still could not go beyond a rough segmentation of the population.[14] The emergence of the Internet brought with it new concepts and abilities such as "the market for one" web page and the "Daily Me" newsletter (that is constructed to include only issues and articles that interest a specific user)[15] which potentially enable personal segmentation and tailoring of the content baskets provided to each specific user. These forms of precise personalization are facilitated by the

[14] On these forms of public segmentation, see David Shenk, DATA SMOG 114 (1997).

[15] This term was coined by Nicholas Negroponte in BEING DIGITAL (1995). Cass Sunstein further elaborates on this issue in REPUBLIC.COM (2001). Also, see Sunstein, at 5, for a description of various forms of these applications.

Internet's digital interface that allows for the customization of every interaction between users and website operators (which could be initiated by configurations set at either side). Such tailoring and segmentation is of course achieved through the analysis and use of personal information previously collected from each user.

Finally, it should be noted that this three-tier process (of collection, analysis and use of personal data) must be viewed as an ongoing cycle that is constantly reassessed and refined, thus forming a feedback loop for the process of providing personally tailored content for every specific user.

3. UNDERSTANDING THE PERILS OF TAILORED CONTENT

After drawing out the characteristics of the personal information flow, we are one step away from grasping the actual privacy concerns. To take this step, I now briefly examine whether the attempts of online content providers to influence their prospective customers, through the use of tailored content, are more effective than other widely-practiced forms of advertising and persuasion, and whether these actions should be deemed problematic and even unacceptable.

The effectiveness of advertisements is a speculative topic that is constantly debated. As decisive empirical data as to the actual effects of online advertising is currently unavailable, this analysis must rely upon theoretical models constructed by social psychologists and advertising experts when attempting to understand and simulate human cognition and various decision making processes. There are many frameworks advertisers use to form their ongoing strategies (which might choose to appeal to the subject's emotions or sense of reason). In this chapter, I apply a commonly-used model which refers to optimal advertisements as those that (1) cross the individuals' barriers of *perception*; (2) capture the individuals' *attention*; and (3) and affect their *comprehension*.[16] Many of these objectives can be met with greater efficiency and success online, as in this context advertisers can provide content that is specifically tailored on the basis of personal data

[16] For additional information as to the various forms of advertising and their underlying strategies, see Kathleen Hall Jamieson and Karlyn Kohrs Campbell, THE INTERPLAY OF INFLUENCE: NEWS ADVERTISING, POLITICS, AND THE MASS MEDIA 195 (4th ed., 1997). Also see Jef Richards, DECEPTIVE ADVERTISING 74 (1990, Lawrence Erlbaum).

(while using data mining analysis to handle the vast amounts of personal data they now control). In the next few paragraphs, I briefly explain why.

First, let us turn to *perception*. Crossing the perception barrier usually refers to having the relevant message optimally enter the individual's sensory register. Information regarding the preferred form of perception will refer to what shapes, colors or sounds are optimally perceived by each individual. Advertisers are constantly trying to construct messages that will be perceived in full by their intended audiences. Psychological testing indicates that even though people share certain sensory traits, other personal attributes vary from one person to another, with each presenting a different threshold for his or her sensory register. Therefore, obtaining the ability to tailor messages in accordance with information concerning a specific individual's perception preferences will prove valuable and move advertisers towards their marketing objectives.

Attention refers to a "higher" cognitive process, in which sensory information is passed on for further analysis. Here, the brain blocks most of the sensory data received so to protect us from being overwhelmed by information. In this context as well, measuring and examining the traits of this "attention barrier" leads to different results for different people; furthermore, results vary when testing the same individual at different times and in different situations. Capturing individuals' attention requires information as to their patterns of interests and preferences as well as additional insights – a somewhat different dataset than the one required for effectively crossing the perception barrier. Thus, marketers aiming to "grasp" a specific user's attention will greatly benefit from any form of information concerning the traits and preferences of that target user.

After obtaining access to an individual's attention, advertisers attempt to cause the subject to *comprehend* a specific point – usually that a consumer should consider purchasing a specific product or service (or, phrased differently, that a specific product is important and essential). The most basic model for promoting the comprehension of ideas relies upon advertisers and marketers merely providing information and descriptive knowledge about their product. Yet in today's competitive market, this is clearly not enough. Therefore, marketers promote their brands by (1) differentiating them from their competitors' products, (2) associating them with other notions and ideas the specific consumer values and (3) inducing consumers to "participate." As the meaning of differentiation is relatively simple, I hereby address the last two stages of this "comprehension" strategy and the way the new forms of personal data flow contribute to their success.

At first, a few words on associating products with the relevant consumer. Here, it is apparent that the greater and better the information available about a specific user, the easier it will be to promote such an association with a specific product. As today's content providers have a great deal of knowledge about each of their users and their values (as reflected in their previous consumer preferences), they can make use of such information by framing and tailoring their marketing pitches in a way that emphasizes specific traits and aspects of products that the targeted individual will most likely associate with. The next factor – inducing a consumer to participate – requires somewhat more thought. Generally, the key to meeting this objective is to change the consumer's attitude about a product into one that will move him or her towards action. Here again personal information about consumers is of great importance; scholars as far back as Aristotle have pointed out that any knowledge as to others' values and attitudes is extremely helpful in framing arguments and successfully persuading them and moving them to action. More specifically, advertisers often attempt to alter a consumer's attitude or to achieve a change in affect without providing actual knowledge about the product, but by appealing to the individual's emotions. For example, shame, guilt, and panic are emotions that are effectively invoked by today's advertisers to persuade consumers to take particular actions. Specific knowledge as to the preferences and personality traits of every specific user can inform advertisers as to which specific *emotional* responses they must try to invoke so to move a consumer to act. Knowledge as to what stimuli will likely induce a desired response may now be available to today's content providers as a result of constantly tracking consumers' ongoing conduct.

As these paragraphs have demonstrated, for a successful advertisement campaign that will have an optimal effect on a consumer's perception, attention, and comprehension, advertisers require *knowledge* about their target consumers, and the ability to *target* them directly. In the offline world, marketers gather this form of information by conducting surveys and focus groups in an attempt to determine the perception and attention barriers of their target audiences. Thereafter, they tailor their content by placing specific billboards in particular areas or by adjusting broadcast commercials for the anticipated audience (after first establishing the age and interests of viewers of specific shows or time slots). Yet these attempts to collect and tailor information are inferior to the online dynamic described above. Offline, the knowledge collected is only partial and in many cases gathered through intermediaries. Furthermore, the tailoring is coarse; though correctly anticipating the characteristics of many consumers, advertisers provide others with advertisements that are outside their realm of interest, while neglecting potential consumers who might be "perfect fits" for the product being promoted. Yet the shift to the digital and online realm brings several changes. Since the dynamics of online advertising offers advantages both in collection and in tailoring, it will prove more effective in meeting advertisers' objectives – achieving influence by crossing the crucial barriers of perception, attention and comprehension (every one of which requires a specific dataset that the online environment makes available). Meeting these objectives will also be made possible by applying the data mining technology illustrated above, which will assist content providers in identifying the preferences, traits and other personal attributes of every user and deducing such valuable information from the vast databases that are now at their disposal.

Arguing that the effectiveness of online persuasion raises different and new policy concerns may be a hard sell. Society has grown accustomed to and accepted other seductive media such as TV and film, which advertisers have learned to master in the past decades. The Internet still lags behind, with its limited abilities to generate sound and video – and appears to be a medium lacking the seductive forces of television and cinema. But, as the technological gap between these media shrinks, the day will come when online advertising moves to center stage, and applies its tailoring abilities to provide a superior advertising product.[17] As broadband deepens its penetration of the household market, I believe that day is not far away.

Having established the effectiveness of online tailored commercial content, I now move to examine its potentially problematic outcomes. Here I argue that an effective message for the advertiser can at times mean an unfair and manipulative message for the consumer/recipient. There are several possible legal paradigms that could be applied to the advertising efforts mentioned above. First, regulators must closely examine whether the personally tailored content is truthful and non-deceptive (a problem that is far from novel or specifically pertaining to the digital environment).[18] Yet even in the event that the tailored content meets these criteria, this issue still requires regulatory scrutiny, for fear that such content will prove to be *manipulative* and impede the relevant consumer's *autonomy*.[19] The premise

[17] Note that tailored advertising need not be limited to the narrow "Internet" context and will also be available at other critical junctures at which content providers can provide customized content – such as to cell phones, digital televisions and additional future applications.

[18] It could be argued that the world of tailored advertising provides more opportunities for advertisements that are deceptive *per se* – as content providers have a greater ability to detect those that might be easier to trick or fool by deceptive campaigns and to act accordingly. I will address this aspect elsewhere.

[19] Autonomy, somewhat like privacy, eludes a clear definition. Raz described it as "... *the vision of people controlling to some degree, their own destiny, fashioning it through successive decisions throughout their lives.*" For a further analysis of this issues, see G. Dworkin, THE THEORY AND PRACTICE OF AUTONOMY 20 (1988); J. Raz, THE MORALITY OF FREEDOM 369 (1986).

of this argument is that individuals – be they acting as citizens or consumers – should be guaranteed the right to act in accordance with their own wills and to control their destinies; in other words, they should be allowed to act *autonomously*. When third parties – in this case advertisers and content providers – are provided with powerful tools of persuasion, the individuals' right of autonomous thought and action might be compromised. In other words, when consumers are bombarded with specially tailored marketing pitches and advertisements that will capitalize on their vulnerabilities and take advantage of their weaknesses, their subsequent actions might not be those that they would have chosen, should they have had the opportunity to reflect on these matters in solitude. Therefore, should empirical evidence prove that the dynamic addressed above (which allows for the delivery of fine-tuned content to specific individuals) leads to extremely influential outcomes, government might be required to step in and protect this aspect of individualistic freedom.

Clearly this argument is both complex and controversial. It is complex because it is extremely difficult to differentiate between a fair attempt to influence and an unfair attempt to manipulate. It is controversial because regulators generally refrain from protecting individuals' autonomy in this way. This of course is not to say that regulators have shied away from the regulation of "unfair" advertising. Laws and regulations do address unfair manipulations aimed at children, as they are considered vulnerable, inexperienced and perhaps gullible.[20] They also address unfair manipulations in the context of harmful (though legal) substances such as tobacco and alcohol. Yet building a case for regulating the manipulation of competent adults beyond these exceptions is quite difficult and rarely successful. Moreover, any attempt to regulate such manipulations and shield individuals and society from their effects can be viewed as an act of paternalism, which is a constraint on autonomy as well.

Nevertheless, I believe that the new reality addressed above will lead to a paradigm shift in the influence of online content providers, and that such influence must be addressed and at times regulated. It is time that regulators recognize that advertising should be construed not only as a means to convey information, but as a tool used to affect consumers' preferences and opinions, while at times treading on individual autonomy. Therefore, I

[20] An example of the FTC's intervention at this juncture is the FTC's complaint against R.J. Reynolds regarding the use of the "Joe Camel" figure in cigarette ads. The FTC investigated allegations that such use was targeting young children in an attempt to convince them to adopt smoking habits – an unfair means of advertising. *See* Press Release, Federal Trade Commission, Joe Camel Advertising Campaign Violates Federal Law, FTC says, (May 28,1997).

suggest that the meaning of "fairness" (in the context of "fair advertising") be broadened to include freedom not only from deception, but from overly influential means of communication as well, especially in the digital context addressed throughout this chapter.[21]

In summation, the use of personal information to provide tailored, manipulative content should be considered a detrimental outcome of today's enhanced flow of personal information. While this danger has yet to manifest itself in its most aggressive form, it should nevertheless be considered at this early juncture. Rules and regulators tend to lag behind technology and business – and therefore the discussion regarding this matter should start now. This issue must be considered as an integral part of today's information privacy policy discourse.

4. SOME THOUGHTS CONCERNING SOLUTIONS

An overall discussion of solutions to the concerns articulated above is clearly beyond the scope of this chapter. In what follows, I draw out a general framework as to these solutions, while acknowledging several objections such solutions will surely confront. In short, the key to mitigating the detriments of tailored content is reliance on two distinct notions: (1) providing consumers with notice as to the tailoring of individualized communications, and the use of personal information within this process; and (2) assuring that users receive a balanced mix of messages. In terms of *notification*, I suggest that online content providers should be required to notify recipients of tailored advertisements that the content they are receiving is tailored, to detail the information used in the tailoring process and to provide some additional insights as to the way in which such personal data has led to the final outcome.[22] When faced with such notification, users could move to guard themselves on both sides of the personal information flow – both by taking actions to avoid the collection of personal information

[21] For an additional analysis of this perspective towards advertising and its regulation, see generally, Iain Ramsay, ADVERTISING, CULTURE AND THE LAW – BEYOND LIES, IGNORANCE AND MANIPULATION (1996).

[22] As I explain elsewhere, a recent settlement reached between several states (led by the State Attorney General of New York) and DoubleClick Inc. includes a similar provision by introducing the "Cookie Viewer" application, which will (among other things) inform users as to how their personal information affects the targeting process; For more on this issues, see Tal Zarsky, *Cookie Viewers and the Undermining of Data-Mining: A Critical Review of the DoubleClick Settlement*, 2002 STAN. TECH. L. REV. P1 (2002).

they might deem harmful (through the use of various forms of "anonymizing" technologies or by opting out of various collection schemes), or by making a mental note of the manipulative nature of the tailored content – and in that way countering its harmful effects. It should be noted that this "notice" requirement differs from today's popular notions of requiring notification at the time personal information is collected[23] as opposed to notification at the time of information use. I hereby advocate the latter, as it allows users to perceive the possible manipulative uses of personal information, rather than merely acknowledging the collection of personal data.

By addressing *balancing*, I suggest that a suitable remedy to unfair, partial and manipulative advertising is assuring an open marketplace of ideas.[24] Therefore, steps must be taken to assure that no single firm or voice will dominantly influence a specific set of consumers. This could be achieved by requiring dominant online players to operate a non-discriminating platform and implementing minimal changes in browser and search engine technologies. In other words, powerful players in the online content market should be required to "deal" with all advertisers and should not be permitted to discriminate among them for strategic reasons alone.[25]

Clearly, much more research is required at this juncture; especially an inquiry as to whether these solutions will constitute excessive impositions on the free speech rights of content providers that will be forced to comply. In doing so, the legal analysis must distinguish among commercial and other forms of speech, and examine whether specific Internet realms should be considered public fora. I leave these important research tasks for future analysis.

[23] The requirement to notify a "data subject" of the collection of personal information is a basic privacy right according to the Fair Information Practices. For more on this issue, see *Fed. Trade Comm'n, Privacy Online: Fair Information Practices in the Electronic Marketplace*, 106th Cong. (2000), *available at* http://www.ftc.gov/os/2000/05/index.htm#22.

[24] In other words, I here argue that a balance of many different advertisements and messages might, to a certain extent, mitigate the detriments of the tailoring process – by allowing competing messages to offset each other.

[25] At this juncture, additional analysis is required to establish the reasons that market mechanisms will be insufficient in dealing with this matter.

5. CODA: THE GMAIL EXAMPLE – OR WHY THE PUBLIC IS USUALLY RIGHT

I conclude this short chapter by addressing a recent turn of events that demonstrates the ideas developed above. In 2004, Google shocked the free email market (as well as the broader Internet arena) by introducing *Gmail* - its free email service. Gmail promised free access to a generously large email account (1 gigabyte of memory)[26] which is supported by a sophisticated and user-friendly interface that incorporates Google's search technology into the user's inbox. But there was a "catch" to this generous offer, as Google revealed that it would also provide users with tailored advertisements which would be posted throughout the email interface. Google further explained that it would "tailor" such advertisements by analyzing the contents of a user's incoming and outgoing email messages.[27] This final aspect of the Gmail service has led to a public uproar and was considered a dangerous breach of users' privacy by several privacy advocates. However, many technologists found this reaction somewhat surprising, for several reasons. First, recent court decisions have held that users have a very limited expectation of privacy within their free email accounts. Furthermore (and as Google executives quickly pointed out) this service was voluntary (not to mention, free) and therefore users could accede to this privacy standard or take their business elsewhere. Finally, Google may have the ability to automatically scan email messages, yet it cannot link the authors and recipients of these messages to the actual "physical" individuals from whom they originated, thus minimizing "traditional" privacy concerns that focus on the transfer of personal information to the government.[28]

Nevertheless, and in view of the analysis provided above, the public's visceral response to Google's new scheme was indeed warranted and reflects a serious concern, even though few have been able to articulate it correctly. This new application brings together all the elements drawn out throughout this chapter, and might lead to the emergence of unfair and manipulative forms of advertisements on the basis of personal information. Therefore, Google is well advised to adopt the measures stated above, those premised

[26] It should be noted that Gmail accounts have recently been expanded to include over 2 gigabytes of memory. Other free email providers have tried to follow suit by providing larger mailboxes to preferred users as well.

[27] See David Pogue, *Google Mail: Virtue Lies In the In-Box*, N. Y. TIMES, May 13, 2004.

[28] This is due to the fact that the accounts may be acquired and registered anonymously – without providing information regarding the user's offline identity during the registration process.

on *notification* and *balancing*, to respond to the public's concern and in that way provide a product that would be fair to the public and profitable to its operators.

Chapter 13

DATA MINING AND ATTENTION CONSUMPTION

ERIC GOLDMAN
Assistant Professor of Law, Marquette University Law School. Email: eric.goldman@marquette.edu. Website: http://www.ericgoldman.org. Interested readers may wish to read my lengthier article on this topic, A Coasian Analysis of Marketing *(forthcoming).*

Abstract: This Essay challenges the prevailing hostility towards data mining and direct marketing. The Essay starts by defining data mining and shows that the only important step is how data is used, not its aggregation or sorting. The Essay then discusses one particular type of data use, the sending of direct marketing. The Essay establishes a model for calculating the private utility experienced by a direct marketing recipient. The model posits that utility is a function of the message's substantive content, the degree of attention consumed, and the recipient's reaction to receiving the message. The Essay concludes with some policy recommendations intended to help conserve recipients' attention while preserving space for direct marketing tailored to minority interests.

Key words: data mining, database, data warehouse, privacy, advertising, marketing, email, spam, telemarketing, direct marketing, direct mail, junk mail, customer relationship management (CRM), economics of attention, economics of marketing, externalities, Coase Theorem

1. INTRODUCTION

The term "data mining" has developed a pejorative taint. Commentators frequently assume, without explication, that data mining is wrong or harmful,[1] as if the harms of data mining are so universally acknowledged that no one would question the assumption.

[1] *See, e.g.*, Andrew J. McClurg, *A Thousand Words Are Worth a Picture: A Privacy Tort Response to Consumer Data Profiling*, 98 NW. U. L. REV. 63

This Essay questions that assumption. First, I question how data mining, without more, creates consequential harm. If defined properly, data mining appears to be merely a prerequisite to possibly objectionable activity. Second, I question the prevailing hostility towards direct marketing assisted by data mining.[2] While direct marketing imposes some negative utility on every recipient by consuming some of the recipient's scarce attention, direct marketing can enhance overall social welfare. Data mining specifically can increase the likelihood that a particular message enhances social welfare.

Direct marketing's effect on attention leads to three policy observations. First, we should not allow attention consumption concerns to foreclose socially beneficial communications between minority interests. Second, we should not discourage marketers from targeting their marketing communications, including using data mining as appropriate. Finally, we should not discourage the display of summary/preview content that recipients can use to make efficient sorting decisions.

2. DATA MINING AS AN INCHOATE ACTIVITY

Although the term "data mining" is often treated as a term of art, it actually has multiple definitions.[3] To understand the term, we need to understand a bit about database operations as illustrated in the following figure:[4]

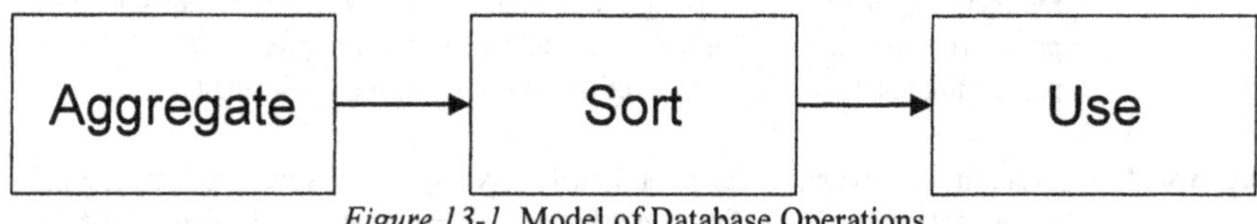

Figure 13-1. Model of Database Operations

To build a database (or "data warehouse"), a data controller first obtains data, either from interactions with data subjects (such as by asking the data subject to volunteer information or by recording interactions between the

(2003) (providing examples of data aggregation and sorting without explaining the specific harms arising from these examples).

[2] *See, e.g.*, ANNE W. BRANSCOMB, WHO OWNS INFORMATION? (1994).

[3] *See* McClurg, *supra* note 1, at 71 & n.50.

[4] *See generally* Tal Z. Zarsky, *Desperately Seeking Solutions: Using Implementation- Based Solutions for the Troubles of Information Privacy in the Age of Data Mining and the Internet Society*, 56 ME. L. REV. 13, 17-31 (2004) (describing an analogous three-stage process).

data subject and the system) or from third party sources, such as data vendors or business partners providing joint services to the data subject.

After obtaining data, the data controller may aggregate the data into one or more databases. In theory, a data controller could aggregate all data collected from or about a data subject into a single database. The reality, of course, is far different. A data controller may obtain data that it does not aggregate into the data warehouse. For example, a data subject may use an "email this page to a friend" web tool where the data subject provides two email addresses (the data subject's and the friend's), but neither address may be added to the data warehouse. Other examples include non-web communications, such as customer support emails, telephone calls or in-person communications, where the data subject may provide valuable data to the data controller but the data controller may lack a technical or operational means to add that data to its data warehouse.

Further, in some cases, the data controller may have multiple independent databases instead of a single unified data warehouse. In these situations, fragmented database architecture may prevent the data controller from "connecting the dots" about that data subject.

Once data is aggregated, the data controller can sort it in a variety of ways, such as (1) using personally identifying information as a criteria (or not), (2) systematically or on an ad hoc basis, and (3) using rudimentary criteria (e.g., provide every mailing address we have) or very sophisticated criteria (e.g., every Wisconsin resident who purchased wool sweaters on a Friday evening during the last month).

After sorting, the data controller or a third party can use the sorted data to take some action or make some decision. For example, sorted data can be used to determine whether a data subject is extended credit, hired (or fired), treated differently from other customers (given enhanced status in a customer loyalty program, for example), or targeted for a marketing communication.

I define "data mining" as data aggregation and sorting done as preparation for some subsequent data use. In doing so, I distinguish data mining from data use. Privacy advocates often consider mere data aggregation and sorting to be harmful, regardless of how the sorted data is used.[5] But how, exactly, does mere data aggregation or sorting cause harm? While data mining may be a predicate to some unwanted use, how do the preparatory activities cause harm by themselves?

[5] *See, e.g.*, A. Michael Froomkin, *Symposium: Cyberspace and Privacy: A New Legal Paradigm?: The Death of Privacy?*, 52 STAN. L. REV. 1461 (2000).

A hypothetical situation illustrates how data aggregation and sorting, without use, lack any meaningful consequences. Assume a data controller aggregates the following data about data subjects into a database: social security numbers, birthdates, addresses, gender, race, sexual orientation and HIV status. The data controller then initiates a query: identify all homosexual Latino males over 40 who live in Texas and have tested positive for HIV, and list their addresses, social security numbers and birth dates. The computer generates a results list. However, the list is not displayed to a human, printed out, archived or further processed by the computer; instead, it is immediately discarded.

This hypothetical is admittedly implausible because this behavior lacks business sense. However, the hypothetical demonstrates the illogic of focusing on aggregation and sorting divorced from usage. How have the data subjects identified on this list been harmed?

Unquestionably, many data subjects would object to public *disclosure* of this information. Being identified on this list might publicize private facts, like sexual orientation or health condition, that could lead to further adverse treatment. Disclosure of the information could allow identity thieves to prey on the individual.

However, based on the hypothethical's parameters, none of these adverse consequences would or could occur. Indeed, no adverse consequence of any sort occurs because the world is exactly the same whether the list is generated or not. The data subject does not experience any change, internally (the data subject never knows that the list was generated) or externally (no one else knows either). This situation brings to mind the ancient Zen parable: if a tree falls in a forest and no one is around to hear it, does it make a sound? Applied to this hypothetical, we might restate the parable: whether the tree makes a sound or not, why do we care?

Some privacy advocates view privacy as a fundamental right[6] or believe that data subjects have the right to control their information.[7] To them, this hypothetical might still be objectionable despite the seeming lack of consequences. By engaging in behavior data subjects might find objectionable if known, a data controller deprives data subjects of control over their circumstances even if the behavior has no perceivable consequences.

[6] *See* Council Directive 95/46/EC of the European Parliament and of the Council of 24 October 1995 on the Protection of Individuals with Regard to the Processing of Personal Data and on the Free Movement of Such Data, art. 1(1), 1995 O.J. (L 281) 31, 38.

[7] *See, e.g.*, Jerry Kang, *Information Privacy in Cyberspace Transactions*, 50 STAN. L. REV. 1193 (1998).

It is, of course, impossible to refute the argument that privacy is a fundamental right. Social scientists cannot empirically prove or disprove the claim, and no single objective source authoritatively classifies what constitutes a fundamental right. Instead, classification of fundamental rights often devolves into an irresolute binary polemic ("Yes it is!" "No it's not!"). I am not attempting to resolve that debate. If people believe that privacy is a fundamental right, this Essay will not convince them otherwise. The rest of us, however, cannot understand how inconsequential data aggregation and sorting is inherently harmful.

While privacy advocates generally reject data mining in all of its forms, many privacy advocates harbor particular animus towards the sale of data by one data controller to another.[8] However, data sales, without more, are indistinguishable from other types of preparatory data mining. We can see this by slightly modifying the prior hypothetical. Assume that data controller #1 generates the results list and electronically sends it to data controller #2 for a fee, but data controller #2 (instead of data controller #1) immediately discards the results list without looking at it or acting on it. In this situation, the data is exposed to a new party (data controller #2), but the data subject still does not experience any consequences from this transfer. It remains as much a non-event to the data subject as the initial hypothetical.

Admittedly, both hypotheticals are highly stylized because the actors' behavior does not make business sense. However, the data controller's motivations for data mining are irrelevant. Even the worst motivations do not make inconsequential behavior consequential. Data mining becomes consequential only when the data mining leads to some impactful use. In other words, all of the preparatory steps prior to data use result in harm only if the usage results in harm. Therefore, to determine the possible harms from data mining, we need to understand how the data is used.[9]

In this respect, data mining concerns are analogous to concerns about the regulation of new technologies. The technology community regularly argues that regulators should outlaw bad technology uses, not the technology itself. I advocate a similar approach to data mining. Data mining is not the problem; the problem is bad uses of mined data, and that is where our focus should lie.

[8] *See, e.g.*, Paul M. Schwartz, *Property, Privacy, and Personal Data*, 117 HARV. L. REV. 2055 (2004).

[9] *See* Zarsky, *supra* note 4.

3. MARKETING COMMUNICATIONS AND ATTENTION COMSUMPTION

To many, using personal information to send direct marketing (including junk mail, telemarketing and spam) is the archetypical bad use of data mining. Recipients passionately hate direct marketing[10] and may transfer negative feelings to data mining by association. However, the academic literature is surprisingly opaque about why consumers hate direct marketing.[11] Why do people object to direct marketing so much; or, stated in economic terms, how does direct marketing create negative utility?

3.1 A utility model of direct marketing

To analyze these questions, we can model an individual recipient's utility from a single direct marketing communication. The total utility contains three discrete components.

Substantive Utility. To the extent that the recipient is substantively exposed to the content (as opposed to discarding the message without being exposed to the substantive content), the recipient derives utility from the communication's content. An individual's response to the communication's contents can vary from highly positive (e.g., contents will lead to a transaction producing significant consumer surplus) to highly negative (e.g., the contents were uninteresting and offensive/objectionable).

Attention Utility. Attention is a scarce resource, both temporally (we have only a certain number of attention-minutes in our lives) and at any one time (we can only pay simultaneous attention to a limited number of focal targets). Each communication consumes some of the recipient's attention.

Reaction Utility. Recipients may derive utility from their reaction to receiving a communication. For example, many recipients are annoyed to

[10] *See, e.g.*, *Consumer Perceptions of Various Advertising Mediums*, Dynamic Logic Beyond The Click (Mar. 2004) (showing that consumer attitudes towards telemarketing, spam and direct mail are significantly more negative than ads delivered through other media), *at* http://www.dynamiclogic.com/na/research/btc/beyond_the_click_mar2004_part2.html (last visited Dec. 29, 2004).

[11] *See* Eric Goldman, *Where's the Beef? Dissecting Spam's Purported Harms,* 22 J. MARSHALL J. COMPUTER & INFO. L. 13 (2003) (discussing possible reasons why recipients hate spam).

receive a telephone call during dinner or a favorite TV show, regardless of the communication's contents.[12]

Equation (1) recaps the model for an individual's utility derived from a single marketing communication:

$$NPU = SU + AU + RU \quad (1)$$

where
NPU = net private utility
SU = substantive utility
AU = attention utility
RU = reaction utility

3.2 Attention consumption and externalities

Compared to SU and RU, the AU component of equation (1) has received little scrutiny from commentators. This is somewhat surprising because all recipients experience negative utility from having their scarce attention consumed. Thus, AU is the only component that is guaranteed to be negative for all recipients (with respect to SU and RU, recipients may derive positive, negative or no utility).

Because direct marketing communications create negative AU for every recipient, some commentators have analogized these communications to negative externalities like pollution.[13] Taking this argument to its logical conclusion, marketers overproduce direct marketing communications because the true social cost (including the negative AU imposed on every recipient) is greater than the marketers' private costs. If so, economically efficient levels of production could be reached by forcing the externality producer (the marketer) to internalize the externalized costs (the attention consumed), such as through a cost-internalizing tax (a "Pigouvian tax").

[12] While typically consumers derive negative utility from their reaction to the interruption/intrusion of direct marketing communications, in some cases this reaction can generate positive utility. *See* Susan Chang & Mariko Morimoto, *An Assessment of Consumer Attitudes toward Direct Marketing Channels: A Comparison between Unsolicited E-Mail and Postal Direct Mail* (Apr. 1, 2003) (quoting one student as getting a "thrill" when he receives mail, even if it is "junk"), *at* http://www.inma.org/subscribers/papers/2003-Chang-Morimoto.doc (last visited Dec. 29, 2004).

[13] *See, e.g.*, Kenneth C. Laudon, *Markets and Privacy*, COMM. OF THE ACM, Sept. 1996, at 92.

Alternatively, some commentators have proposed schemes that would allow consumers to shift costs to marketers.[14]

Unfortunately, the analogy between pollution and direct marketing does not survive critical scrutiny. Pollution constitutes a negative externality because it imposes negative utility on everyone. By focusing solely on the AU component of equation (1), direct marketing looks analogous to pollution because it too imposes negative utility on all recipients. However, it is analytically inaccurate to isolate a single component of equation (1) to assess the social utility of direct marketing. While pollution uniformly generates negative utility on all affected individuals, direct marketing recipients can experience positive NPU from the communication even though the AU component may be negative.

In addition, direct marketing may facilitate marketplace competition.[15] As a result, non-recipients of a marketing communication may benefit from lower prices or better products caused by the marketing communication. This non-recipient effect may constitute a *positive* externality from the communication.

In theory, we could calculate the social welfare impact of a direct marketing communication by adding together all recipients' NPUs, the marketer's net private utility, and any externalities (positive or negative) generated by the marketing communication.

In practice, of course, this calculation cannot be made on an *ex ante* basis because the recipients' interests are heterogeneous but undisclosed. No one – not the government, not the marketer, perhaps not even recipients themselves – precisely knows the recipients' substantive interests, tolerance of attention consumption or reaction to receiving a communication.

Indeed, a recipient's utility may vary from day to day. Consider the following example. A marketer delivers to Jane a coupon offering a $100 discount on a new Dell computer. In scenario 1, the coupon arrives after Jane has already decided to buy a Dell but before she has made the purchase. In this case, the coupon may generate significant positive utility for Jane. In scenario 2, the coupon arrives immediately after Jane has made her purchase and cannot take advantage of the coupon. In this case, the coupon may be irrelevant; indeed, it could be upsetting because Jane may develop buyer's remorse (because she thinks she overpaid). As this example illustrates, the

[14] *See* Ian Ayres & Matthew Funk, *Marketing Privacy*, 20 YALE J. ON REG. 77 (2003); Laudon, *supra* note 13. According to the Coase Theorem, it is unclear how these cost-shifting efforts would change the outcome. *See* Ronald Coase, *The Problem of Social Cost*, 3 J.L. & ECON. 1 (1960).

[15] *See* Lee Benham, *The Effect of Advertising on the Price of Eyeglasses*, 15 J.L. & ECON. 337 (1972).

utility generated by a particular marketing communication varies dynamically, making reliable computations impossible.

3.3 Policy implications

While the social welfare effects of direct marketing communications may be indeterminate, the prior discussion suggests three policy observations.

3.3.1 Observation #1: Everyone must tolerate some communications of interest only to a minority of recipients

Because the social welfare effects from a particular direct marketing communication are unknown (and unknowable) *ex ante*, we cannot assume the message has negative net social welfare. Indeed, even if most recipients experience negative utility from a communication, the communication could still create net positive social utility if the remaining recipients (or non-recipients benefiting from positive externalities) experience more positive NPU than the negative NPU experienced by the majority recipients. This scenario most likely arises when a communication is extremely useful to a minority community and the majority of recipients are simply uninterested. Preferably marketers will do more targeting (discussed in Observation #2) to avoid unnecessary impositions on uninterested recipients, but this is not always possible.

Over-responding to majority interests regarding messages that are broadly unpopular but extremely useful to minority recipients could reduce social welfare. Ultimately, members of the majority must have some attention consumed by unwanted or irrelevant messages as the "price tag" of allowing members of the minority to communicate with each other.[16]

[16] Everyone has some interests that are minority in nature. Our obligation to tolerate – and defend – minority interests that do not coincide with our own interests reminds me of the poem *First They Came...* by Reverend Martin Niemöller:

> First they came for the communists, and I did not speak out—
> because I was not a communist;
> Then they came for the socialists, and I did not speak out—
> because I was not a socialist;
> Then they came for the trade unionists, and I did not speak out—
> because I was not a trade unionist;
> Then they came for the Jews, and I did not speak out—
> because I was not a Jew;
> Then they came for me—
> and there was no one left to speak out for me.

3.3.2 Observation #2: Facilitate marketer targeting

While we each must tolerate some unwanted messages catering to minority interests, social welfare would improve if marketers did a better job of targeting their messages. In an ideal world, marketers would communicate only with recipients who derive positive SU from the message, and recipients would receive only communications that create positive SU.[17] While we may never reach this idyllic state, we would nevertheless benefit by encouraging marketers to do more targeting.

Data mining can help marketers with this targeting, which in turn could increase social welfare. Therefore, it would be counterproductive to set up a regulatory scheme that discourages marketers from engaging in data mining.

Beyond data mining, marketers are developing other technologies to infer consumers' undisclosed and latent interests. An example of this technology is "adware," which monitors an individual's online behavior and, based on the activity, generates marketing communications that reflect the adware vendor's prediction of the individual's interests. These technologies may allow marketers to significantly improve message targeting.

However, a regulatory assault on relevancy-improving technologies threatens their development. For example, in March 2004, Utah enacted a law prohibiting adware vendors from delivering certain types of contextually relevant advertising *even if the recipient wanted it*.[18] Although this law was preliminarily enjoined in June 2004,[19] it represents an all-too-common regulatory paranoia about marketers engaging in social welfare-enhancing targeting efforts.

In part, this paranoia reflects the inherent tension between "privacy" and relevancy targeting. To provide consumers with more highly targeted marketing, marketers must know more about individuals' interests – including latent interests that individuals may not be able to articulate even if asked. However, the more data that marketers capture about individuals, the greater the privacy concerns. As discussed in Part 2 *supra*, this tension can be resolved only by focusing on bad uses of data, not by prophylactically inhibiting data mining or other data aggregation/sorting techniques. When

[17] *See* JOHN HAGEL III & MARC SINGER, NET WORTH: SHAPING MARKETS WHEN CUSTOMERS MAKE THE RULES (Harv. Bus. Sch. Press 1999) (arguing that "infomediaries" would mediate communications between marketers and consumers to improve marketing targeting for both groups).

[18] Utah Spyware Control Act, H.B. 323, 2004 General Session, Part 2 (Utah 2004).

[19] WhenU.com, Inc. v. Utah, Civil No. 0407097578 (Utah Dist. Ct. June 22, 2004), *available at* http://www.benedelman.org/spyware/whenu-utah/pi-ruling-transcript.pdf.

regulating bad data uses, we must be very careful not to mischaracterize socially beneficial targeting as a bad data use.

3.3.3 Observation #3: Facilitate recipients' ability to make predictive relevancy judgments

Each marketing communication necessarily consumes some of the recipient's attention for sorting purposes. Ordinarily, a recipient will scan a marketing communication to make a predictive judgment about the message's relevancy to the recipient's interests.[20] If the recipient initially deems the message irrelevant, the recipient will usually discard the message. On the other hand, if the recipient initially makes a predictive judgment that the communication may be relevant, the recipient usually will then do a more careful review to make an evaluative judgment of relevancy.[21]

This two-stage review process suggests that social welfare can increase if recipients can make more efficient predictive judgments, thereby lessening the need to make more time-consuming evaluative judgments. In other words, each recipient will experience less negative utility due to attention consumption if the recipient can sort the communication faster.

Recipients can make quicker predictive judgments if the marketing communication provides easy-to-scan content that summarizes or previews the marketing communication's contents. I refer to this summary or preview content as "filtering content." An email's subject line is an example of filtering content. If the subject line contains useful predictive information, many uninterested recipients can delete the email without having to open the email and scan its contents. While reviewing email subject lines still consumes some of the recipient's attention, the quick sorting can reduce the negative utility.

Regulators recognize the importance of filtering content inconsistently. In some situations, regulations mandate filtering content. For example, certain emails now must be labeled "advertising" or "sexually explicit"[22] and

[20] *See* Soo Young Rieh, *Judgment of Information Quality and Cognitive Authority in the Web*, 53 J. AM. SOC'Y FOR INFO. SCI. & TECH. 145, 150 (2002).

[21] *Id.*

[22] *See* 15 U.S.C. § 7704(a)(5)(A)(i) (commercial electronic mail messages must provide "clear and conspicuous identification that the message is an advertisement or solicitation"); 16 C.F.R. § 316.1 (commercial electronic mail messages that include sexually oriented material must "include in the subject heading the phrase 'SEXUALLY-EXPLICIT:' in capital letters as the first nineteen (19) characters at the beginning of the subject line").

telemarketers must display their name and phone number to recipients using caller identification devices.[23]

However, regulations do not mandate or encourage filtering content in comparable circumstances. For example, senders of direct mail are not required to include any filtering content on the envelope's exterior – no name, return address or other predictive content of any type. As a result, direct mail recipients often must open the envelope to make any predictive judgments.

In yet other situations, the regulatory scheme discourages filtering content. For example, some Internet trademark cases have imposed liability on intermediaries merely for displaying filtering content to searchers, irrespective of the filtering content's usefulness in making predictive judgments.[24] This type of liability may curtail the display of filtering content seen by searchers, thus unnecessarily consuming more attention.

While regulations should not inhibit the provision of accurate filtering content, the converse proposition – regulators should require marketers to provide accurate filtering content – does not always hold true. Specifying that recipients be exposed to filtering content structures how the recipients' attention will be consumed, which may require more attention from recipients than would have been required without the regulation. If so, the additional benefits of the mandatory filtering content need to be weighed against the implicit attention consumption costs.[25] Mandatory filtering content could create social welfare, but regulators rarely or never balance the attention consumption costs they are imposing – and they should.

Furthermore, even when mandatory filtering content creates positive net social welfare, the filtering content should reflect how individuals process information. Usually, regulators determine what mandatory filtering content to require based on intuitive assumptions about individuals' informational needs. However, regulators often have no training in human cognitive processes and thus may require unhelpful filtering content (or fail to require useful filtering content). To the extent that regulators determine that recipients should get filtering content, regulators should rely upon experts,

[23] *See* 16 C.F.R. § 310.4(a)(7) (defining as an abusive telemarketing practice the failure "to transmit or cause to be transmitted the telephone number, and, when made available by the telemarketer's carrier, the name of the telemarketer, to any caller identification service in use by a recipient of a telemarketing call").

[24] *See, e.g.*, 1-800 Contacts, Inc. v. WhenU.com, Inc., 309 F. Supp. 2d 467 (S.D.N.Y. 2003).

[25] There may be other costs to consider, such as the marketer's production and compliance costs.

such as information scientists, to structure the filtering content in a useful and efficient manner.

4. CONCLUSION

People who obsess over data mining are misdirecting their energies. Data mining is not the problem; at worst, it is just a preparatory step towards a problem. In the case of direct marketing, data mining is a *beneficial* preparatory step, and social welfare would improve if direct marketers used it more often.

The obsession with data mining as a standalone harm has also masked the important role of attention consumption in direct marketing. Properly isolated, we can see the value of trying to conserve attention where it is conservable; but we also recognize that everyone must sacrifice some attention to preserve breathing room for communications catering to minority interests.

Chapter 14

IS PRIVACY REGULATION THE ENVIRONMENTAL LAW OF THE INFORMATION AGE?

DENNIS D. HIRSCH
Capital University Law School

Abstract: This chapter argues that information-based businesses injure personal privacy in much the same way that smokestack industries damage the environment, and that this analogy can teach us something about how to preserve privacy better in the information age. The chapter shows that two of the principal constructs that have been used to understand environmental damage – the negative externality, and the tragedy of the commons – apply equally well to privacy injuries. Thus, a common conceptual structure can be used to understand both environmental damage and privacy injuries. Can environmental law and policy serve as a model for the nascent field of privacy protection? The chapter examines the evolution of environmental regulation from first generation, "command-and-control" methods to more flexible, second generation strategies. It argues that first generation approaches, while appropriate for addressing some social ills, are not a good fit for the regulation of fast-changing information businesses. Second generation strategies, which demand meaningful results while also providing flexibility and reducing regulatory costs, will work better in the privacy context. The chapter concludes that environmental covenants and environmental management systems can be adapted to protect personal information. An emission fee approach can be used to combat spam effectively.

Key words: information revolution, privacy, data privacy, online privacy, privacy regulation, spam, e-mail, personal information, data mining, electronic commerce, e-commerce, privacy policy, environmental law, environmental policy, externality, tragedy of the commons, environmental covenant, environmental management system, emission fee

Just over a century ago, Western nations experienced an Industrial Revolution.[1] It was based on new technologies and forms of production, and gave rise to many new businesses and products.[2] It also generated environmental pollution at unprecedented levels.[3] Over time, a new form of law – environmental law – emerged to address these injuries.[4] During the past fifteen years, the world has experienced a second industrial revolution, the Information Revolution.[5] This revolution, too, is premised on new technologies and is giving rise to new forms of business. It is also generating new types of harms. This time, however, the harms are not to the external environment, but to the internal one. They are harms to privacy.

As with the environmental harms of a century ago, the law is now seeking to respond to these new injuries to privacy. Legal scholars have drawn on property law,[6] contract law,[7] and even trade secret law[8] to develop strategies for protecting privacy. However, they have largely ignored environmental law.[9] This chapter argues that environmental law can make a large contribution to the protection of privacy. On the level of theory, two of the main constructs that have been used to understand environmental injuries – the negative externality and the tragedy of the commons – can be effectively applied to privacy harms. One the level of policy, the regulatory instruments that have, through much trial and error, been developed in the environmental field can provide useful models for regulation to protect privacy.

[1] *See generally,* T.K. Derry & Trevor I. Williams, A SHORT HISTORY OF TECHNOLOGY: FROM THE EARLIEST TIMES TO A.D. 1900 (1960).

[2] *Id.*

[3] Eric Pearson, ENVIRONMENTAL AND NATURAL RESOURCES LAW 1 (2002).

[4] *See generally* Philip Shabecoff, A FIERCE GREEN FIRE: THE AMERICAN ENVIRONMENTAL MOVEMENT (1993) (describing the evolution of American environmental law).

[5] Howard Isenberg, *The Second Industrial Revolution: The Impact of the Information Explosion*, 27 IND. ENG. 14 (March 1, 1995).

[6] *See e.g.*, Jacqueline Lipton, *Information Property: Rights and Responsibilities,* 56 FLA. L. REV. 135 (2004); Laurence Lessig, *The Law of the Horse: What Cyberlaw Might Teach*, 113 HARV. L. REV. 501, 520-22 (1999).

[7] Paul M. Schwartz, *Privacy and the Economics of Personal Health Care Information*, 76 TEX. L. REV. 1 (1997).

[8] Pamela Samuelson, *Privacy as Intellectual Property*, 52 STAN L. REV. 1125 (2000).

[9] One scholar has noted the value of the environmental movement for the field of intellectual property. *See* James Boyle, *A Politics of Intellectual Property: Environmentalism for the Net?*, 47 DUKE L.J. 87 (1997) (arguing for a politics of intellectual property modeled on the environmental movement This chapter heads in a related, but different, direction by applying environmental pollution law and policy to the regulation of privacy.)

1. HARMS TO PRIVACY

The Information Revolution injures personal privacy in at least two major ways. First, as many have recognized,[10] the increased generation, availability and use of personal information decreases an individual's ability to control access to such information. This impairs "informational privacy." Second, unsolicited "spam" e-mail invades privacy. While many have analyzed the problem of spam, few have framed it as a privacy issue.[11] This chapter will argue that spam, by invading our virtual space, infringes on our privacy.

1.1 Harms to informational privacy

In their foundational law review article, published in 1890, Louis Brandeis and Samuel Warren conceived of privacy as "the right to be let alone."[12] They argued that this right protects an individual against unwarranted intrusion into her personal realm and, if violated, gives rise to an action in tort.[13] In the years that have followed, this right has taken strong root in the law and has branched out in multiple directions. One of these has been the recognition of "informational privacy."[14] This branch responds to modern society's tendency to generate and make available increased quantities of personal information. It treats this tendency as an invasion of the personal realm. It provides individuals with certain rights over the "collection, use and disclosure" of their personal information.[15]

The Information Revolution threatens informational privacy in a number of ways.[16] The increasing digitization of personal records, and the relative ease and speed with which they can now be searched, have made it possible to put together "computer profiles" of individuals that offer snapshots of their purchases, health issues, financial situation, habits and the like.[17] The rise of the Internet, e-commerce and "cookies" have made it possible to track

[10] *See, e.g.* Daniel J. Solove & Mark Rotenberg, INFORMATION PRIVACY LAW 1 (2003).

[11] For two who have, see Robert H. Hahn & Anne Layne-Farrar, *The Benefits and Costs of Online Privacy Legislation*, Brookings Joint Center for Regulatory Studies Working Paper 01-14, 2 (October 2001).

[12] Louis Brandeis & Samuel Warren, *The Right to Privacy*, 4 HARV. L. REV. 193 (1890).

[13] *Id.*

[14] *See generally* SOLOVE & ROTENBERG, *supra* note 10 (describing the history and contours of information privacy).

[15] *Id.* at 1.

[16] *See, e.g., id.*

[17] Richard C. Turkington & Anita L. Allen, PRIVACY LAW: CASES AND MATERIALS 398 (2nd ed. 2002).

an individual's travels through cyberspace and so to learn even more about that person.[18] Finally, data mining uncovers, and brings to light, the latent information that can be gleaned from the relationships between disparate pieces of data about an individual.[19] In these ways, and others, the Information Revolution has harmed informational privacy.[20]

1.2 Harms to spatial privacy

A second important offshoot of Brandeis and Warren's arguments for a right to privacy has been the rise of legal doctrines that protect an individual against invasions of her personal spaces. For example, the Restatement (Second) of Torts recognizes the tort of "intrusion upon seclusion" when one "intentionally intrudes, physically or otherwise, upon the solitude or seclusion of another or his private affairs or concerns."[21] Unauthorized entry into another's private home violates this right.[22] So do "telephone calls [that] are repeated with such persistence and frequency as to amount to a course of hounding the plaintiff, that it becomes a substantial burden on his existence."[23]

Today, a person's in-box is as important a personal space as her living room or her phone line. The spammer's daily invasion of this virtual, personal space is as intrusive as the unwanted visitor to one's living areas, or the harassing telephone caller. If privacy is the "right to be let alone," as Brandeis and Warren defined it, then the incessant hounding by spammers should be recognized as an invasion of that right.[24] This chapter will accordingly discuss two ways in which the information revolution harms personal privacy: the capturing, use and mining of our personal information; and the intrusion into our personal space through spam.

[18] *See* Jerry Kang, *Information Privacy in Cyberspace Transactions*, 50 STAN. L. REV. 1193 (1998).

[19] *See* Joseph Fulda, *Data Mining and Privacy,* 11 ALB. L.J. SCI. & TECH. 105 (2000).

[20] *See* SOLOVE & ROTENBERG, *supra* note 10, at 1.

[21] RESTATEMENT (SECOND) OF TORTS § 652B (1977).

[22] *Id.*, comment (b).

[23] *Id.*, comment (d).

[24] *See* Hahn and Layne-Farrar, *supra* note 11, at 2 (interpreting the right to be left alone as including "the right to refuse unwanted e-mail or online solicitations").

2. PRIVACY HARMS ARE THE ENVIRONMENTAL HARMS OF THE INFORMATION AGE

Privacy harms are conceptually similar to environmental injuries. Two of the principal constructs that have been employed to understand environmental injuries – the negative externality, and the tragedy of the commons – can usefully be applied to privacy harms as well.

Negative externalities are costs of an activity that are borne, not by the actor herself, but by others in society.[25] For example, consider a company that uses steel to make a product. The company must pay for this commodity; it must bear the cost of using up that amount of steel. If the company's factory emits air pollutants from its smokestack, thereby causing respiratory problems and other illnesses in the surrounding populace, this too creates a social cost. But it is one that, in the absence of regulation or a successful tort suit, the company does not have to bear. It is able to "externalize" this cost onto the surrounding community. Since it does not have to bear the cost of this pollution, the company has little incentive to minimize it. Instead, it will wastefully "use up" the clean air resource in a way that it would never consume a resource, such as steel, for which it had to pay.

The "tragedy of the commons," another core construct used to understand environmental harms, applies to commonly owned resources. The classic example, drawn from an essay by Garrett Hardin, is of cattle herders who graze their animals on a grass field that is open to all.[26] A cattle herder gets the full benefit of adding another animal to the pasture. However, the herder shares the cost of adding that additional animal, in terms of the exploitation of the grass field, with all others who have rights to the field. From an individual perspective, it is rational for the herder to keep bringing more and more cattle to graze in the field. The same is true for all the other herders. This dynamic leads to the field being so overgrazed that it cannot regenerate itself and becomes useless for grazing purposes. All the cattle herders thus lose this valuable resource. What was individually rational turns out to be collectively destructive.

[25] *See* James Salzman & Barton H. Thompson Jr., ENVIRONMENTAL LAW AND POLICY 17 (2003) (defining "negative externality").

[26] Garrett Hardin, *The Tragedy of the Commons*, 168 SCIENCE 1243 (1968).

2.1 Spam as an environmental harm

These constructs, often used with reference to environmental injuries, can also help to explain the phenomenon of spam. Spam imposes social costs. These include the time spent downloading and deleting spam messages, the cost of phone line usage while performing this activity (the majority of Americans still use dial-up modems), the cost of installing filters to screen out unwanted messages, and costs due to the loss of desired messages that are mistakenly deleted as a result of the filtering process.[27] These costs are not borne by the spammer. Instead, they are externalized onto the recipients of the spam. Spam spewed from a computer thus creates negative externalities in much the same way that emissions from a smokestack do.

If left unchecked, spam will also generate a tragedy of the commons. Here, the commons is the in-box (or, more properly, all in-boxes), since almost anyone who learns of an e-mail address can enter the corresponding in-box.[28] When a spammer sends a message, she gets all the benefit of that communication in terms of selling or advertising a product. By contrast, she bears only a small fraction of the cost of that message, most of it having been externalized onto the recipient as described above.[29] Much as the cattle herders in Garrett Hardin's essay have an incentive to keep adding more head of cattle until the common area is overgrazed and destroyed,[30] so each spammer has incentive to continue sending more and more spam messages in the hope of securing one more customer. This is, in fact, what we see: 140 billion spam messages sent in 2001, 260 billion in 2002, and nearly 2 trillion in 2003, with no end in sight.[31] The tragedy, which is already beginning to occur, will be when excessive spam makes e-mail unattractive

[27] *See* Senate Report on the CAN-SPAM Act of 2003, S. Rep. 108-102, 2004 U.S.C.C.A.N. 2348, 2353 [hereinafter, Senate Report] (describing these harms).

[28] An in-box is not a perfect commons. Individuals may protect it by refusing to reveal the e-mail address. Few people do this and this measure, itself, seriously constrains the usefulness of the in-box and may be thought of as a cost of spam. In addition, Internet Service Providers and/or organizations or individuals themselves may install filters to keep out unwanted e-mail messages. Experience has shown that such filters are imperfect. Spammers respond by increasing the volume and variety of spam, thereby finding ways to avoid the censors. The addition of filters may be partially responsible for the exponential increase in the volume of spam.

[29] The spammer bears only the marginal cost of sending the spam message coupled with the cost of having her own in-box filled up with spam (assuming that she has a valid e-mail address of her own that is subject to spamming).

[30] HARDIN, *supra* note 26.

[31] Senate Report, *supra* note 27, at 2349.

to the user and people seek alternatives to this communications medium.[32] This will destroy e-mail for all of us, including the spammers. Like the cattle herders who overgraze the common pasture,[33] spammers will have destroyed the very resource on which they themselves rely.

2.2 Invasion of informational privacy as an environmental harm

These concepts also apply to privacy harms that result from the excessive use of personal information. When a web merchant or data miner collects and uses someone's personal information, it imposes costs on that person by taking away some of her personal privacy. Marketers and data miners do not bear these costs. Instead, they externalize them onto the individuals whose personal information they are using. The cost thereby qualifies as a negative externality.[34] Whereas smokestack pollution affects the world around us, the injuries here are to the inner environment, the sense of privacy. One might even refer to them – if the phrase is not too glib – as "*internal* externalities."[35]

Once again, tragedy lurks around the corner. Here, the commons is the collective willingness of individuals to share their own personal information on the web. Businesses that use this information claim most of the benefits of doing so and bear only a small fraction of the costs.[36] Each such business accordingly has an incentive to use more and more personal information. This will eventually lead to a point at which individuals feel so insecure about sharing their information on the web that they turn away from e-

[32] *See, id.*, at 2352 ("Left unchecked at its present rate of increase, spam may soon undermine the usefulness and efficiency of e-mail as a communications tool.")

[33] *See* Hardin, *supra* note 26.

[34] Hahn & Anne Layne Farrar, *supra* note 11, at 16 ("In economic terms, the companies collecting personal information impose a negative externality on consumers.") Of course, businesses that use personal information often also create positive externalities for the individuals concerned. For example, a website's arrangement to share information about an individual customer with a web marketing firm may result in that person receiving targeted marketing that she finds valuable. The goal of regulation in this area should not be to eliminate all use of personal information. Rather, it should be to try to preserve the positive externalities associated with such use while, simultaneously, minimizing the negative ones.

[35] I must give credit to my friend, Professor Craig Nard, for helping me to coin this felicitous phrase.

[36] These are the costs of actually manipulating the information, as well as the costs to the individual web marketers or data miners of having their own personal information exploited by others in their field.

commerce. This dynamic is already beginning to occur.[37] If it continues, the willingness of individuals to provide personal information on the web – a resource on which e-commerce depends – will have been over-exploited and used up.[38] All of us will lose, including the website owners, web marketers and data miners themselves. Much like the cattle herders, they too will have destroyed the resource on which their livelihood depends.

3. ENVIRONMENTAL REGULATION AS A MODEL FOR PRIVACY REGULATION.

The conceptual similarity between environmental and privacy harms raises a tantalizing possibility. Might the regulatory strategies developed over the past decades to address the environmental injuries of the smokestack era be similarly applied to the privacy harms of the information age? The answer is a qualified "yes." In particular, the imposition of emission fees, a market-based approach that has been used to address a variety of environmental problems, could be adapted for use in controlling spam. Two other contemporary environmental policy instruments – environmental covenants and government promotion of environmental management systems – could be converted into means for better protecting personal information.

3.1 An emission fee system for spam

Under an emission fee approach, the government requires each polluter to pay a fee equal to the cost that its emissions are imposing on society and the environment.[39] If the fee is set accurately, it should force the polluter to bear ("internalize") the costs that it would otherwise be externalizing onto others.[40] This gives the polluter an incentive to reduce its emissions if the costs of doing so are less than the amount of the fee that it would otherwise have to pay. Under such a system, companies that are able to reduce their

[37] Robert E. Litan, *Law and Policy in the Age of the Internet*, 50 DUKE L. J. 1045, 1058 (2001) (opinion polls demonstrate that "an overwhelming majority of Americans consistently report that they are deterred from using the Internet more than they currently do because of privacy-related fears.")

[38] *See* Neil Weinstock Netanel, *Cyberspace Self-Governance: A Skeptical View from Liberal Democratic Theory*, 88 CAL. L. REV. 395, 474 (2000) ("consumers will not use the Internet for electronic commerce unless they are assured about personal privacy protection.")

[39] James Salzman & Barton H. Thompson, Jr., ENVIRONMENTAL LAW AND POLICY 45 (2003).

[40] Stephen M. Johnson, ECONOMICS, EQUITY AND THE ENVIRONMENT 17 (2004).

emissions relatively cheaply will tend to do so. Businesses for which pollution reduction would be extremely costly will either pay the fee, or shut down. The emission fee approach thus creates a framework in which those who can most cheaply reduce pollution tend to undertake the bulk of the reductions.[41] It also gives a competitive advantage to those firms that can reduce their emissions for less than the cost of paying the fee.

In the spam context, the emissions-fee approach would consist of a government-imposed charge for each e-mail sent. To send an e-mail, one would first have to pay the small fee.[42] Further empirical research will be needed to determine the level at which this fee should be set. For present purposes, assume that a federally-required, uniform fee were set at a tenth of a cent per e-mail. This would dramatically change the incentives that spammers face. A seller of Viagra or "fine Rolex watches" who has been sending a million messages each day would now have to pay $1000 per day, or $365,000 per year, for a privilege that previously had been all but free. This would give such a business a strong incentive to target its messages to individuals with a higher likelihood of interest in the product, rather than spreading them with abandon to all with a valid e-mail address. If we treat spam as a form of pollution, this more refined targeting of messages could be thought of as a kind of pollution prevention. Recent research suggests that it should be technically feasible to implement a fee-per-e-mail system.[43]

Three objections to this idea come immediately to mind. Such a fee may chill beneficial communication *within* organizations. It may prevent non-profit, educational and religious groups from sending out mass e-mails to

[41] Adam Chase, *The Efficiency Benefits of "Green Taxes," A Tribute to Senator John Heinz*, 11 UCLA J. ENVTL. L. & POL'Y 1, 11-12 (1992).

[42] In a January, 2003, speech, Bill Gates, CEO of Microsoft, publicly endorsed having ISP's charge for e-mail as a way to eliminate spam. *Make 'em Pay*, THE ECONOMIST, Feb. 14, 2004 (U.S. Edition). Six months later, after being accused of trying to establish such a system in order to enrich Microsoft, Gates retracted the proposal and assured the public of Microsoft's "firm[] belie[f] that monetary charges would be inappropriate and contrary to the fundamental purpose of the Internet as an . . . inexpensive medium for communications." Kevin Murphy, *Gates Backs Away From Postage Stamps Idea in Spam Vision*, COMPUTERWIRE (June 29, 2004). The present proposal, as explained in more detail below, differs from Gates's in various ways. Most importantly, it contemplates that the federal government, not a private ISP, would establish the fee and receive the proceeds. This would allow the government to spend the monies collected in socially beneficial ways, as outlined below. Such a system would not be subject to the criticism that it is being implemented to enrich a private company.

[43] *See* Martin Abadi, Andrew Birrell, Mike Burrows, Frank Dabek & Ted Wobber, *Bankable Postage for Network Services* (Springer-Verlag 2003) (available at www.springerlink.com) (describing the operation of a "ticket server" that could ensure that senders of e-mail pay a fee in order to have their messages delivered).

members, or to the general public. Finally, it may disrupt socially useful communication among friends, family members and business associates who have grown accustomed to free e-mail.

These concerns, while important, should be solvable. The system could apply only to e-mail that travels through an Internet Service Provider (ISP) with facilities in the United States. The ISP's could be required not to deliver any e-mail for which the sender had not paid the fee. This would address the problem of *intra*-organizational communication, since messages that travel via an organization's own server would remain free. The concerns about non-profit, educational, and religious institutions could largely be addressed by exempting those organizations whose 501(c)(3) status already exempts them from other federal taxes. Finally, a small fee should not deter most individual e-mail users. Consider the person who sends 100 personal e-mails a day. A fee of a tenth of a cent per e-mail would amount to 10 cents per day, or roughly $36.50 per year. This is about the same as the cost of a month of broadband Internet service. To reduce the burden on individuals almost to zero the federal government could create an income tax credit, to be claimed on the individual's annual income tax form, equal to the amount of e-mail fees that the person paid that year. Designed in this way, the system should not deter individual use of e-mail.

As to the fees collected from commercial entities, they would go to the federal treasury. Part could be spent enforcing the fee system and tracking down those who try to undermine it. Another portion could be spent on providing computers and computer training, including instruction on how to use e-mail, to low-income people without e-mail access. This investment would generate more users of e-mail and so would replenish the commons on which the spammers themselves subsist.

3.2 Using covenants and management systems to protect personal information

Regulatory instruments developed in the environmental field can provide strategies for protecting informational privacy as well. The current policy climate with respect to protection of personal information is something like the following: There is a broad recognition that advances in information technology are significantly compromising informational privacy, and that steps must be taken to offer better protection. Congress has taken action to safeguard personal information in some limited areas and is contemplating further legislation.[44] At the same time, there is a strong feeling among many in government, industry and academia that the government should not intrude too forcefully into this area.[45] Information industries are too varied and are changing too rapidly to be governed effectively by centralized decision-makers.[46] Moreover, information industries are central to the nation's economic future, and ill-conceived governmental intervention could do great damage. Thus, there is at once a recognition of the need for additional regulation and a reluctance to impose rules from the top.

The environmental field has faced this very situation and has developed responses to it. Traditionally, environmental regulation has consisted of centrally-mandated technology and/or performance specifications that industries subject to the rule must meet. This method is often referred as "command-and-control" regulation.[47] Regulation of this sort is not appropriate for all situations. Where industries are rapidly changing and centralized decision-makers have a hard time keeping up, environmental policymakers have sought to use other tools.[48] One of these "second generation" strategies is the environmental covenant.

An environmental covenant is an agreement among government, the regulated industry and, sometimes, interested NGOs that defines the steps that industry will take to achieve environmental goals.[49] Government initiates the process and usually offers that the agreement, if reached, will take the place of command-and-control regulation. In contrast to traditional regulation, covenants give the regulated industry a direct say in what the regulatory requirements will look like. They frequently also provide long lead times and flexible methods for achieving the agreed-upon social ends.[50] Government regulators and interested non-profits also must sign the covenant. Before they will do so, they often ask that the industry agree to ambitious environmental performance goals.

[44] Hahn & Layne-Farrar, *supra* note 11, at 29-50 (surveying existing and proposed legislation).

[45] Litan, *supra* note 37, at 1055.

[46] *Id.* at 1063.

[47] This term carries with it unfortunate pejorative connotations. Traditional regulation of this sort, as required by the Clean Air Act, Clean Water Act, and Resource Conservation and Recovery Act (RCRA), is responsible for many of the environmental gains that have been achieved in the United States over the past thirty years.

[48] *See* OFFICE OF TECHNOLOGY ASSESSMENT, ENVIRONMENTAL POLICY TOOLS: A USER'S GUIDE 27 (1995).

[49] *See* Richard Stewart, *A New Generation of Environmental Regulation?*, 29 CAPITAL L. REV. 21, 81 (2001) (analysis of environmental covenants).

[50] *Id.*

The covenant approach thus follows Coasian bargaining principles. If all the relevant parties agree to forego command-and-control regulation and replace it with the covenant, then the negotiated approach must be presumed to be superior to traditional regulation from all perspectives.[51] In theory, government and public benefit by obtaining steeper reductions in pollution than they would have obtained through traditional mechanisms.[52] Industry benefits by being given direct input into how the standards are shaped, longer lead times and implementation flexibility. The United States is still experimenting with environmental covenants.[53] However, the Netherlands, the European Union, and Japan have all made wide use of this tool.[54]

The covenant approach might prove to be an effective regulatory instrument for protecting personal information. The growing public concern about information privacy, and the increasing pressure on the government to take action, should give relevant industries an incentive to come to the negotiating table. Concerns about top-down intervention should motivate government officials to try a non-traditional approach. Finally, the United States is home to several strong and knowledgeable privacy-related NGOs. These organizations would have the expertise and resources to participate in such a process and to provide it with greater accountability and legitimacy. The stage is set for the federal government to negotiate a covenant with industry trade associations and relevant NGOs that will offer strong protection of personal information, while providing industry with the flexibility and extended time frame necessary to meet these objectives in a smart way.

There is even a rough model on which to build. In 1998, members of the Online Privacy Alliance, a group of major companies that utilize personal information in their on-line operations, were faced with a Federal Trade Commission demand that they take steps to protect privacy better or face regulation. The group voluntarily agreed to a set of guidelines to protect informational privacy.[55] While these guidelines were not formally negotiated with the government or NGO's, they can be viewed as a proto-covenant and

[51] *Id.* at 61.

[52] *Id.*

[53] *See* Brad Mank, *The Environmental Protection Agency's Project XL and Other Regulatory Reform Initiatives: The Need for Legislative Authorization,* 1998 ECOLOGY L. Q. 1, 14 (describing the Common Sense Initiative, an EPA program that experimented with industry covenants).

[54] Stewart, *supra* note 49, at 80-81, 84; Paul de Jongh, *The Netherlands' Approach to Environmental Policy Integration* (Center for Strategic and International Studies, Washington, D.C.) 1996.

[55] *See* Litan, *supra* note 37, at 1059. The guidelines themselves are available at http://www.privacyalliance.org/resources/ppguidelines.shtml.

illustrate the potential of the covenant approach. Future efforts might build on this initial one.

The precise contours of a covenant can be spelled out only by the parties themselves. That said, the environmental experience suggests that negotiators of such an agreement should consider requiring companies in the relevant industries to implement the equivalent of an environmental management system, one of the latest, and potentially most effective, tools being used in environmental management today.[56]

An environmental management system (EMS) is a company-initiated, systems approach to managing the environmental side of a business. It consists of an organizational plan for: assessing the company's environmental impacts and auditing its compliance status; setting goals for improved performance; systematically generating ideas about how to meet these goals; monitoring progress towards the objectives; and committing to continuous improvement in performance and compliance over time.[57] These steps are relatively common in business management. The innovative part has been the decision to apply such a management system, or any affirmative management approach for that matter, to the environmental side of the business rather than simply reacting to government demands. Environmental management systems thus represent a major leap forward on this front. The Environmental Protection Agency (EPA) has credited them with achieving better environmental performance, compliance and pollution prevention at the source.[58] The agency has developed programs that support, and give incentives for, the adoption of EMSs by private entities and governmental organizations.[59]

The EMS approach could usefully be applied to the protection of personal information. Most companies that use personal information offer privacy policies to protect their customers. Such policies typically consist of a statement, available on a company's website or in written materials, that discloses how the organization plans to use its customers' personal information and the circumstances under which it will share this information

[56] *See* Christopher L. Bell, *The ISO 14001 Environmental Management Systems Standard: A Modest Perspective*, 27 ENVTL. L. REP. 10,622 (Dec. 1997) (describing environmental management systems).

[57] *Id.*

[58] U.S. EPA, *Draft EMS Action Plan* 16 (Dec. 20, 1999).

[59] For example, the EPA has developed a national database of information on how companies across the country have fared with EMS implementation. *Id.* at 6. The agency has also offered to waive gravity-based penalties for those companies that uncover a violation through a compliance audit conducted as part of a bona fide EMS. U.S. EPA, Incentives for Self-Policing: Discovery, Disclosure, Correction and Prevention of Violations, 65 Fed. Reg. 19,618, 19,620-21 (Apr. 11, 2000).

with other commercial entities. While such a policy may stipulate to useful limits on the spreading or use of such information, it hardly constitutes an affirmative and continually improving system for protecting it. The privacy equivalent of an EMS – it might be called a Personal Information Management System, or PIMS – would offer greater protection. A company, through its PIMS, would commit to assessing the impact of its activities on personal privacy, setting ambitious goals for protecting personal information, systematically developing strategies to meet these goals, monitoring progress, and continually improving in its protection of personal information. The government or a private body might develop a general standard against which such a PIMS could be evaluated.[60] A customer who knows that the company adheres to a Personal Information Management System that meets such a standard would feel more comfortable sharing personal information than one who is merely provided with a privacy policy that she will likely not take the time to read. Government regulators could create incentives for adopting a PIMS, and could provide information and technical assistance to those companies interested in doing so. In addition, the negotiated covenant described above could require adoption of a PIMS as one of its terms.

Environmental management systems have a major blind spot. They merely represent a process for setting and attaining an organization's environmental goals. They have nothing to say about whether these goals are ambitious, or lenient. The risk is that a company might set extremely modest goals, implement an EMS to achieve them, and appear as worthy a company that has set out to achieve much better performance. A PIMS, much like an EMS, will be just a management framework devoid of substantive content. In the environmental field, regulatory requirements often provide the substance that fills out the EMS. But such requirements do not yet exist for many businesses that impact personal information privacy. Where can the substantive content of the PIMS be found?

Perhaps in the covenant. This government-industry-NGO agreement could set standards for the protection of personal information that an individual company could, in turn, incorporate into its PIMS. The covenant could further, as one of its terms, require a signatory business to adopt a PIMS as a means of ensuring achievement of the substantive goals set

[60] In the environmental field, the International Standards Organization, after lengthy negotiations with stakeholders, established the ISO 14001 standard for environmental management systems. *See generally* Bell, *supra* note 56 (describing ISO 14001). The European Union has adopted the Eco-Management and Audit Scheme (EMAS) standards for environmental management, auditing and reporting. *See* Eric Orts, *Reflexive Environmental Law*, 89 NW. U. L. REV. 1227 (1995) (describing the EMAS approach).

elsewhere in the agreement. In this way, the covenant and management system approaches would work together to protect personal privacy better, and so to preserve people's willingness to share their personal information on the Web.

4. CONCLUSION

The privacy law of today shares much with the environmental law of thirty years ago. Injuries caused by current business practices are just coming into the public eye, and pressure is growing on governments to protect their citizens against them. The theory and practice of environmental law, developed through hard experience over the past three decades, provides a resource on which emerging privacy regulation can draw. In particular, this chapter has suggested that emission fees, regulatory covenants, and the promotion of management systems may be transferable from the environmental to the privacy arena. If implemented, these strategies should protect individual privacy better and so prevent the tragedy of the commons that threatens both e-mail and e-commerce. In so doing, such regulatory measures will help to preserve these highly useful activities for future generations. Such an approach may contribute to the "sustainable development" of the information age economy.

Chapter 15

DOCUMENT SANITIZATION IN THE AGE OF DATA MINING

Dilek Hakkani-Tür,[1] Gokhan Tur,[2] Yücel Saygin,[3] and Min Tang[4]
[1]AT&T Labs Research, Florham Park, NJ 07932 dtur@research.att.com, [2]AT&T Labs Research, Florham Park, NJ 07932, gtur@research.att.com, [3]Sabanci University, Istanbul, Turkey, ysaygin@sabanciuniv.edu, [4]Center for Spoken Language Research, University of Colorado, Boulder, CO, tangm@cslr.colorado.edu

Abstract: The volume of data collected about people and their activities has increased over the years, especially with the widespread use of the internet. Data collection efforts coupled with powerful querying and data mining tools have raised concerns among people regarding their privacy. Recently the issue of privacy has been investigated in the context of databases and data mining to develop privacy preserving technologies. In this work, we concentrate on textual data and discuss methods for preserving privacy in text documents.

Key words: privacy, text mining, named entity extraction

1. INTRODUCTION

Data mining, which aims to turn heaps of data into useful knowledge, has found many applications both in government and in commercial organizations. Data mining techniques have been used mostly for better customer relationship management and decision-making in commercial organizations.

Government agencies are also using data mining techniques to track down suspicious behavior. For that reason, data collection initiatives such as CAPPS II (Computer Assisted Passenger Prescreening System) have been proposed to collect passenger and itinerary information. Data collection efforts by government agencies and enterprises have raised a lot of concerns among people about their privacy. In fact, the Total Information Awareness project – which aimed to build a centralized database to store the credit card transactions, e-mails, web site visits, and flight details of Americans – was not funded by the U.S. Congress due to privacy concerns. Privacy concerns have recently been addressed in the context of data mining, in work initiated by Rakesh Agrawal from IBM Almaden, and Chris Clifton from Purdue University, for example. However, privacy preserving data mining research deals mostly with the problem of mining the data without actually seeing confidential information. Another aspect of privacy preserving data mining is to be able to protect data "against" data mining techniques — since privacy risks increase even more when we consider powerful data mining tools. This aspect is important when the data itself is not confidential but the implicit knowledge in the data that could be extracted by data mining tools is confidential. The following quote from a NY Times article supports our point: "The Pentagon has released a study that recommends the government pursue specific technologies as potential safeguards against the misuse of data-mining systems similar to those now being considered by the government to track civilian activities electronically in the United States and abroad."[1]

In this work, we concentrate on privacy leaks in text, since text is one of the main widely available data sources, especially on the Internet, where text is available in newsgroups, e-mail, technical papers, and in many other formats.

2. MOTIVATING SCENARIO

Text data sources are of great interest to data mining researchers. Text mining techniques have been developed to classify a given text into predefined categories, or to cluster a given set of documents into groups for improving document retrieval. Techniques for finding patterns in text, such as term associations, have also been developed. Using text classification techniques, one can identify the author of a text easily. For example, authorship identification techniques have been used to resolve claims about some works of Shakespeare. However, authors of documents would not like to be public in all cases. Consider the reviews of papers (or books) published on the Internet. If someone could collect a sufficient number of

[1] J. Markoff, *Study Seeks Technology Safeguards for Privacy*, N.Y. TIMES, December 19, 2002.

documents related to a set of authors, this set could be used to train a classification model to identify the author(s) of a given document.

Consider the following scenario: Some database conferences are now doing blind review, but, the database community being finite, an exhaustive list of authors could easily be obtained from the internet (for example, by querying DBLP bibliography for authors of published papers in databases; http://www.informatik.unitrier.de/ley/db/). A curious reviewer could then employ some of his or her graduate students to collect a set of these authors' papers and construct a classification model to identify the authors of anonymous papers. Maybe curiosity is not a very convincing motivation for a reviewer to want to know the authors of the papers he or she is reviewing, but anger could be. Consider now an author whose paper has been rejected. Included with the rejection letter are a few paragraphs of anonymous reviews. Knowing that reviewers are mostly from the Program Committee of the conference, it is not difficult to obtain a full list of possible reviewers from the conference web site. A set of documents by these reviewers could also be easily obtained by crawling from the Internet to form a training set for authorship identification. Privacy may be even more important in the case of a declaration in the news that needs to be anonymous for security reasons.

In addition to the implicit information – such as the author of a document – that could be discovered by text mining tools, one can also discover private information from textual data through simple searching. For example, people's names, phone numbers, and addresses might be obtained. Such information may be inherently private, depending on the context in which it appears, or it can be linkable with other public data sources to infer private information. In this work, we assume a large volume of textual data such that it would not be practical to remove private information by hand. We address Named Entity Extraction (NE) as a means to locate such private information and sanitize text data in order to remove it. This step can be used before text documents are published. It could also be employed as a form of data preprocessing before data mining.

3. PRIVATE INFORMATION IN DOCUMENTS

Privacy issues in document repositories arise from the fact that the text may contain private information, which can be discovered by adversaries who are curious to know more about privacy-conscious individuals. Adversaries could turn this information to their own advantage, for example by publishing it in a tabloid.

We have previously provided a taxonomy of privacy threats in document databases.[2] Since our main concern is document repositories, the main elements we are going to consider are documents as information sources. There are also what we call Privacy Conscious Entities (PCEs) whose privacy may be jeopardized by the release of the documents. A PCE could be a person, a company, or an organization. Privacy concerns of PCEs may require that the identifying information of the PCE (person name, company name, etc) not be seen in a document, since the document is related to a sensitive topic. A PCE may not want to be seen as the author of the document, or appear in a document in a certain context, such as being a criminal or being in debt. A PCE may not want to be associated with another entity, such as being a friend of a convict. Links between documents, such as references or hyperlinks to a document, should also be taken into account.

The private information in a text can be grouped into two classes, namely explicit and implicit information. Explicit information might be the name, salary, or address of a person that could be viewed by text editors and browsers, and can be found by search engines. Implicit information might be the characteristics of a document, such as its author; or its statistical properties such as the frequencies of some words and punctuation marks; or its usage of particular phrases that can be identified with an individual. Data mining, natural language processing, and machine learning tools are needed to extract such information.

The tools that can be used to protect privacy depend on the private information type and the type of threat. We classify privacy threats as direct and indirect. Direct threats occur due to the existence and availability of explicit information in the data. Names of individuals, phone numbers, and salary amounts are just a few examples of pieces of information that may form a direct threat to privacy when they are revealed to a third party. Upon disclosure of the text data, users can see the contents using a browser or an editor. Indirect threats can be of two kinds: one is due to data integration and the other is caused by data analysis. The former type of indirect threats stem from the integration of different documents in order to infer private information that cannot be revealed by each individual document when considered alone. The integration can be achieved by finding those documents that are considered "similar" based on similarity measures

[2] Yucel Saygin, Dilek Hakkani-Tür, and Gokhan Tur, *Sanitization and Anonymization of Document Repositories, in* WEB AND INFORMATION SECURITY (Elena Ferrari & Bhavani Thuraisingham, eds. 2005).

defined in the context of information retrieval.[3] Indirect threats via data analysis, on the other hand, are due to the application of machine learning and data mining techniques to the available data. New data mining tools, especially tools for text classification, can be used with a training database (which is easy to construct from news groups and so forth) to infer private information such as the author of a document. In a preliminary study we collected a set of ten authors of newspaper articles. Using very simple features such as word and punctuation frequency counts, we were able to train a Support Vector Machine model to identify the authors of the newspaper articles with more than 90% accuracy. This was a very limited domain, with a limited set of authors, which resulted in high prediction accuracy. However, we believe that with some domain knowledge, the possible authors can often be reduced to a limited set (as in the case of the conference review example) to increase the accuracy in more general cases.

4. DOCUMENT SANITIZATION WITH NAMED ENTITY EXTRACTION

The aim of sanitization is to hide personal information to meet certain privacy requirements in order to disable the extraction of personal or other private business-related information from spoken language databases. Sanitization can be considered as a step towards privacy preserving text mining. The security implications of data mining and threats to privacy that can occur through data mining have been described by others.[4] Methods to preserve privacy in data mining are well studied for numeric and categorical values.[5] However, preserving privacy for natural language databases has not previously been studied to the best of the authors' knowledge.

The success of sanitization depends on the corresponding task to be performed with the data. We need to make sure that the data quality is still preserved after sanitization. Data quality could be measured in terms of readability and of the ability to use the sanitized text for the corresponding task. For example, if the data is going to be used for text classification, it is

[3] William W. Cohen, Data Integration Using Similarity Joins and a Word-Based Information Representation Language, 18 ACM TRANSACTIONS ON INFO. SYS. (July 2000).

[4] *See* B. Thuraisingham, *Data Mining, National Security, Privacy and Civil Liberties*, 4 SIGKDD EXPLORATIONS 1-5 (2002) and C. Clifton and D. Marks, *Security and Privacy Implications of Data Mining, in* 1996 PROC. ACM SIGMOD CONF. WORKSHOP ON RES. ISSUES IN DATA MINING AND KNOWLEDGE DISCOVERY

[5] R. Agrawal and R. Srikant, *Privacy-Preserving Data* Mining, in 2000 PROC. ACM SIGMOD CONF. ON MANAGEMENT OF DATA 439-450.

necessary to perform sanitization without diminishing the classification accuracy. If the task is information retrieval, the sanitization methods should not interfere with the indexing and document matching methods.

We will discuss data sanitization approaches to protecting the privacy of the speakers of natural language utterances. Our method is based on detecting named entities — such as persons, locations and organization names — and numeric values — such as dates, and credit card numbers — in the spoken utterances. We then sanitize them using various means with the purpose of hiding the personal information. The challenge in the sanitization of these utterances is ensuring that the accuracy of the spoken dialog system models trained using the new sanitized data and used for data mining techniques such as automatic speech recognition and spoken language understanding is as good as the accuracy before the sanitization. We show that, by hiding task-dependent named entities, we can preserve the privacy of the speakers, and still achieve a comparable accuracy.

4.1 Main Approach

The first step in text anonymization is to remove any explicit personally identifying information. Named entity extraction (NE) techniques can be used to find and remove personally identifying information from text. NE techniques have been previously shown by Latanya Sweeney from Carnegie Mellon University[6] to be useful for sanitizing medical documents. In recent work, we have also addressed how documents could be sanitized to remove explicit private information, or personally identifying information. However full text anonymization needs to consider data mining techniques for authorship identification of anonymous documents. This is a difficult task since there are many features that could be used by text classification techniques. Another concern is preserving the quality of the released text (comprehensibility, and readability) after anonymization.

We have identified k-anonymity as the basic privacy metric that has been previously proposed for statistical disclosure control. It was proposed as a tool for disclosure control of tabular data by Pierangela Samarati from the University of Milan and Latanya Sweeney from Carnegie Mellon University.[7] In a recent study we used k-anonymity to anonymize a given

[6] L. Sweeney, *Replacing Personally-Identifying Information in Medical Records, the Scrub System*, *in* 1996 PROC. AMER. MED. INFORMATICS ASS'N FALL SYMP.

[7] Pierangela Samarati and Latanya Sweeney, *Protecting Privacy when Disclosing Information: k-Anonymity and Its Enforcement through Generalization and Suppression*, SRI TECH. REP. (1998).

set of documents so that the author of a document in a given set of documents cannot be distinguished from k-1 other authors.[8] We also proposed heuristics based on updating the documents to reduce the significance of important features used for document classification. Updating the word counts without disturbing the readability of text can be achieved by replacing words with their synonyms. This will homogenize the word frequencies in different documents so that word frequencies cannot be used in text classification. This initial work, which identified data mining tools as a threat to anonymity of documents, will hopefully lead to more research to ensure full anonymity.

As we previously stated, the first step of the sanitization process is to extract personally identifying information, such as the name or SSN of a person, or a company name if we would like to protect the privacy of a company. However, spotting the personally identifying information may not be enough. We also need to find the quasi-identifier named entities that could be used to identify individuals by linking to other documents, such as the ZIP Code, birth date, and gender. Sanitization also depends on the corresponding task.

There are three known methods for partial access to databases[9] which can also be used for sanitization: value distortion, value dissociation, and value-class membership. Value distortion alters the confidential values to be hidden with random values. Value dissociation keeps these values but dissociates them from their actual occurrences. This can be achieved, for example, by exchanging the values across sentences. Value-class membership exchanges the individual values for disjoint, mutually exhaustive classes. For example, all the proper names can be changed to a single token, "Name."

The simplest form of sanitization is modifying the values of named entities or replacing them with generic tokens, as in the value-class membership approach. If the named entities are not already marked using XML tags, we can utilize automatic named entity extraction methods, which are well studied in the computational linguistics community. The concept of k-anonymity can be assured for text sanitization while determining the generic tokens. For example, people's names can be generalized until they map to at least k people. For the case of numeric values such as salary, a concept hierarchy can be exploited. The salary can be mapped to a more

[8] *See* Saygin *et al.*, *supra* note 2.

[9] R. Conway and D. Strip, *Selective Partial Access to a Database, in* 1976 PROC. ANN. ACM CONF.

generic value, which refers to at least k people – (*e.g.*, “low,” “average,” “high,” and “astronomic” linguistic hedges in the concept hierarchy – even when quasi-identifier information is used. For the case of addresses, we can ensure that the address maps to k different people in the company or to a district for which at least k distinct addresses exist.

Generic tokens can also preserve non-sensitive information to ensure readability of the text. For example, the gender or identification of an individual can be marked in the token for the resolution of further (pronominal) references (*i.e.*, < *PERSON* > versus < *PERSON, GENDER* = *MALE* >). An even harder task would be associating references during sanitization, as, for example, where <*DATE2* > is extended as <*DATE2= DATE1+3days*>.

- **System:** How may I help you?
- **User:** Hello. This is John Smith. My phone number is area code 973 1239684. I wish to have my long distance bill sent to my Discover card for payment.
- **System:** OK, I can help you with that. What is your credit card number?
- **User:** My Discover card number is 28743617891257 hundred and it expires on the first month of next year.
- **System:** . . .

Figure 15-1. A sample natural language dialog[10]

4.2 Case Studies

Goal-oriented spoken dialog systems aim to identify the intentions of humans, expressed in natural language, and to take actions accordingly, to satisfy their requests. In a spoken dialog system, typically, first the speaker’s utterance is recognized using an automatic speech recognizer (ASR). Then, the intent of the speaker is identified from the recognized sequence, using a spoken language understanding (SLU) component. This

[10] All names and numbers in the dialog have been made up for purposes of illustration.

step can be framed as a classification problem for goal-oriented call routing systems.[11] Figure 15-1 presents a sample dialog between an automated call center agent and a user. As is clear from this example, these calls may include very sensitive information about the callers, such as names and credit card and phone numbers.

State-of-the-art data-driven ASR and SLU systems are trained using large amounts of task data which is usually transcribed and then labeled by humans, a very expensive and laborious process. In the customer care domain, "labeling" means assigning one or more of the predefined intents (call-types) to each utterance. As an example, consider the utterance *I would like to pay my bill*, in a customer care application. Assuming that the utterance is recognized correctly, the corresponding intent or call-type would be *Pay(Bill)* and the action would be learning the caller's account number and credit card number and fulfilling the request. The transcribed and labeled data is then used to train automatic speech recognition and call classification models.

The bottleneck in building an accurate statistical system is the time spent to produce high quality labeled data. Sharing of this data has extreme importance for machine learning, data mining, information extraction and retrieval, and natural language processing research. Reuse of the data from one application, while building another application, is also crucial in reducing development time and making the process scalable. However, preserving privacy while sharing data is important since such data may contain confidential information. Outsourcing of tasks requiring private data is another example of information sharing that may jeopardize the privacy of speakers. It is also possible to mine these natural language databases using statistical methods to gather aggregate information, which may be confidential or sensitive. For example, in an application from the medical domain, using the caller utterances, one can extract statistical information such as the following:

y% of US doctors prescribe <DRUG1> instead of <DRUG2>

which may also be need to be kept private for business-related reasons. However, the medical reports of patients, if made publicly available, can significantly facilitate medical research. Ruch *et al.* have worked on hiding

[11] Narendra Gupta, Gokhan Tur, Dilek Hakkani-Tür, Srinivas Bangalore, Giuseppe Riccardi, Mazin Rahim, *The AT&T Spoken Language Understanding System,* IEEE TRANS. ON SPEECH AND AUDIO PROCESSING (forthcoming).

names in formatted medical reports.[12] In the literature, word and sentence accuracies are used to evaluate ASR performance. Sentence accuracy is the percentage of sentences that are recognized correctly (*i.e.*, they exactly match the transcriptions). Word accuracy is computed as follows:[13]

$$\text{Word Accuracy} = 1 - \frac{\text{\# of insertions} + \text{\# of substitutions} + \text{\# of deletions}}{\text{Total \# of words in the correct transcript}}$$

In the baseline system, the language model has a vocabulary of 5,345 words and was trained from the transcribed training set with all the personal information untouched. We tried three approaches to hide personal information in the training set as described in Section 4, namely value distortion, value dissociation and value-class membership.

Table 15-1. ASR Experiments

System	Word Accuracy	Sentence Accuracy
Baseline	73.8%	50.9%
Value Distortion	73.6%	50.6%
Value Dissociation	73.7%	50.7%
Value-Class Membership with uniform distribution	72.6%	50.3%
Value-Class Membership with keeping priors	72.9%	50.4%

As can be seen from Table 15-1, after finding the sensitive named entities in the training set and hiding them by using the value distortion or value dissociation approach, the ASR accuracy did not degrade significantly. The value dissociation approach has a better ASR accuracy than value distortion because it preserves the distribution of each value. The result also shows that with the assumption that class members are uniformly distributed, the value-class membership approach degraded ASR word accuracy from 73.8% to 72.6%. When the class member priors were kept in the class-based language models, the ASR word accuracy of this approach went up from 72.6% to 72.9%.

[12] P. Ruch *et al.*, *Medical Document Anonymization with a Semantic Lexicon, in* 2000 PROC. AMER. MED. INFORMATICS ASS'N SYMP. 729-33.

[13] D. Jurafsky and J. H. Martin, SPEECH AND LANGUAGE PROCESSING (2000).

4.3 Experiments on Real Data Sets

The SLU experiments were conducted using different combinations of ASR output or transcription of the training set and ASR output of the test set, and with or without sanitization. The total number of classes in this application is 97. The experimental results are shown in Table 15-2. In the table, "Top CE" stands for top scoring class error rate, which is the fraction of examples in which the call-type with maximum probability assigned by the classifier was not one of the true call-types. The baseline classification system does not use any named entities and sanitization. "General" is the system that extracts named entities by using regular grammars and then sanitizes the data using the value-class membership approach. "General+PN" does the same thing as "General" except that person names ("PNs") are extracted by using the heuristic scheme.

Table 15-2. SLU Experiments (Test set is from ASR Output).

Training set	*Sanitization*		*Top CE*	*F-Measure*
Transcribed	Baseline		0.2628	0.7252
	Value-Class Membership	General	0.2619	0.7290
		General + PN	0.2612	0.7270
	Value Distortion		0.2657	0.7268
	Value Dissociation		0.2633	0.7253
ASR output	Baseline		0.2664	0.7181
	Value-Class Membership	General	0.2658	0.7190
		General + PN	0.2640	0.7189
	Value Distortion		0.2678	0.7158
	Value Dissociation		0.2633	0.7193

As can be seen from Table 15-2, no matter whether the transcribed training set or the ASR output training set is used, the classification systems which sanitize data using the value-class membership approach (replacing personal information by named entities) achieved lower top scoring class error rate (the lower the better) and higher F-Measure (the higher the better). This may be due to the fact that there are stronger associations between call-types and named entities than between call-types and values of named entities. Also there is no significant change of SLU performance after using the value distortion approach or the value dissociation approach. Therefore we are able to maintain or even slightly improve the SLU performance while hiding personal information in spoken language utterances.

5. SUMMARY

Text data sources are very common and can be easily collected through the internet and by other means. However, such data sources are also a concern from the privacy perspective since they may contain sensitive information about people. In this work, we addressed privacy concerns about text documents and presented a sanitization approach to preserving privacy in spoken language databases. We have shown that by hiding task-dependent named entities we can preserve the privacy of the speakers, and still apply data mining techniques such as automatic speech recognition, and spoken language understanding.

V

Implications of Technology for Anonymity and Identification

Chapter 16

NYMITY, P2P & ISPS

Lessons from BMG Canada Inc. v. John Doe[1]

IAN KERR[1] and ALEX CAMERON[2]
[1]Canada Research Chair in Ethics, Law & Technology, Faculty of Law, University of Ottawa (iankerr@uottawa.ca); [2]LL.D. (Law & Technology) Candidate, University of Ottawa (acameron@uottawa.ca)

Abstract: This chapter provides an exploration of the reasons why a Canadian Federal Court refused to compel five Internet service providers to disclose the identities of twenty nine ISP subscribers alleged to have been engaged in P2P file-sharing. The authors argue that there are important lessons to be learned from the decision, particularly in the area of online privacy, including the possibility that the decision may lead to powerful though unintended consequences. At the intersection of digital copyright enforcement and privacy, the Court's decision could have the ironic effect of encouraging more powerful private-sector surveillance of our online activities, which would likely result in a technological backlash by some to ensure that Internet users have even more impenetrable anonymous places to roam. Consequently, the authors encourage the Court to further develop its analysis of how, when and why the compelled disclosure of identity by third party intermediaries should be ordered by including as an element in the analysis a broader-based public interest in privacy.

Key words: privacy, anonymity, compelled disclosure of identity, Internet service providers, peer-to-peer, copyright, cybercrime

Some people go online to share music – to explore the limits of their imaginations, to sample, to up and download songs from various musical genres and to feel the beat of previous generations. In the U.S., sharing

[1] BMG Canada Inc. v. John Doe, 2004 FC 488, *available at* http://decisions.fct-cf.gc.ca/fct/2004/2004fc488.shtml.

music across some peer-to-peer (P2P) networks is illegal.[2] In Canada, it is not.[3] Not yet.[4]

Some people go online to construct nyms – to engage in a social process of self-discovery by testing the plasticity of their identities and the social norms from which they are constituted. In the U.S., this form of personal exploration has been compromised by litigation campaigns that have successfully sought to compel Internet service providers (ISPs) to disclose their customers' offline identities.[5] In Canada, such campaigns have not enjoyed the same success. Not yet.

Why did a Canadian court refuse to compel the disclosure of the identities of twenty-nine P2P file-sharers whom the Canadian Recording Industry Association (CRIA) wished to sue for copyright infringement? Ought this decision to be upheld on appeal? What can be learned from this decision?

This chapter aims to address the above questions and to reinforce the motif that we must tread carefully at the intersection between the procedures and policies supporting digital copyright enforcement and online privacy.[6]

[2] A&M Records, Inc. v. Napster, Inc., 239 F.3d 1004 (9th Cir. 2001).

[3] *See* BMG v. Doe, *supra* note 1.

[4] BMG v. Doe, 2004 FC 488, *appeal filed*, No. A-T-292-04 (F.C.A. Apr. 13, 2004), http://www.cippic.ca/en/projects-cases/file-sharing-lawsuits/criaappealnotice.pdf.

[5] *See, e.g.*, Keith J Winstein, *MIT Names Student as Alleged Infringer*, THE TECH, Sept. 9, 2003, *at* http://www-tech.mit.edu/V123/N38/38riaa.38n.html; John Borland, *RIAA Targets Students in New File-Swapping Suits*, CNet News.com, Oct. 28, 2004, *at* http://news.com.com/2102-1027_3-5431231.html?tag=st.util.print; *Electronic Frontier Foundation Defends Alleged Filesharer*, Electronic Frontier Foundation, Oct. 14, 2003, http://www.eff.org/IP/P2P/20031014_eff_pr.php;
Katie Dean, *RIAA Hits Students Where it Hurts*, WIRED NEWS, Apr. 5, 2003, http://www.wired.com/news/digiwood/0,1412,58351,00.html. Not every attempt to compel subscriber identities from ISPs in the United States has proven successful. Some courts have shown that subscriber identities should not be handed over too easily. *See, e.g.*, Recording Indus. Ass'n of Am., Inc. v. Verizon Internet Servs., 351 F.3d 1229 (D.C. Cir. 2003).

[6] This motif, though it is not novel among legal academic circles, has not yet enjoyed general recognition outside of a relatively small community of experts. *See, e.g.* Michael Geist, *Web Privacy vs. Identifying Infringers*, THE TORONTO STAR, Oct. 6, 2003, *available at* http://www.michaelgeist.ca/resc/html_bkup/oct62003.html; Alex Cameron, *Digital Rights Management: Where Copyright and Privacy Collide*, 2 CANADIAN PRIVACY LAW REV. 14 (2004),
http://anonequity.org/files/a_comeron-Where_Copyright_and_Privacy%20Collide.pdf;
Alice Kao, *RIAA V. Verizon: Applying the Subpoena Provision of the DMCA*, 19 BERKELEY TECH. L.J. 405 (2004); Robert J. Delchin, *Musical Copyright Law: Past, Present and Future of Online Music Distribution*, 22 CARDOZO ARTS & ENT. L.J. 343 (2004).

1. NYMITY

As many scholars have pointed out, there is little consensus as to whether our ability to disconnect our actions from our identities is, on balance, a good thing.[7] Anonymity is like the Duke's toad – ugly and venomous, and yet it wears a precious jewel in its head.[8]

Ugly and venomous, because it disables accountability and enables wrongdoing. In the P2P context, an inability to ascertain the real-life identities of *geekboy@KaZaA, mr_socks@KaZaA, chickiepoo25@KaZaA*[9] and other file-sharers facilitates their ability to copy and disseminate music *en masse,* carefree and without a trace. Without knowing their identities, CRIA and other such organizations cannot sue these individuals and consequently cannot test the claim that file-sharers are engaging in illegal conduct. This could be a serious problem because, if anonymous P2P networks were undefeatable, copyright industries would have no means of legal recourse. As Professor Lawrence Lessig once remarked, in its broader context, "[p]erfect anonymity makes perfect crime possible."[10] While illegal copying of MP3s is unlikely to unravel civilization as we know it, a more generalized ability to commit perfect crime might. There are good reasons to fear a society in which people are able to act with impunity. Consequently, there are good reasons to fear anonymous P2P networks.

Though dangerous, anonymity is at the same time precious. It is Plato's *pharmakon;*[11] a drug that is both poison and remedy. As Derrida might have described it: "[t]his charm, this spellbinding virtue, this power of fascination, can be - alternately or simultaneously - beneficent or maleficent."[12] The ability to use "nyms" – alternative identifiers that can encourage social experimentation and role playing – is "an important part of the rich fabric of human culture."[13] Anonymity facilitates the flow of information and communication on public issues, safeguards personal reputation and lends voice to individual speakers who might otherwise be silenced by fear of

[7] *See, e.g.,* A. Michael Froomkin, *Anonymity in the Balance, in* DIGITAL ANONYMITY AND THE LAW (C. Nicoll et al. eds., 2003). *See generally* G.T. Marx, *What's in a Name? Some Reflections on the Sociology of Anonymity,* 15(2) INFO. SOC'Y 99, 99-112 (1998).

[8] William Shakespeare, *As You Like It,* act 2, sc. 1.

[9] These are some of the nyms at issue in BMG v. Doe, *supra* note 1.

[10] L. Lessig, *The Path of Cyberlaw*, 104 YALE L.J. 1743, 1750 (1995). *See also* A. Michael Froomkin, *Anonymity and Its Enmities*, 1995 J. ONLINE L. art. 4 (June 1995), para. 46.

[11] J. Derrida, *Plato's Pharmacy, in* DISSEMINATION 95 (B. Johnson trans., 1981).

[12] *Id.* at 70.

[13] Roger Clark, *Famous Nyms* (Aug. 31, 2004), http://www.anu.edu.au/people/Roger.Clarke/DV/FamousNyms.html.

retribution.[14] Nyms can be used to enhance privacy by controlling the collection, use and disclosure of personal information. Anonymity can also be used to protect people from unnecessary or unwanted intrusions and to "encourage attention to the content of a message or behavior rather than to the nominal characteristics of the messenger."[15]

It is not our aim in this short chapter to resolve the conflicting value sets generated by the possibility of perfect anonymity, nor to make a case for some intermediate solution such as pseudonymous or traceable transactions. Although there are a number of technological applications seeking to create both such states of affairs,[16] the typical uses of online nyms are much more leaky. That is, one nym can usually be associated with another or with other information to create a personal profile that enables identification.

For example, "*geekboy@KaZaA*" is one kind of nym; "24.84.179.98" is another. The latter, sometimes referred to as an IP address, is a numeric identifier assigned to computers or devices on TCP/IP networks. IP addresses are easily discovered and observed as people transact online. The particular IP address referred to above is alleged to belong to the network device used by an individual whose KaZaA pseudonym is *geekboy@KaZaA*. In the context of the recording industry's campaign against P2P file-sharers, finding out the IP address of a device is currently the best first step in uncovering the identity of an individual file-sharer. But the IP address is not enough. In order to sue, it is necessary to tie the device's IP address to a legal name. This is not always easy to do; at least not without help from a third party intermediary. In this case, the ISPs were the targeted intermediaries and they will be the focus of discussion in this chapter. However, the information might have been available from any intermediary, including from the operators of KaZaA or other P2P networks.

In our information society, ISPs have increasingly become trusted holders of and gatekeepers to our personal information. ISPs uniquely hold information about many online activities, including information which ties

[14] Marx, *supra* note 7.

[15] *Id.*

[16] *See* Peter Biddle et al., *The Darknet and the Future of Content Distribution, in* PROC. ACM WORKSHOP ON DIGITAL RIGHTS MANAGEMENT (2002), *available at* http://crypto.stanford.edu/DRM2002/darknet5.doc; Ian Clarke, *The Philosophy Behind Freenet*,http://freenetproject.org/index.php?page=philosophy&PHPSESSID=fca0b9ec8c97a4797456a0c20a26097a; George F. du Pont, *The Time has Come for Limited Liability for Operators of True Anonymity Remailers in Cyberspace: An Examination of the Possibilities and Perils*, 6 FALL J. TECH. L. & POL'Y 3 (2001); Tal Z. Zarsky, *Thinking Outside the Box: Considering Transparency, Anonymity, and Pseudonymity as Overall Solutions to the Problems of Information Privacy in the Internet Society*, 58 U. MIAMI L. REV. 991 (2004).

individuals' pseudonymous surfing and downloading activities to their 'real-world' identities. In this context, individuals trust and are dependent on ISPs to safeguard sensitive personal information and communications. Indeed, given the relationship between some ISPs and their subscribers, it is possible that the conditions for the imposition of fiduciary duties on ISPs might exist in some cases.[17] Canadian courts, including the Supreme Court of Canada, continue to recognize the importance of maintaining a degree of privacy or confidentiality with respect to the personal information held by ISPs. Especially so when it comes to linking legal names or other common identifiers to particular IP addresses. As one member of the Supreme Court of Canada recently held:

> [an individual's surfing and downloading activities] tend to reveal core biographical information about a person. Privacy interests of individuals will be directly implicated where owners of copyrighted works or their collective societies attempt to retrieve data from Internet Service Providers about an end user's downloading of copyrighted works. We should therefore be chary of adopting a test that may encourage such monitoring.[18]

In *BMG v. Doe*, the Federal Court of Canada was forced to confront the conflict between copyright enforcement and privacy, head-on, when CRIA commenced a litigation campaign against P2P file-sharers 'in parallel' with the one commenced by the Recording Industry Association of America (RIAA). As *BMG v. Doe* ascends through the appellate process, it promises to be an important comparative IP and cyberlaw case, forcing the courts to craft a judicial test for determining when ISPs should be compelled to disclose their customers' identities to copyright owners.

2. *BMG V. DOE*

On March 31, 2004, the Federal Court of Canada issued a widely-publicized ruling in *BMG v. Doe*. This decision propelled Canada into the

[17] Ian Kerr, *Personal Relationships in the Year 2000: Me and My ISP, in* PERSONAL RELATIONSHIPS OF DEPENDENCE AND INTERDEPENDENCE IN LAW (Law Commission of Canada ed., 2002) 78, 110-11; Ian Kerr, *The Legal Relationship Between Online Service Providers and Users*, 35 CAN. BUS. L.J. 40 (2001) [hereinafter Kerr, *Legal Relationship*].

[18] Society of Composers, Authors and Music Publishers of Canada v. Canadian Association of Internet Providers, 2004 SCC 45, at para. 155 (LeBel, J., dissenting) [hereinafter *SOCAN v. CAIP*]. *See also* Irwin Toy Ltd. v. Doe 2000 O.J. No. 3318 (QL) (Ont. S.C.J.) at paras. 10-11.

international spotlight because of the Court's statements regarding the legality of sharing music files on P2P networks.[19] The media coverage tended to obfuscate the other issue central to the decision, which focused on whether the privacy concerns in the case outweighed the interest of a private party in obtaining discovery in civil litigation. *BMG v. Doe* is significant because it may have set the threshold test for future cases in Canada and perhaps elsewhere, in which courts are asked to compel ISPs to link individuals' online nyms – specifically their IP addresses – to their offline identities.

2.1 Nature of the case

BMG v. Doe involved an impressive matrix of fifty-three organizations and individuals, divided into four categories as follows:

Plaintiffs: seventeen music recording companies who were members of CRIA (collectively "CRIA");

Defendants: twenty-nine unnamed individuals identified only by their P2P pseudonyms and IP addresses;

Non-party respondents: five of Canada's largest telecommunications and cable ISPs: Shaw Communications Inc., Telus Inc., Rogers Cable Communications Inc., Bell Sympatico, Vidéotron Ltée. (collectively the "ISPs"); and

Interveners: Canadian Internet Policy and Public Interest Clinic (CIPPIC) and Electronic Frontier Canada (collectively the "Interveners").[20]

The case began in February 2004 when CRIA commenced a copyright infringement lawsuit against the Defendants, alleging that the Defendants

[19] *See, e.g.*, Electronic Privacy Information Center, *Canadian Court OKs peer to peer sharing*, EPIC ALERT, Apr. 8, 2004 *at* http://www.epic.org/alert/EPIC_Alert_11.07.html; John Borland, *Judge: File Sharing Legal in Canada*, Cnet News.com, Mar. 31, 2004, *at* http://news.com.com/2102-1027_3-5182641.html?tag=st.util.print; *Keep on Swapping! Cdn File Sharers Told*, p2pnet.net News, Mar. 31, 2004, *at* http://p2pnet.net/story/1118; Tony Smith, *File Sharers Not Guilty of Copyright Infringement – Canadian Judge*, THE REGISTER, Mar.31, 2004, http://www.theregister.co.uk/2004/03/31/file_sharers_not_guilty/; Gene J. Koprowski, *Canada Feds Rule Song Swapping Legal*, TECHNEWSWORLD, Apr. 1, 2004, *at* http://www.technewsworld.com/story/33290.html.

[20] It should be noted that co-author of this chapter, Alex Cameron, was also co-counsel for CIPPIC (http://www.cippic.ca) in *BMG v. Doe*, *supra* note 1.

had unlawfully shared copyrighted music files on P2P networks. CRIA could only identify the Defendants by their P2P pseudonyms and IP addresses.

CRIA immediately brought a preliminary motion (the "Motion") seeking to compel the ISPs to release the identities of the twenty-nine unknown subscribers. The initial reactions of the ISPs differed widely, with Shaw taking the strongest stand to protect its subscribers' privacy. With the exception of Vidéotron, all of the ISPs opposed the Motion in Federal Court. The Interveners also opposed the Motion.

2.2 Evidence

2.2.1 CRIA's evidence

The bulk of CRIA's evidence in support of the Motion came from Mr. Millin, the President of MediaSentry. MediaSentry is a New York company that CRIA had hired to gather evidence of copyright infringement on P2P networks.

Millin explained that his company had searched P2P networks for files corresponding to CRIA's copyrighted sound recordings and then randomly downloaded such files from each Defendant between October and December 2003. He claimed that MediaSentry was able to determine the IP address of each Defendant at the time MediaSentry downloaded the files. Using the American Registry for Internet Numbers,[21] MediaSentry was then able to determine to which ISPs the IP addresses had been assigned at the relevant times. This allowed MediaSentry to determine the ISP through which each Defendant had been sharing the files.

During cross-examination, Millin admitted that he had not listened to the files that MediaSentry downloaded. He also acknowledged that linking an IP address to a subscriber account would identify only the ISP subscriber, not necessarily the P2P user engaged in file-sharing. For example, Millin admitted that an ISP account may have hundreds of users on a local area network or that a wireless router might be used by any number of authorized and unauthorized users to engage in file-sharing.

Finally, Millin explained that MediaSentry used files called "MediaDecoys" as part of its work with CRIA. "MediaDecoys" are files that appear, based on their filenames, to be copyrighted songs. However, once a P2P user downloads and opens such a file, the user discovers that the

[21] This is a non-profit organization that assigns IP addresses to ISPs. See http://www.arin.net for a description of this organization.

file is actually inoperative. Such measures are designed to reduce the attractiveness of P2P networks by frustrating P2P users. Because Millin did not listen to any of the files downloaded by MediaSentry, he admitted that he did not know whether any of those files were in fact MediaDecoy files, thus rendering impossible a determination in any given instance whether CRIA-owned content was in fact being shared.

2.2.2 ISPs' evidence

Three ISPs – Shaw, Telus and Rogers – were the only parties to file evidence opposing the Motion. Neither Bell nor Vidéotron filed evidence and, by order of the Court, the Interveners were not permitted to file evidence.

Shaw and Telus gave evidence that they almost always assigned IP addresses to their subscribers "dynamically." This means that each time a subscriber went online, the subscriber would be randomly assigned a new IP address for that session. Shaw stated that it did not keep historical records of which IP addresses were assigned to particular subscribers at particular times. For this and other technical reasons, Shaw's evidence indicated that it could not, with the degree of certainly required, provide the personal information sought by CRIA. This was a point of difference between the ISPs which is important to bear in mind for the discussion of *Lawful Access* under Part 3 below. Shaw also registered its concern about potential legal liability in fulfilling CRIA's request; for example, the liability that might arise if it incorrectly matched an IP address to a subscriber, even through no fault of its own.

Telus gave evidence that it did not have any records of the information sought by CRIA and that it had no commercial reason to maintain those kinds of records. Multiple databases would have to be cross-referenced in order to search for and produce the information sought by CRIA. Further, Telus stated that the longer the delay between an event and Telus' search, the less reliable the information would become. This turned out to be an important evidentiary point since MediaSentry had gathered CRIA's evidence as early as October 2003, roughly six months before the court hearing. Finally, Telus provided evidence about how responding to CRIA requests would be costly and disruptive to Telus' operations, particularly if such requests were made in significant numbers in the future.

Rogers provided evidence indicating that it had some information about eight of the nine Rogers subscribers targeted by CRIA and that it had sent notice of the CRIA lawsuit to almost all of those subscribers. Rogers indicated that it generally retained the kind of information sought by CRIA for a period of six days.

Although Bell did not file evidence, Bell's counsel advised the Court that Bell had already identified and was holding information about all of the targeted Bell customers. Bell's counsel also echoed concerns raised by the other ISPs about compensation for ISPs' costs to comply with a disclosure order.

2.3 Privacy arguments in *BMG v. Doe*[22]

2.3.1 CRIA's arguments

CRIA argued that the following seven-part test should be applied by the Court in deciding whether to compel disclosure of the identities of the ISP subscribers:

1. Is there a *prima facie*, or *bona fide* case, at least, of copyright infringement?
2. Is there a relationship between the ISPs and the alleged infringers?
3. Do the ISPs have information about the identities of the alleged infringers?
4. Are the ISPs the only practical source of the information sought?
5. Is the information necessary for CRIA to proceed with its lawsuit?
6. Would the information sought be compellable at trial and useful to CRIA's case?
7. Is there any interest, privacy or otherwise, that would outweigh the ISP's duty to disclose the identity of the alleged infringers?

Privacy concerns figure into the first and last elements of this test. They arise under the first element in the sense that privacy concerns might justify a higher evidentiary threshold at the preliminary stage of the lawsuit. For example, the requirement of proving a *prima facie* case of infringement would be a higher threshold than proving a mere *bona fide* (good faith) case. CRIA did not draw a distinction between these evidentiary thresholds, arguing, in any event, that it had satisfied either threshold.

Privacy might arise under the last element of the test as a factor which could prevent disclosure outright. With respect to this element, CRIA argued that there were no privacy concerns at issue in the Motion that would

[22] The written arguments in the case can be accessed online at http://www.cippic.ca/file-sharing-lawsuit-docs. A blog of the oral arguments is also available at http://www.cippic.ca/file-sharing-lawsuits.

outweigh CRIA's interest in having disclosure in order to sue the alleged infringers.

CRIA asserted that Canadian privacy law did not prevent disclosure because the law expressly permitted disclosure without an individuals' consent where required by an order of a court.[23] CRIA also argued that the ISP subscribers had already consented to disclosure in the circumstances (where violation of a legal right was at issue) by agreeing to such provisions in their ISPs' "acceptable use" agreements, upon subscribing to the ISPs' services.

CRIA further argued that many of the privacy concerns raised by the other parties to the Motion were diminished by virtue of the fact that there was little likelihood that the Defendants' Internet activities at large would be associated with their actual identities on the basis of merely providing CRIA with the link between their P2P usernames, IP addresses and their legal names, as sought by the order. Finally, in response to the ISP's evidence and arguments, CRIA claimed that the ISPs were able to identify the subscribers because, for example, Shaw and Rogers admitted that they had done so in response to police or other requests on numerous occasions in the past.

2.3.2 ISPs' arguments

Shaw sought to protect its customers' personal information in accordance with Canadian privacy law, in part because it could be held accountable to its customers or to Canada's Federal Privacy Commissioner. Shaw expressly adopted parts of CIPPIC's argument and asserted that there were substantial privacy interests at stake which required the Court to impose a high standard – a "strong *prima facie* case" – on CRIA before ordering disclosure. Shaw argued that the CRIA request amounted to a civil search warrant in circumstances where there was no legal authority for such a warrant and where there were no privacy protections for the targets of the inquiry.

In terms of whether the test had been met, Shaw claimed that CRIA had not made out a *prima facie* case of copyright infringement. For example, Shaw asserted that there was no evidence as to how CRIA linked the P2P pseudonyms to the IP addresses and no evidence that anyone at CRIA had listened to the downloaded songs.

[23] Personal Information Protection and Electronic Documents Act, S.C. 2000, c. 5, ss. 7(3)(c).

Telus stated that it had no documents sought by CRIA and characterized the Motion as a mandatory order conscripting Telus to conduct investigations for CRIA and to create documents for CRIA without concern for the impact it would have on Telus and without concern for the reliability of the information produced. This was a time-consuming and costly process which would be disruptive to Telus' ordinary course of business. It was also something for which Telus argued it might face liability to its customers. Telus suggested that CRIA should have asked KaZaA and other P2P companies for the information sought before coming to the ISPs.

Rogers made brief arguments, asserting that the order sought by CRIA was extraordinary and that CRIA should be required to produce evidence commensurate with the nature of the order sought. Rogers also asserted that the form of order sought by CRIA should be more restrictive. For example, Rogers submitted that if the order were granted, Rogers should only be required to produce the last known name and address of the account holders at issue.

Finally, Bell took a relatively neutral approach in its argument by highlighting issues and questions that the Court should consider. Bell submitted that the Court should only make an order for disclosure of personal information where the moving party has made out a *prima facie* case based on admissible evidence. Bell asserted that there was no evidence as to how the IP addresses of the alleged Defendants were linked to the pseudonyms and that the affidavits filed by CRIA were not based on personal knowledge.

2.3.3 CIPPIC's arguments

CIPPIC filed a substantial written brief regarding privacy and copyright issues. Drawing on a number of Supreme Court of Canada search and seizure cases and Canada's recently enacted private-sector privacy laws, CIPPIC asserted that there were fundamental privacy values at stake in the case, demanding that a high threshold test be applied before identity should be disclosed. These values included protection of informational privacy which the Supreme Court had expressly recognized in Canada.[24] In explaining why the threshold test was critical from a privacy perspective, CIPPIC pointed to *R. v. Dyment* where the Supreme Court of Canada stated that "if privacy of the individual is to be protected, we cannot afford to wait to vindicate it only after it has been violated."[25]

[24] R. v. Dyment, 1988 2 S.C.R. 417 at 427-30.

[25] *Id.* at 429-30.

Further justifying a high threshold test, CIPPIC advanced arguments regarding the particular importance of online privacy and anonymity:

> The Internet provides an unprecedented forum for freedom of expression and democracy. The ability to engage in anonymous communications adds significantly to the Internet's value as a forum for free expression. Anonymity permits speakers to communicate unpopular or unconventional ideas without fear of retaliation, harassment, or discrimination. It allows people to explore unconventional ideas and to pursue research on sensitive personal topics without fear of embarrassment.
>
> If the Plaintiffs are able, by virtue of a court order, to link an IP address (*e.g.*, 66.51.0.34) and a KaZaA user name to a presumptive "real world" person (*e.g.*, John Smith), and thus commence an action against that person, the action could connect information about John Smith to the world (with consequences beyond the scope of the allegation). For example, John Smith might have visited a Web site on sexually-transmitted diseases, posted or shared documents criticizing the government or his employer, discussed his religious beliefs using a pseudonym in a chat room, or virtually any other type of expression. John Smith would likely hold an assumption that he was and would remain anonymous in many of these activities. The effect of the Court order in this case would shatter that anonymity and potentially cause significant embarrassment and irreparable harm to John Smith, independent of and prior to a determination of his culpability. It would have a corresponding chilling effect on free speech and online activity generally.[26]

During oral argument, CIPPIC expanded on this hypothetical in response to a question from the Justice von Finckenstein. CIPPIC pointed out that P2P systems can be used to share virtually any kind of document, software, music, video or other types of files. In fact, CIPPIC was able to point the Court to actual examples of documents and pictures being shared by some of the Defendants. CIPPIC argued that this sharing had been done on an assumption of anonymity and that to reveal the identity of those sharing files would effectively shatter their anonymity much more broadly.

[26] Memorandum of Argument of the Intervener CIPPIC at para. 17-18, BMG v. Doe, 2004 FC 488 (No. T-292-04), *available at* http://www.cippic.ca/en/projects-cases/file-sharing-lawsuits/memorandum_fctd_final_12pt.pdf.

CIPPIC asserted that CRIA should have to provide clear evidence of the alleged infringement, clear evidence of copyright ownership and clear evidence that they have identified the correct defendants. CIPPIC also pointed out that where a case is unlikely to proceed to trial after an interlocutory order is made, courts will and should engage in a more extensive review of the merits of plaintiffs' claims. CIPPIC and Shaw argued that this was important because if disclosure was ordered, CRIA would likely follow the aggressive approach adopted by the RIAA in the US, which pressed defendants to immediately engage in 'settlement discussions' – a potentially problematic practice when one considers the vast inequality in bargaining power between the plaintiff and defendants. CIPPIC argued that a more extensive review of CRIA's case was similarly justified because disclosure of the Defendants' identities could lead to seizure of computers and consequent loss of privacy and the ability to work.

2.4 The decision

The Court began its decision with a cursory review of the facts and then adopted the description of how P2P systems work set forth in *Metro-Goldwyn-Mayer Studios Inc. v. Grokster*.[27] In terms of the legal issues, the Court framed the case by raising three questions, each of which involves balancing privacy against other considerations: (i) "What legal test should the Court apply before ordering disclosure?"; (ii) "Have the Plaintiffs met the test?"; and (iii) "If an order is issued, what should be the scope and terms of such order?"

2.4.1 What legal test should the Court apply before ordering disclosure?

Following largely on the factors proposed by CRIA, Justice von Finckenstein of the Federal Court held that the following five criteria must be satisfied before a disclosure order would be made:

a) The Plaintiff must establish a *prima facie* case against the Defendants;
b) The ISPs must be in some way involved in the matter under dispute (*i.e.*, the ISPs must be more than innocent bystanders);
c) The ISPs must be the only practical source of the information;

[27] 259 F. Supp 2d 1029 (C.D. Cal. 2003).

d) The ISPs must be reasonably compensated for their expenses arising out of compliance with the order, in addition to their legal costs; and

e) The public interests in favour of disclosure must outweigh legitimate privacy concerns.

These five elements comprise the threshold test established in this case. Although not all of these factors bear on privacy in an obvious way, it is important to consider the Court's findings with respect to each factor because they could have an impact, on what we characterize as *a broader public interest in privacy*, as discussed below in part 3.

2.4.2 Have the Plaintiffs met the test?

The Court concluded that CRIA did not meet the test for disclosure because: (i) CRIA did not make out a *prima facie* case; (ii) CRIA did not establish that the ISPs were the only practical source of the information; and (iii) the privacy concerns in the case outweighed the public interest in disclosure. The Court's analysis followed each factor of the threshold test as follows.[28]

Factor 1: CRIA must establish a prima facie case against the Defendants. The Court found that there were three deficiencies in the *prima facie* copyright infringement case advanced by CRIA. First, the Millin affidavit was deficient because it was not based on personal knowledge and gave no reason for his beliefs. The Court also remarked that Millin had not listened to any of the downloaded files and in particular did not know if they were MediaDecoy files. On this basis, it concluded that there was "no evidence before the Court as to whether or not the files offered for uploading are infringed files of the Plaintiffs."[29]

Second, the Court noted that Millin had not explained how MediaSentry linked the P2P pseudonyms to specific IP addresses. Therefore, the Court concluded that it would be "irresponsible" for the Court to order disclosure:

> There is no evidence explaining how the pseudonym "Geekboy@KaZaA" was linked to IP address 24.84.179.98 in the first place. Without any evidence at all as to how IP address 24.84.179.98 has been traced to Geekboy@KaZaA, and without being satisfied that such evidence is reliable, it would be irresponsible for the Court to order the

[28] The Court offered most of its analysis between paras. 10-42.

[29] BMG v. Doe, *supra* note 1, at para. 19.

disclosure of the name of the account holder of IP address 24.84.179.98 and expose this individual to a law suit by the plaintiffs.[30]

Finally, Justice von Finckenstein found that CRIA had not provided any evidence that copyright infringement had taken place under Canadian law. The Court rejected each of CRIA's infringement claims, noting *inter alia* that "[n]o evidence was presented that the alleged infringers either distributed or authorized the reproduction of sound recordings. They merely placed personal copies into their shared directories which were accessible by other computer users via a P2P service."[31] In part, the Court relied on the landmark Supreme Court of Canada decision in *CCH Canada Ltd. v. Law Society of Upper Canada,*[32] holding that providing facilities to copy does not by itself amount to authorizing infringement. Justice von Finckenstein also held that distribution requires a positive act by the owner of a shared directory, beyond merely placing a file in a shared directory.

Factor 2: The ISPs must be more than innocent bystanders. The Court found that the ISPs were not mere bystanders because they are the means by which file-sharers access the Internet and connect with one another. Although the Court did not specifically say so, its recognition that ISPs play the role of gatekeeper is consistent with the view that there is a legal relationship between ISPs and those who use their services; a relationship which may create privacy-related obligations that are not applicable to innocent bystanders.

Factor 3: The ISPs must be the only practical source of the information. The Court found that it could not make a determination on this issue because CRIA had not described the entities that operate P2P systems, where they are located or whether the names corresponding to the pseudonyms could be obtained from the P2P operators. For example, Telus' evidence suggested that CRIA may be able to obtain the identities from KaZaA in cases where users had signed up for 'KaZaA Plus' and would therefore be billed by KaZaA.

Factor 4: The ISPs must be reasonably compensated for their expenses. The Court concluded that the process sought to be imposed by CRIA would be costly and divert the ISPs' resources from other tasks. The Court held

[30] *Id.* at para. 20.

[31] *Id.* at para. 26.

[32] 2004 SCC 13. *See generally* Michael Geist, *Banner year for digital decisions,* TORONTO STAR, Dec. 20, 2004 (hailing *CCH Canada Ltd. v. Law Society of Upper Canada* as "the most important copyright case of the year"), *available at* http://www.thestar.com/NASApp/cs/ContentServer?pagename=thestar/Layout/Article_PrintFriendly&c=Article&cid=1103496608921&call_pageid=968350072197.

that ISPs would need to be compensated for their reasonable costs as well as their legal costs of responding to the Motion.

Factor 5: The public interests in favour of disclosure must outweigh legitimate privacy concerns. The Court began the heart of its privacy analysis by noting that "it is unquestionable but that the protection of privacy is of utmost importance to Canadian society."[33] The Court cited with approval passages from *Irwin Toy v. Doe,* which articulated the value of privacy on the Internet:

> In keeping with the protocol or etiquette developed in the usage of the internet, some degree of privacy or confidentiality with respect to the identity of the internet protocol address of the originator of a message has significant safety value and is in keeping with what should be perceived as being good public policy. As far as I am aware, there is no duty or obligation upon the internet service provider to voluntarily disclose the identity of an internet protocol address, or to provide that information upon request.[34]

The Court in *BMG v. Doe* noted, however, that privacy is not absolute and cannot be used to insulate anonymous persons from civil or criminal liability. Because courts are required to balance one individual's privacy rights against the rights of other individuals and the public interest, the Court recognized that CRIA had legitimate copyrights in their works and were entitled to protect them against infringement. Thus, the Court recognized that CRIA had an interest in compelling disclosure of the identities of the peer-to-peer file-sharers. The Court also implied that there was a public interest favouring disclosure in litigation so that parties are not denied the ability to bring and try their legal claims merely because they cannot identify the alleged wrongdoers.[35] Consequently, it held that the privacy concerns in the case must be balanced against CRIA's interest and the broader interest that it stands for.

In its analysis of the privacy concerns, the Court held that the reliability and scope of the personal information sought by CRIA were the most significant factors to consider. The information sought must be reliable and ought not to exceed the minimum that would be necessary in order to permit CRIA to identify the alleged wrongdoers.[36] Here, the Court held that CRIA

[33] BMG v. Doe, *supra* note 1, at para. 36.

[34] *Id.* at para. 37 (quoting Irwin Toy Ltd. v. Doe (2000), 12 C.P.C. (5th) 103 (Ont. Sup. Ct.) at paras. 10-11).

[35] *Id.* at para. 42.

[36] *Id.*

had sought too much information from the ISPs and that the information sought was not sufficiently reliable to justify disclosure:

> In this case the evidence was gathered in October, November and December 2003. However, the notice of motion requesting disclosure by the ISPs was not filed until February 11, 2004. This clearly makes the information more difficult to obtain, if it can be obtained at all, and decreases its reliability. No explanation was given by the plaintiffs as to why they did not move earlier than February 2004. *Under these circumstances, given the age of the data, its unreliability and the serious possibility of an innocent account holder being identified, this Court is of the view that the privacy concerns outweigh the public interest concerns in favour of disclosure.*[37]

In the above passage, the Court expressly mentions the age of the data as contributing to its unreliability. Perhaps even more importantly, its reference to the "serious possibility of an innocent account holder being identified" ought to be understood in reference to the lack of an evidentiary link between P2P pseudonyms and IP addresses. Even on the assumption that a given ISP is able to accurately link IP addresses to its customers' legal names, without being able to prove the connection between online pseudonyms and IP addresses, the Court determined that CRIA is unable to ensure that it is seeking to compel disclosure of the identities of the appropriate individuals. As a result of these weighty privacy concerns, the Court refused to compel disclosure.

2.4.3 If an order is issued, what should be the scope and terms of such order?

Although the Court did not order disclosure in this case, it did propose a privacy-protective framework for orders that might be granted in future cases. The Court noted that if an order for disclosure had been made, certain restrictions would have been needed to protect the privacy of the Defendants because "the invasion of privacy should always be as limited as possible."[38]

First, the use of subscriber names by CRIA would be strictly limited to substituting the John Doe and Jane Doe names in the lawsuit. Second, the P2P pseudonyms would be used as proxies for the legal names for the Defendants on the Statement of Claim. This would protect the names of the subscribers from public disclosure, at least initially. An annex (protected by

[37] *Id.* (emphasis added).
[38] *Id.* at para. 44.

a confidentiality order) would be added to the Statement of Claim relating each P2P pseudonym to the legal name and address of a particular ISP account holder. Finally, the ISPs would be required to disclose only the name and last known address of each account holder. These kinds of protections would provide the information CRIA needed to proceed with a given claim while, at the same time, providing a measure of privacy protection to Defendants.

3. LESSONS FROM *BMG v. DOE*

The decision in *BMG v. Doe* has precipitated two significant events in Canada. First, CRIA commenced an appeal of the decision to Canada's Federal Court of Appeal. That appeal was heard April 20-21, 2005. We await the court's decision. Second, CRIA has continued its lobbying efforts to persuade the Government of Canada to ratify the *WIPO Copyright Treaty* and *WIPO Performances and Phonograms Treaty*.[39] Though implementation of the treaties has not yet happened, it is likely imminent.[40] These copyright wars, pitting our cultural industries against various segments of the general population, have received much attention. However, the possible ramifications of these battles for online privacy have received considerably less airplay.

[39] World Intellectual Property Organisation Copyright Treaty, Dec. 20, 1996, S. Treaty Doc. No.105-17 at 1 (1997), 36 I.L.M. 65; World Intellectual Property Organisation Performances and Phonograms Treaty, Dec. 20, 1996, S. Treaty Doc. No.105-17 at 18 (1997), 36 I.L.M. 76 (providing "adequate legal protection and effective legal remedies against the circumvention of effective technological measures that are used by authors [performers or producers of phonogram] in connection with the exercise of their rights under [those] Treaties"). Interestingly, the lobbying has proceeded in both directions. In addition to CRIA lobbying the Canadian government, Canadian politicians made public promises to the recording industry – just prior to Canada's most recent Federal election – that the Government would respond to the decision through legislation. *See, e.g.*, *Scherrer vows to crack down on file sharers*, CBC NEWS ONLINE, Apr. 13, 2004, *at* http://www.cbc.ca/arts/stories/scherrer20040413; Press Release, CRIA, The Canadian recording industry calls for adoption of Heritage Committee copyright report recommendations (May 12, 2004), *available at* http://www.cria.ca/news/120504a_n.php.

[40] *See, e.g.*, Press Release, Government of Canada, *The Government of Canada Announces Upcoming Amendments to the Copyright Act* (Mar. 24, 2005), *available at* http://strategis.ic.gc.ca/epic/internet/incrp-prda.nsf/en/rp01140e.html; Michael Geist, *'TPMs': A Perfect Storm For Consumers*, THE TORONTO STAR, Jan. 31, 2005, *available at* http://geistcanadiandmca.notlong.com; Ian Kerr, Alana Maurushat, & Christian S. Tacit, *Technical Protection Measures: Part II – The Legal Protection of TPMs* (2002), *available at* http://www.pch.gc.ca/progs/ac-ca/progs/pda-cpb/pubs/protectionII/index_e.cfm.

On the one hand, *BMG v. Doe* sends a clear message to future plaintiffs – they should come to court with solid evidence of the alleged wrongdoing as well as solid evidence of a reliable link between the alleged activity and specific individuals. Without this kind of evidence, privacy concerns may militate against disclosure, as they did in this case. Further, even where privacy concerns do not justify refusing disclosure outright, the Court also sent a message to future courts that any invasion of privacy should be limited and minimized by protective measures in the disclosure order. For these reasons, *BMG v. Doe* must unquestionably be read as a victory for privacy and as an endorsement for preserving online anonymity unless there are strong reasons to justify compelling the disclosure of identity.

On the other hand, a number of the Court's findings in *BMG v. Doe* may quite unintentionally diminish Internet privacy in the future. Recall that the result in *BMG v. Doe* turned on the inadequate evidence provided by CRIA. The decision openly invites CRIA to come back to court with better evidence of wrongdoing in a future case. Such an invitation may well result in even closer scrutiny of Internet users targeted by CRIA, both to establish a reliable link between their pseudonyms and their IP address and to carefully document the kinds of activities that the individuals were engaged in for the purpose of attempting to show a *prima facie* copyright violation.[41] It could also motivate the development of even more powerful, more invasive, surreptitious technological means of tracking people online. This increased surveillance might be seen as necessary by potential litigants in any number of situations where one party to an action, seeking to compel disclosure of identity information from an ISP, is motivated to spy on the other, set traps and perhaps even create new nyms in order to impersonate other peer-to-peer file-sharers with the hope of frustrating them, intimidating them, or building a strong *prima facie* case against them.

Still, there are good reasons in favour of upholding the decision in *BMG v. Doe* on appeal. Independent of the copyright claims, the serious deficiencies in CRIA's evidence – particularly the lack of a link between the pseudonyms and the IP addresses – is itself a sufficient reason to reject disclosure, even if privacy protections were built into an order. In its appeal factum, CIPPIC elaborates on the reasons why the evidentiary issues are so important:

[41] While monitoring the activities of peer-to-peer file sharers may achieve these objectives, surveillance can also be used as a broader means of social manipulation or control. *See, e.g.*, James Boyle, *Foucault in Cyberspace: Surveillance, Sovereignty, and Hardwired Censors*, 66 U. CIN. L. REV. 177 (1997); Oscar H. Gandy, Jr., THE PANOPTIC SORT: A POLITICAL ECONOMY OF PERSONAL INFORMATION (1993); SURVEILLANCE AS SOCIAL SORTING: PRIVACY, RISK AND AUTOMATED DISCRIMINATION (David Lyon ed., 2002).

> The Appellants [CRIA] have relied upon automated computer systems to gather and generate evidence in support of their motion. They have not disclosed the details of how these systems work. Without explaining how the error was made, the Appellants admit that they made an error in one of the IP addresses at issue - rather than 64.231.255.184, one of the targeted IP addresses should be 64.231.254.117.
>
> ...
>
> When dealing with this kind of evidence in support of such extraordinary *ex parte* relief, the court should be presented with frank and full disclosure. For example, full disclosure might include an explanation from an independent expert as to how the P2P pseudonyms are linked to IP addresses (along with a solid documentary backup to put the explanation beyond doubt in every case). One incorrect number in an IP address means all the difference to the innocent person that would be exposed by the order sought.[42]

The real challenge for the Federal Court of Appeal in the *BMG v. Doe* case will not simply be the determination of whether or not to grant a disclosure order. The real challenge will be to formulate and then clearly articulate general principles about how to account for privacy and other interests in a way that accommodates the many concerns expressed above. In so doing, it will be crucial for the Court to recognize that any such exercise does not merely involve weighing the privacy interests of the individual defendants against CRIA and the public interest in permitting parties to proceed with lawsuits. *There is a broader public interest in privacy that must also be considered.*

This broader public interest in privacy on the Internet has been hinted at in other Canadian cases.[43] To the extent that the Court's order in *BMG v.*

[42] Memorandum of Fact and Law of the Intervener CIPPIC at paras. 7 and 36, BMG v. Doe, *supra* note 1, *appeal filed*, No. A-T-292-04 (F.C.A. Apr. 13, 2004), *available at* http://www.cippic.ca/en/projects-cases/file-sharing-lawsuits/CIPPIC%20FINAL%20Factum%20Aug%2010%202004.pdf

[43] *See, e.g.*, Irwin Toy Ltd. v. Doe 2000 O.J. No. 3318 (QL) (Ont. S.C.J.) at para. 10 ("In keeping with the protocol or etiquette developed in the usage of the internet, some degree of privacy or confidentiality with respect to the identity of the internet protocol address of the originator of a message has significant safety value and is in keeping with what should be perceived as being good public policy"); *SOCAN v. CAIP*, *supra* note 18, at para. 155 ("Privacy interests of individuals will be directly implicated where owners of copyrighted works or their collective societies attempt to retrieve data from Internet Service Providers about an end user's downloading of copyrighted works") (LeBel, J., dissenting).

Doe may result in privacy invasions through increased monitoring and surreptitious surveillance, this broader-based public interest in privacy must be taken into account in the analysis of how, when and why disclosure should be ordered or rejected. As one Supreme Court of Canada Justice recently acknowledged, courts considering the intersection of copyright and privacy should "be chary of adopting a test that may encourage [the monitoring of an individual's surfing and downloading activities]."[44]

One final lesson to be learned from *BMG v. Doe* is that the view of the ISP as the trusted guardian of its customers' privacy may soon be relegated to the past.[45] In the early days of the World Wide Web, most commercial ISPs put a sincere premium on their customers' privacy and, at that time, were in a plausible position to do so.[46] More recently, ISPs have faced reputational pressure to protect privacy. This pressure is particularly present where one major ISP breaks from the pack and indicates that it will protect its subscribers' privacy, thereby creating intense pressure for other ISPs to follow suit.

However, in the time that has passed since those heady days, the ISP-customer relationship has become more complex.[47] Although some of the ISPs involved in *BMG v. Doe* continue to play a role in advocating their customers' privacy, perhaps partly as a result of the pressure imposed by Shaw's strong lead to protect privacy in the case, others have chosen to play a lesser role, despite indications in their corporate privacy policies that claim a "longstanding commitment to safeguarding [subscribers'] right to

[44] *SOCAN v. CAIP*, *supra* note 18, at para. 155 (LeBel, J., dissenting). Although Justice LeBel was discussing a test for jurisdiction, his rationale in that context seems to apply even more so to the issue of the *BMG v. Doe* threshold for compelling the disclosure of identity.

[45] For a better understanding of those heady days, see generally Kerr, *Legal Relationship*, *supra* note 17.

[46] From time to time one still hears this. *See, e.g.*, Declan McCullagh, *Verizon appeals RIAA subpoena win*, CNET NEWS.COM, Jan. 30, 2003 (in the context of the RIAA lawsuits in the United States, Verizon stated that it would "use every legal means to protect its subscribers' privacy"), *at* http://news.com.com/2100-1023-982809.html. One may wonder whether such litigation is motivated more by user privacy or by the administrative cost to ISPs in complying with disclosure demands.

[47] *See generally Alex Cameron, Pipefitting for Privacy: Internet service providers, privacy and DRM*, Presentation to the 5th Annual Center for Intellectual Property Law and Information Technology Symposium at DePaul University College of Law: Privacy and Identity: The Promise and Perils of a Technological Age (Oct. 14, 2004).

privacy"[48] and an "eager[ness] to ensure protection of information carried over the Internet and respect for [subscribers'] privacy."[49]

One reason for this may be that the business of ISPs is no longer merely that of providing Internet access. For example, some Canadian ISPs have entered into the music downloading business.[50] Bell offers its customers music downloading through a service called Puretracks.[51] Vidéotron, on the other hand, is wholly owned by Quebecor Media Inc.[52] which provides its own music downloading service through another subsidiary company, Archambault Group Inc.[53] It should come as no surprise, therefore, that Vidéotron did not oppose CRIA's motion in *BMG v. Doe*. In fact, on appeal, Vidéotron has actually *supported* CRIA's position on the copyright issues, leaving little doubt about where it stands on the issues in the case: "[Vidéotron] agrees to protect its clients' privacy. [Vidéotron] does not agree to protect its clients' piracy."[54] As the ISP industry continues to evolve, it will be interesting to see whether other ISPs might follow Vidéotron's example.

Another reason why ISPs are no longer the trusted guardians of privacy they once were is the increasing role that ISPs are being forced to play in aiding international law enforcement and the fight against cybercrime and terrorism.[55] Recognizing that ISPs are not only the pipeline of online communication but also the reservoirs of their customers' personal information and private communications, cybercrime legislation proposed or enacted in many jurisdictions – including legislative reforms currently under

[48] Bell, Customer Privacy Policy, http://www.bell.ca/shop/en/jsp/content/cust_care/docs/bccpp.pdf.

[49] Vidéotron, Legal Notes, *at* http://www.videotron.com/services/en/legal/0_4.jsp.

[50] *See, e.g.*, Press Release, BCE, Bell Canada Launches the Sympatico Music Store (May 13, 2004), *available at* http://www.bce.ca/en/news/releases/bc/2004/05/13/71214.html.

[51] *Id.* Another Canadian ISP, Telus, is also offering music downloads in conjunction with Puretracks which can be accessed at http://telus.puretracks.com/. In addition to possibly helping to attract and retain customers, such services provide ISPs with an alternative revenue stream outside of charging a flat fee for access to the Internet.

[52] For a description of Vidéotron's relationship with Quebecor Media Inc., see http://www.videotron.com/services/en/videotron/9.jsp.

[53] See http://www.archambault.ca for a description of this company and its services.

[54] Memorandum of Fact and Law of the Third Party Respondent Vidéotron Ltée at para.7, BMG v. Doe, *supra* note 1, *appeal filed*, No. A-T-292-04 (F.C.A. Apr. 13, 2004), *at* http://www.cippic.ca/en/projects-cases/file-sharing-lawsuits/videotron_factum.pdf.

[55] *See, e.g.*, Convention on Cybercrime, Nov. 23, 2001, Europ. T.S. No. 185, *available at* http://conventions.coe.int/Treaty/Commun/QueVoulezVous.asp?NT=185&CL=ENG. *See also* Ian R. Kerr & Daphne Gilbert, *The Changing Role of ISPs in the Investigation of Cybercrime*, *in* INFORMATION ETHICS IN AN ELECTRONIC AGE: CURRENT ISSUES IN AFRICA AND THE WORLD (Thomas Mendina & Johannes Brtiz eds., 2004).

consideration in Canada[56] – will be used to expedite criminal investigations by substantially reducing the threshold tests required to obtain a judicial order for various forms of state surveillance. In other words, ISPs will be compelled to disclose identity information in a number of circumstances without anyone having to come before a judge and prove that the privacy rights of the individual under investigation and the broader public interest in protecting personal privacy are outweighed by the public interest in investigating cybercrime.

Canada's *Lawful Access*[57] agenda is one example of where ISPs may be forced to play a bigger role in aiding law enforcement. Under the current system, each major ISP in Canada has a different network architecture. The differences can be particularly significant between telecommunications and cable ISPs. These differences have important implications for privacy protection because the abilities of various ISPs to capture and retain information relating to the identity of their subscribers can differ greatly. This difference is reflected in the different ISP responses in the *BMG v. Doe* case – for example, Shaw and Telus claimed that they had none of the information sought by CRIA and Bell claimed that it had all of the information. Under *Lawful Access*, this will change as certain providers will be forced to completely re-engineer their networks to provide law enforcement with easy access to data.

As Professor Michael Geist has noted, *Lawful Access* will create a baseline standard for all ISPs' data retention and network configurations that will make it far easier for identity information to be obtained from them. This easier access to identity information will undoubtedly spill over from the law enforcement context to the civil actions. At the very least, ISPs will no longer be able to argue that they do not have the information sought. The reputational pressure on ISPs to protect privacy may also become negligible.

Finally, cost is another reason why ISPs may no longer be trusted guardians of privacy. When law enforcement or private parties have knocked on ISPs' doors seeking identity information to date, the first concern of the ISPs has often been "Who is going to pay for this?" Provided

[56] Dept. of Justice et al., *Lawful Access: Consultation Document* (Aug. 25, 2002), *available at* http://canada.justice.gc.ca/en/cons/la_al/law_access.pdf.

[57] "Lawful Access" is the euphemism designated to describe the Government of Canada's attempt to modernize law enforcement by expediting various forms of investigatory procedures, including procedures by which law enforcement agencies are able to more easily obtain identity information and personal communications from ISPs (in some cases without going to court). The Department of Justice is currently in the midst of an extensive "Lawful Access Consultation" process. For a description of this process and documents related to it, see http://canada.justice.gc.ca/en/cons/la_al/.

that ISPs are reimbursed for their costs of providing the information, ISPs will likely put up little privacy-based resistance to initiatives like *Lawful Access* and notice-and-takedown, or in civil actions like *BMG v. Doe*. Cost was a central issue for the ISPs in *BMG v. Doe*. At times during the hearing, it seemed as though the ISPs and CRIA were effectively engaged in a negotiation, mediated by the court, about the price for which subscribers' privacy rights could be bought.

As is the case in private sector disputes such as *BMG v. Doe*, some members of Canadian civil society[58] are also intervening in public sector hearings in order to ensure that the legal thresholds for compelling ISPs to disclose their customers' personal identity information are not diminished in the public law context as a result of Canada's *Lawful Access* agenda.

4. CONCLUSION

As *BMG v. Doe* ascends through the appellate process, it is uncertain whether the *privacy values* articulated by the Federal Court – in a case that will ultimately become known and remembered as *a-case-about-copyright* – will be affirmed and instantiated on appeal. We live in interesting times. At the same time that CRIA and other powerful private sector entities continuously intensify their growing arsenals of powerful new surveillance technologies, governments are seeking to pass laws which make it easier to obtain and make use of the personal information and communications that those private sector surveillance technologies are able to collect in their ever-growing databases, often without the consent of those about whom the information is being collected, used or disclosed to others.

The progression of *BMG v. Doe* through the courts runs 'in parallel' to the development of cybercrime legislation in government. Both of these private and public sector decision-making processes run the risk of diminishing online privacy in favour of an alleged public interest said to conflict with it.

One of the pioneers of the Internet, Stewart Brand, famously said that: "[o]nce a new technology rolls over you, if you're not part of the steamroller, you're part of the road."[59] This unseemly prospect is so powerful, so compelling that it paves an attitude in some of those in

[58] Canadian civil society groups include: On the Identity Trail (http://www.anonequity.org), CIPPIC (http://www.cippic.ca), Public Interest Advocacy Center (http://www.piac.org), and the B.C. Civil Liberties Association (http://www.bccla.org/).

[59] Stewart Brand, THE MEDIA LAB: INVENTING THE FUTURE AT MIT (1987) at 9.

opposition to the value set underlying the dominant technology to develop bigger steamrollers and steer them in the opposite direction. There is little doubt that for a small subset of those who cherish P2P file-sharing, the answer to CRIA and surveillance technologies like those used by MediaSentry will be the development of an extremely potent anonymous P2P network.[60] Such systems would enable Lessig's horrific vision: "[p]erfect anonymity makes perfect crime possible."[61]

For the many netizens who see social value in the ability to exist online and off in various states of nymity, and who abhor those others who intentionally exploit anonymity as nothing more that a means of escaping accountability for immoral or illegal acts, the steamroller mentality is not a promising road. Our courts and legislatures need desperately to pave other paths.

ACKNOWLEDGEMENTS

Ian Kerr wishes to extend his gratitude to the Social Sciences and Humanities Research Council, to the Canada Research Chair program, to Bell, Canada and to the Ontario Research Network in Electronic Commerce for all of their generous contributions to the funding of the research project from which this paper derives: *On the Identity Trail: Understanding the Importance and Impact of Anonymity and Authentication in a Networked Society*" (www.anonequity.org). Alex Cameron wishes to thank Philippa Lawson for the privilege of representing CIPPIC (www.cippic.ca) in *BMG v. Doe* and Ian Kerr for his steadfast encouragement and support. Both authors are grateful to Todd Mandel for his very capable research support and his outstanding contributions to this project. The authors also wish to express their sincere thanks to Professors Jane Bailey, Michael Geist, and Philippa Lawson for their very helpful comments, which resulted in a much better chapter. Finally, the authors wish to congratulate Katherine Strandburg and Daniela Stan Raicu for organizing CIPLIT®'s successful 2004 symposium

[60] Biddle et al., *supra* note 16. *See also* John Borland, *Covering tracks: New privacy hope for P2P*, CNET NEWS.COM, Feb. 24, 2004, *at* http://news.com.com/2100-1027-5164413.html.

[61] Lessig, *supra* note 10, at 1750.

"Privacy and Identity: The Promise and Perils of a Technological Age" at DePaul University College of Law.

Chapter 17

FOURTH AMENDMENT LIMITS ON NATIONAL IDENTITY CARDS

Daniel J. Steinbock
Harold A. Anderson Professor of Law and Values, University of Toledo College of Law

Abstract: In the past three years there have been serious calls for a national identity system whose centerpiece would be some form of national identity card. This chapter analyzes the Fourth Amendment issues raised by two major features of any likely national identity system: requests or demands that individuals present their identity cards; and governmental collection, retention, and use of personal information to be used in identity checks. These issues are evaluated in several different contexts in which they might plausibly arise. The chapter concludes that, while the Fourth Amendment might bar certain practices and block others depending on their purposes, it would be possible to have a constitutional national identity card system of a fairly comprehensive type. Even where an identity system would not strictly run afoul of the Fourth Amendment, however, an analysis of the interests that the Amendment is designed to protect provides an insight into the price in privacy and liberty a national identity system would exact. The chapter also indicates how these effects might be mitigated somewhat in the system's design. This chapter thus aims to illuminate not only what kind of national identity system the U.S. lawfully could have, but how it might be devised, and, implicitly, whether we want to have one at all.

Key words: identity card, national identity system, identity check, checkpoints, Fourth Amendment, stops, seizure, search, privacy, records, database

1. INTRODUCTION

The United States is currently slouching toward a national identity card. The past three years have seen the most serious and detailed consideration of a national identity system ever,[1] and we are probably just one domestic terror attack away from its implementation in some form. For almost twenty years, everyone, citizen or not, has needed to present specified documents to prove work authorization.[2] The already ubiquitous social security number is being demanded in an increasing number of contexts.[3] The recently enacted REAL ID Act effectively preempts state laws regarding drivers' licenses with detailed federal standards set by the Act and ensuing regulations.[4] It specifies what documents a state must require before issuing a driver's license, as well the data that the license would have to contain.[5] A license meeting the criteria, or an equivalent card issued to nondrivers for identification purposes, will be needed to board commercial aircraft and for whatever other purposes the Secretary of Homeland Security designates.[6] Licensing agencies must confirm lawful immigration and social security status, using database links. If the Act, which takes effect in 2008, does not actually establish a *de facto* national identity card it certainly makes the prospect of one increasingly likely.

The basic function of a national identity system would be "to link a stream of data with a person. . . . [H]uman identification is the association

[1] *See, e.g.*, Alan M. Dershowitz, WHY TERRORISM WORKS: UNDERSTANDING THE THREAT, RESPONDING TO THE CHALLENGE 200-01 (2002) (arguing that a national identity card is "an issue that deserves careful consideration"); David Frum & Richard Perle, AN END TO EVIL: HOW TO WIN THE WAR ON TERROR 69-73 (2003) (advocating a national identity card with biometric data showing citizenship or immigration status); Tova Wang, The Debate Over a National Identity Card, 1 (2002), *available at* http://www.tcf.org/4L/4LMain.asp?SubjectID=1&TopicID= 0&ArticleID=284 (last accessed Apr. 22, 2004) ("[P]roposals for creating a national identification card system have gained new attention."); Donald A. Dripps, *Terror and Tolerance: Criminal Justice for the New Age of Anxiety*, 1 OHIO ST. J. OF CRIM. L. 9, 36 (2003) ("If security professionals believe national biometric identity cards would contribute substantially to the prevention of terrorism, then we need to proceed promptly in that direction."); Editorial, *A National ID*, N.Y. TIMES, May 31, 2004, at 16 (urging establishment of a national commission to study identity cards); Nicholas D. Kristof, *May I See Your ID?*, N.Y. TIMES, Mar. 17, 2004, at A25; *see also* Wang, *supra*, at 8-9 (listing proposals for enhancements to identity documents and identification requirements.

[2] For a detailed specification of documents establishing employment authorization and identity for purposes of verifying authorization to work in the U.S., see Immigration and Nationality Act §274A(b)(1), 8 U.S.C.A. §1324a(b)(1) (identity established by U.S. passport, resident alien card, driver's license).

[3] *See e.g.*, Rick S. Lear & Jefferson D. Reynolds, *Your Social Security Number or Your Life: Disclosure of Personal Identification Information by Military Personnel and the Compromise of Privacy and National Security*, 21 B. U. INTL. L. J. 1, 13-14 (2003).

[4] Pub. L. No. 109-13, Title II, 119 Stat. 311 (May 11,2005).

[5] *Id.* at §202 (b), (c).

[6] *Id.* at §202 (a).

of data with a particular human being."[7] Once that connection is made, official reaction can take a variety of forms, depending, of course, on what the data show and the legal consequences of that knowledge. What data to collect, who would have access to the data, and what uses would be made of it are all important issues in the system's design. Any such system, however, depends on two major features: the database (or databases) containing information about particular individuals and the means to connect a given person with that information.[8]

Connecting an individual with a certain identity, and hence to a body of data, necessitates that people identify themselves at some point or points in time, by way of codes, tokens, biometrics, or some other means.[9] A national identity system would almost certainly produce new interactions between the populace and law enforcement personnel – a subject regulated by the Fourth Amendment when interaction amounts to a seizure of the person. One Fourth Amendment question, then, concerns which identification opportunities constitute governmental seizures and under what circumstances state agents may demand to see a person's identity card.

Each identification encounter would also be an occasion to add information to the central database, facilitating government surveillance of movement and activity. This possibly raises a second set of Fourth Amendment issues, particularly when it involves governmental collection of data generated in circumstances in which there might otherwise be some legitimate expectation of privacy, such as information provided to health

[7] Roger Clarke, *Human Identification in Information Systems: Management Challenges and Public Policy Issues*, 7 Info. Tech. & People, (No. 4) 6, 8 (1994), *available at* http://www.anu.edu.au/people/Roger.Clarke/DV/HumanID.html (last visited Mar. 28, 2004) (emphasis omitted).

[8] This is no easy task. For an indication of how much work would need to be done to construct a standard national database, *see* U.S. Gen. Accounting Office, *Information Technology Terrorist Watch Lists Should be Consolidated to Promote Better Integration and Sharing*, GAO-03-322 at 1-2 (Apr. 15, 2003).

[9] Roger Clarke, *Human Identification in Information Systems: Management Challenges and Public Policy Issues*, 7 INFO. TECHNOLOGY & PEOPLE 6, 10 (1994) gives an interesting account and capsule history of each. He also gives the following shorthand summary:
appearance – or how the person looks;
social behavior – or how the person interacts with others;
names – or what the person is called by other people;
codes – or what the person is called by an organization;
knowledge – or what the person knows;
tokens – or what the person has;
bio-dynamics – or what the person does;
natural physiography – or what the person is;
imposed physical characteristics – or what the person is now.

care or educational institutions or in other registration procedures. Would such data collection be a search under the Fourth Amendment and, if so, would it be a reasonable one?

Although the possible Fourth Amendment ramifications of a national identity system have often been acknowledged, they have not been explored in depth.[10] While the impact on Fourth Amendment rights would of course depend on the particular features of the system, because of their centrality to any likely national identity system it is worth examining these issues even in the absence of a concrete proposal.[11] This chapter attempts to sketch out some of the issues and their possible resolution.

2. IDENTITY CARD REQUESTS AND DEMANDS

The Fourth Amendment bars "unreasonable searches and seizures," including seizures of the person. Evaluating the effect of the Fourth Amendment on demands for identification involves determining first whether the demand involves a "seizure." If not, the Fourth Amendment inquiry is at an end. If the encounter does entail a "seizure" the second question is whether it is one that is reasonable.

2.1 Consensual Interactions

Consensual interactions between individuals and governmental personnel are not seizures under the Fourth Amendment. The Supreme Court has refused to find a seizure in governmental questioning and identification requests of people already in confining circumstances.[12] These would include demands for identification during what may be described as registration procedures, such as those for driver's licenses, medical services, schools, and flights, as well as employment eligibility verification. Airport check-in, which now necessitates government-issued photo identification,

[10] Computer Science and Telecommunications Board, National Research Council, IDs – NOT THAT EASY 7 (2001) ("Clearly, an examination of the legal . . . framework surround identity systems . . . would be essential.")

[11] *Id.*, at 29 ("The constitutional limitations on an agent's ability to require presentation of IDs . . . should be explored before any such enactment to avert the costs of imposing the system and then having to revise or abandon it in the face of its unconstitutionality, to say nothing of its effects on civil liberties.").

[12] INS v. Delgado, 466 U.S. 210, 212-13 (1984); Florida v. Bostick, 501 U.S. 429, 436 (1991); United States. v. Drayton, 536 U.S.194, 201-02 ((2002).

might be considered the paradigm. All of these tasks require the customer to stop to register, and while the display of identification tokens may slightly lengthen the process, it is unlikely that those additional moments would convert the registration process to a state-mandated seizure. Further, this precedent strongly suggests that a national identity system could mandate the display of identification at other registration points, such as hotel or car rental check-ins, without creating any seizure of the person, so long as the registration would, by itself, ordinarily restrict movement.

Moreover, similar reasoning allows law enforcement agents to approach people and ask for identification even when their freedom of movement is not already limited by their own actions or decisions; in other words, as they go about their business in public. As long as a reasonable person would feel free to "terminate the encounter" in the circumstances the person has not been seized.[13] This doctrine has important implications for any national identity system. Law enforcement agents could approach an individual on the street and ask to see a national identity card at any time, without any prior suspicion of criminality or other illegality. Because the Fourth Amendment generally puts little restraint on racial, national, and ethnic profiling in such police law enforcement actions, these factors could legally be the basis for such identification "requests."

If a true cross-section of the American population was routinely asked by government agents to show their identity cards, the incidence of "consensual" compliance might decline drastically. A national identity system that depended on voluntary responses to requests for identification would thus run the risk of perfectly legal, and possibly organized, civil disobedience.[14] It is unlikely, therefore, that a national identity system could rely solely, or mainly, on consensual compliance as a means of identity verification.

2.2 Investigative Stops

Though the real life factual differences between them can be quite small, an investigative stop is conceptually quite different from a consensual encounter. Under *Terry v. Ohio*[15] and its numerous Supreme Court

[13] *Bostick,* 501 U.S. at 434-36.

[14] By way of analogy, as of March 16, 2004, 269 communities in thirty-eight states had passed anti-Patriot Act resolutions. ACLU, *List of Communities That Have Passed Resolutions*, *at* http://www.aclu.org/SafeandFree/SafeandFree.cfm?ID=11294&c=207 (last visited Apr. 12, 2004). This includes three statewide resolutions. *Id.*

[15] 392 U.S. 1 (1968).

progeny,[16] such stops are Fourth Amendment seizures and can be conducted only on "specific and articulable facts which, taken together with rational inferences from those facts, reasonably warrant [the] intrusion."[17] These facts and inferences can relate to an ongoing crime or to a past felony.[18] The level of evidence need for a lawful stop is often called "reasonable" or "articulable" suspicion, which can be distinguished from the "inarticulable hunches"[19] or "inchoate and unparticularized suspicion"[20] that are not sufficient to compel a person to halt for investigation. Under this body of law, persons could not involuntarily be seized on a random or individual basis for identity checks in the absence of reasonable suspicion of criminality.[21] The Supreme Court similarly has barred the suspicionless stopping of motor vehicles to check license and registration[22] or for questioning about a traveler's citizenship.[23] The Fourth Amendment thus bars random or suspicionless *demands* that a person "show her papers."

But when reasonable suspicion is present the suspect may be compelled to identify himself, as the Supreme Court recently held in *Hiibel v. Sixth Judicial District Court of Nevada*.[24] Said a 5-4 majority of the Court, "The request for identity has an immediate relationship to the purpose, rationale and practical demands of a *Terry* stop."[25] Because it would make many investigative stops more efficient and effective, by implication a national identity card could also be required in investigative stops.[26]

The Court in *Hiibel* stated, however, that under established stop and frisk principles, "an officer may not arrest a suspect for failure to identify himself if the identification request is not reasonably related to the circumstances

16 *See, e.g.*, Florida v. J.L., 529 U.S. 266 (2000); Illinois v. Wardlow, 528 U.S. 119, 124-25 (2000); United States v. Sokolow, 490 U.S. 1, 7-8 (1989); United States v. Cortez, 449 U.S. 411, 417-18 (1981).

17 *Terry*, 392 U.S. at 21.

18 United States v. Hensley, 469 U.S. 221, 229 (1985).

19 *Terry*, 392 U.S. at 22.

20 *Terry*, 392 U.S. at 27.

21 Brown v. Texas, 443 U.S. 47, 52 (1979).

22 Delaware v. Prouse, 440 U.S. 648, 661 (1979).

23 United States v. Brignoni-Ponce, 422 U.S. 873 (1975). This holding does not apply at the border or its functional equivalents. *Id.* at 876.

24 124 S. Ct. 2451 (2004).

25 *Id.* at 2459.

26 As the Court took pains to point out, the Nevada statute at issue in *Hiibel* had been interpreted to require that the suspect disclose his name, but not necessarily provide his driver's license or any other document. If this distinction was important to the holding, it would obviously undermine *Hiibel*'s support for insistence on identity card presentation. It seems likely, however, that this language is simply the usual confining of a case to its facts rather than a suggested prohibition on document demands.

justifying the stop."[27] Is there, then, a category of stops in which a demand for identification would be constitutionally unreasonable? The Ninth Circuit has said yes, and found criminal punishment for refusal to give identification during a *Terry* stop to violate the Fourth Amendment when identification is not needed in the investigation.[28] There is not much reason, though, to follow this line of cases. While there could be some circumstances in which requiring the suspect's identity card would not advance the officer's inquiry, merely asking for or reading it would not seem to add appreciably to the stop's intrusiveness. Moreover, it seems silly and unworkable to ask the officer on the street to distinguish between stops for which identification would further the investigation and those for which it would not. Rather, this would seem an area where a bright-line rule would make good sense.

2.3 Arrests and Citations

The reasons for permitting law enforcement personnel to demand identification from those stopped for investigation apply even more forcefully to persons being arrested or given a traffic citation. These people have already been seized, so a demand for identification imposes no additional physical restraint. Identification is even more necessary for the processing of the arrest or traffic citation than for an investigative stop.[29]

Traffic stops present a particularly fertile field for identity checking. The Supreme Court held in *U.S. v. Whren* that the officer's motivation for a traffic stop is irrelevant to its reasonableness under the Fourth Amendment; in other words, a driver has no constitutional objection to being pulled over on the ground that it is a "pretext" for the officers to perform one of the many investigation measures attendant on a traffic stop.[30] Because identification verification is always part of traffic law enforcement,[31] police

[27] *Hiibel*, 124 S. Ct. at 2459.

[28] Martinelli v. City of Beaumont, 820 F.2d 1491, 1494 (9th Cir. 1987); Lawson v. Kolender, 658 F.2d 1362, 1364-69 (9th Cir. 1981).

[29] Illinois v. Lafayette, 462 U.S. 646, 640 (1983) (noting the importance of ascertaining or verifying the arrestee's identity); Smith v. United States, 324 F.2d 879, 882 (D.C. Cir. 1963) (holding that routine identification processes are part of custodial arrest).

[30] 517 U.S. 806 (1996); Arkansas v. Sullivan, 532 U.S. 769, 771-72 (2001) (valid traffic stop not rendered unreasonable by officers' subjective aim to search for evidence of crime.). *See also* David A. Harris, *The Stories, the Statistics, and the Law: Why "Driving While Black" Matters*, 84 MINN. L. REV. 265, 311-18 (1999).

[31] David A. Harris, *Car Wars: The Fourth Amendment's Death on the Highway*, 66 GEO. WASH. L. REV. 556, 568 (1998) ("The traffic stop gives the officer the opportunity to walk to the driver's side window and . . . request[] license and registration . . . ."); Wayne R.

officers could, consistent with Supreme Court precedent, pull drivers over for the very purpose of checking that identification, as long as they first observed a traffic violation.

Traffic stops, therefore, present a powerful and dangerous tool for any national identity system. They are powerful because with enough law enforcement desire and effort, virtually any driver could be made to stop and present an identity card; they are dangerous because of the unlimited official discretion this possibility allows. For this reason, statutory restraint on the use of traffic stops as a means of identity checking should be considered as part of any national identity scheme, though it must be acknowledged that such limitations would be difficult to define and enforce. The temptation to stop vehicles for minor offenses in order to demand drivers' identity cards thus counts as a major disadvantage of any extensive identity system.

2.4 Checkpoints

Checkpoints entail stopping all persons or vehicles (or a pre-designated subset) passing a particular location. They are a potentially important method of identity determination for any national identity system, both because they could reach large numbers of people and because they could be placed at or around sensitive locations. They are quite efficient in that they do not rest on the voluntary compliance of consensual encounters, nor do they depend upon the reasonable suspicion of illegal behavior required for investigative stops. Checkpoints, in short, are both compulsory and encompassing in their coverage. Vehicle checkpoints have already been employed for immigration enforcement, driver sobriety checks, drug interdiction, witness identification, and other national security and law enforcement objectives. In dicta, the Court has indicated approval of roadblock-type stops for highway license and registration checks,[32] at government buildings or airports,[33] or "to thwart an imminent terrorist attack or to catch a dangerous criminal who is likely to flee by way of a particular route."[34] If a national identity system did employ checkpoints, they would likely be applied to pedestrians as well, a use not yet specifically addressed by the Supreme Court.

Lafave, *The "Routine Traffic Stop" from Start to Finish: Too Much "Routine," Not Enough Fourth Amendment,* 102 MICH. L. REV. 1843, 1875 (2004) (running a computer check on the driver's license and registration incident to a "routine traffic stop," "is well established as a part of the 'routine,' and has consistently been approved and upheld by both federal and state courts.")

[32] Delaware v. Prouse, 440 U.S. 648, 663 (1979).

[33] *Edmond,* 531 U.S. at 48-49.

[34] 428 U.S. 543, 551 (1976).

For checkpoints the primary purpose of which is not general law enforcement, the Supreme Court requires that reasonableness (and thus compliance with the Fourth Amendment) be determined using a three-pronged balancing test. The Court weighs 1) the government's interest in preventing the relevant harm, 2) the extent to which the checkpoint system can be said to advance that interest, and 3) the degree of intrusion on those persons who are stopped.[35]

With respect to the first factor, a national identity system directed at a real and present danger of terrorist attacks or the illegal presence of millions of non-citizens would almost certainly suffice. Avoidance of terrorist attack needs no justification as a substantial governmental interest. As for immigration enforcement, what the Court said in 1976 in *United States v. Martinez-Fuerte* is still true: despite (or perhaps because of) the national policy to limit immigration, "large numbers of aliens seek illegally to enter or to remain in the U.S." [36]

Checkpoint effectiveness can be measured in several different ways: 1) the *absolute number* of suspects apprehended, 2) the *rate* of apprehensions (the number of suspects divided by the number of individuals stopped); or 3) their *relative effectiveness* compared to other methods of prevention and enforcement. The Supreme Court has considered all three measures in evaluating checkpoints, although not in any systematic way. Existing checkpoints impact an enormous number of vehicles and the resulting success rate is relatively low. In *Martinez-Fuerte* the Court found that .12% (or 12 of every 10,000) of detained vehicles contained deportable aliens. In *Sitz* it approved a 1.6% success rate for sobriety checkpoints.[37] More general use of checkpoints would probably test even those numerical limits. On the other hand, the Court in *Sitz* explicitly cautioned the judiciary against "a searching examination" of checkpoint effectiveness.[38] With *relative effectiveness*, too, the Court urges deference to governmental preferences.[39]

[35] Mich. Dep't of State Police v. Sitz, 496 U.S. 444, 455 (1990); *see also Martinez-Fuerte*, 428 U.S. at 555; Brown v. Texas, 443 U.S. 47, 50-51 (1979).

[36] *Martinez-Fuerte*, 428 U.S. at 551. The Court mentioned a possible 10-12 million aliens illegally in the country. *Id.*

[37] Michigan Dept. of State Police v. Sitz, 496 U.S. 444, 455 (1990) (noting that nationally sobriety checkpoints result in drunk driving arrests of 1% of motorists stopped). *But see id.* at 461 (0.3% success rate at 125 Maryland checkpoints).

[38] *Sitz*, 496 U.S. at 454 (Effectiveness evaluation is "not meant to transfer from politically accountable officials to the courts the decision as to which among reasonable alternative law enforcement techniques should be employed to deal with a serious public danger.") *Id.* at 453.

[39] Id. at 453-54.

On balance, an expanded use of immigration checkpoints in conjunction with a national identity card would seem to pass the "effectiveness" threshold of current Fourth Amendment balancing. To the extent that checkpoints would be employed to identify potential "terrorists" rather than undocumented aliens, the various measures of efficacy come out differently. Presumably, checkpoints would be used either to catch individuals already identified as terrorism suspects or to single out previously unknown persons who met some kind of "terrorist" profile. Preventing terrorism is certainly a powerful national goal, but it seems likely that the absolute number of suspects apprehended through extensive identification checkpoints would be very small and the success rate would be miniscule.[40] The number of terrorism suspects seems to be tiny, and the group of those who are not also in the U.S. illegally is even smaller. Moreover, given what a crude instrument terrorist "profiling" is, the likelihood of mistaken "hits" is substantial.

The third factor in the Supreme Court's assessment of roadblocks has been the degree of intrusion experienced by the affected motorists. The Court speaks in terms of "objective" and "subjective" intrusions, with "objective" referring to physical interference with movement and "subjective" referring to concern or fright caused to those stopped by the roadblock.[41] In the cases involving immigration and sobriety checkpoints, the Court has not found either form of intrusion particularly weighty.

The "objective" intrusion obviously depends on how a checkpoint is operated. Hypothetical checkpoints under a national identity system would most likely involve stopping each motorist or pedestrian, a demand for presentation of her identity card, and a "swipe" of the card through a reader linked with computerized records. The individual effectively would be detained both while this occurred as well as pending the electronic reply from the database. If this sequence of events took no longer than an ordinary credit card authorization, it is doubtful the Court would find the objective intrusion to differ significantly from that of immigration or

[40] *See* Stephen J. Ellmann, *Racial Profiling and Terrorism*, 22 N.Y.L.S. L. REV. 675, 699-700, n. 65-70 (2003), for an attempt to estimate the number of Al Qaeda terrorist within the United States and the relative hit rate of a program that tried to profile them. Recognizing that we "simply do not know what the true number [of such terrorists] is" and that no profile is completely accurate, he assumes that a hit rate of even 1 in 10,000 would be a "plausible basis for action." *Id.* at n. 65-66. He also notes that a profiling program's effect in deterring attacks adds to the measure of its usefulness. *Id.* at n. 63. On racial profiling's effectiveness, *see* David A. Harris, PROFILES IN INJUSTICE: WHY RACIAL PROFILING CANNOT WORK 73-87 (2002).

[41] Michigan Dept. of State Police v. Sitz, 496 U.S. 444, 451-53 (1990); Martinez-Fuerte v. United States, 428 U.S. 543, 558-59 (1976).

sobriety checkpoints, which the Court has characterized as "quite limited"[42] and "brief."[43] The average delay occasioned by the sobriety checkpoints in *Sitz* was approximately 25 seconds,[44] but checkpoints requiring an individual to remove an identity card from her pocket or purse, to have it read, and then to await clearance could easily take considerably longer, beyond what the Supreme Court has approved thus far.

In one way, the "subjective" intrusion of an identification checkpoint would be less than for other kinds, because the checking would be limited to asking for a clearly designated piece of identification, a much less open-ended inquiry than looking for signs of intoxication, for example, at sobriety checkpoints. Fear of facing unknown questions on unknown topics would thereby be reduced as long as the officers stationed at the checkpoints restricted themselves to inspecting the identification card. Although the Court has yet to consider pedestrian checkpoints, it seems unlikely that it would find them more invasive than vehicle checkpoints as a general matter.[45] Being checked against a database is probably much more frightening, though, than the current immigration or sobriety checkpoints – even for an innocent person.[46] The individual has no way of knowing the contents of the database against which her identification is being run, whether they are accurate or not, or what further impositions might be triggered by the information linked to her identity card. This uncertainty will turn every identification demand into cause for apprehension. Identification checkpoints, it may be argued, have an additional subjective

[42] *Martinez-Fuerte*, 428 U.S. at 557-58 ("The stop does intrude to a limited extent on motorists' right to 'free passage without interruption, and arguably on their right to personal security. But it involves only a brief detention of travelers during which all that is required of the vehicle's occupants is a response to a brief question or two and possibly the production of a document evidencing a right to be in the U.S.") (internal quotations and citations omitted)

[43] Illinois v. Lidster, 540 U.S. 419, 425 (2004); *Sitz*, 496 U.S. at 451.

[44] *Id.* at 448.

[45] At one point in *Martinez-Fuerte*, the Court adverted to the general principle of lesser expectation of privacy in automobiles, 428 U.S. at 561, but in the context of the entire case the holding does not seem to turn on this factor. In *Lidster*, 540 U.S. at 428-29, the Court recently intimated that informational pedestrian checkpoints, at least, are *less* intrusive than vehicular ones in that the former can be conducted in a voluntary, consensual manner while the latter necessarily involve a seizure under the Fourth Amendment.

[46] Normally the Supreme Court only accounts for the reaction of a reasonable *innocent* person when determining Fourth Amendment issues. *See, e.g.*, *Florida v. Bostick*, 501 US at 438.

effect on a grand scale: the psychic harm to a free people of having to "show your papers," even if only at certain designated locations.[47]

On balance, these considerations would probably result in upholding immigration-focused identity card checkpoints against constitutional challenge. Apart from immigration enforcement, checkpoints directed at catching potential terrorists would, and should, mainly fail to pass muster under the Fourth Amendment. Even if suspects' names were known, huge numbers of people would need to be stopped in the hopes of locating a very small collection of individuals unless the search were geographically limited.[48] Moreover, profiling potential terrorists is at a very rudimentary stage and coupling it with identification checkpoints would in all likelihood yield a low rate of success and a large number of wrongful investigative detentions. Despite the obvious importance of preventing further domestic attacks, these likely results should stand in the way of finding such general anti-terrorism identity card checkpoints to be reasonable seizures under the Fourth Amendment. The outcome would probably be different, however, at especially sensitive locations like airports, monuments, public buildings, or so-called national special security events.[49]

3. DATA, GENERATION, COLLECTION, AND RETENTION

At the same time an identity check taps into existing databases, it can also generate new data by inputting the location and activity of the person whose identification card is being read. National identity cards could thus be the fulcrum for a massive exercise in governmental data collection. For example, everyone is now required to present some form of identity token in the process of taking a flight. In addition to indicating whether that person was a flight risk (or perhaps an undocumented alien or wanted criminal), the same readable token could easily create new data, such as the location, time,

[47] *See e.g.*, Edmond v. Goldsmith, 183 F. 3d 659, 662 (7th Cir. 1999) (referring to checkpoints as "methods of policing associated with totalitarian nations") (Posner, J.).

[48] For an argument in favor of "seizing groups" as defined by place and time (not, for example, race), *see* William J. Stuntz, *Local Policing After the Terror*, 111 YALE L. J. 2137, 2163-69 (2002).

[49] When an event is designated a National Special Security Event, the Secret Service becomes the lead agency for the design and implementation of the operational security plan. http://www.secretservice.gov/nsse.shtml. Florida v. J.L., 529 U.S. 266 (2000) (Fourth Amendment expectations of privacy diminished in certain places, including airports).

and date the token was read, and even the flight and itinerary. The database would now contain new information about the person's location and activities. This information potentially could be used not only in subsequent identity checks but also for more general law enforcement or other purposes.[50]

There are a surprising number of registration occasions in modern American life.[51] In addition, private businesses collect and retain "staggering amounts of personal information about American citizens and compil[e] it into electronic dossiers designed to predict the way people think and behave."[52] If a national identity token were used in these procedures, the data could end up in a database linked to that identity and be accessible for government use. This, in fact, is true already of a great deal of personal information through its association with social security numbers. A centralized national database would only sharpen the picture of a person's life.

Deciding whether government access to such information is a "search" under the Fourth Amendment, therefore requiring a warrant, probable cause, or some other evidentiary basis, would put to the test a series of cases in which the Court has applied a concept of assumption of the risk of surveillance. The Court has held in a variety of contexts that a person "takes

[50] *See* ABA Standards for Criminal Justice, Electronic Surveillance (3d ed. 1999), Section B: Technologically-Assisted Physical Surveillance, Commentary to Standard 2-9.1(d) (vii) (p. 43) ("[T]he results of tracking operations . . . can be preserved well after the surveillance ends, in theory indeterminately. This capability raises the specter of extensive libraries that retain information on vast numbers of individuals in perpetuity."). The ACLU characterizes the possibilities this way:

> When a police officer or security guard scans your ID card with his pocket bar-code reader, for example, will a permanent record be created of that check, including the time and your location? How long before office buildings, doctors' offices, gas stations, highway tolls, subways and buses incorporate the ID card into their security or payment systems for greater efficiency? The end result could be a nation where citizens' movements inside their own country are monitored and recorded through these "internal passports."

American Civil Liberties Union, 5 Problems with National ID Cards, http://www.aclu.org/Privacy/Privacy.cfm?ID=13501&c=39 (February 13, 2004).

[51] Andrew J. McClurg, *A Thousand Words Are Worth a Picture: A Privacy Tort Response to Consumer Data Profiling*, 98 Nw. U. L. Rev. 63, 65 (2003). ("Consumer data profiles may contain information pertaining to some of the most intimate aspects of human life: religion, health, finances, political affiliation, leisure activities, and much more.") *Id.* at 70.

[52] *Id.* at 65.

the risk, in revealing his affairs to another, that the information will be conveyed by that person to the Government."[53] It has stated, "This Court has held repeatedly that the Fourth Amendment does not prohibit the obtaining of information revealed to a third party and conveyed by him to Government authorities, even if the information is revealed on the assumption that it will be used only for a limited purpose and the confidence placed in the third party will not be betrayed."[54] The Court has upheld the installation of pen registers, which record telephone numbers dialed, for the same reason: the caller assumes the risk that the phone company will divulge these numbers, and therefore has no expectation of privacy in that information.[55] This approach allows unregulated government access to just about all information "voluntarily" conveyed to third parties in the course of one's activities, including much of the contents of commercial databases.[56]

The assumption of risk collapses under its own weight, however, when applied to information *required* to be passed to the government, as would be the case if a national identity card were mandated for certain registrations and the information was then, by statutory or regulatory mandate, transmitted to the system's computers. The two requirements that information be linked to a national identity token and transmitted to the government for inclusion in its identity system database make all the difference, even if the information itself would have been generated anyway in the registration transaction. The risk of disclosure would be imposed, not assumed, and would not be a risk at all but a certainty.[57] In addition, many occasions for registration involve activities that are central to a free and full

[53] United States v. Miller, 425 U.S. 435, 443 (1976).

[54] *Id.*, citing cases involving government informants being sent, without judicial authorization, to converse with suspects and report, record, or transmit the suspect's statements.

[55] Smith v. Maryland, 442 U.S. 735, 743 (1979).

[56] Certain statutory protections now exist for some personal records, Right to Financial Privacy Act, 12 U.S.C. §3401 (1978) (permitting depositors to challenge subpoenas for their financial records except where notifying them would "seriously jeopardize the investigation"), and Electronic Communication Privacy Act, 18 U.S.C. §3121 (1987) (court approval required for government access to telephone records). It is not clear whether the existence of such statutes would be found to create an expectation of privacy under the Fourth Amendment. In the context of a national identity system with data-gathering powers this would likely be a moot point, however. By establishing federal access to certain records, the enabling legislation would thereby eradicate any inconsistent prior protections, destroying any expectation of privacy that might have been present.

[57] This would be analogous to Justice Marshall's hypothetical of an official announcement that henceforth mail or private phone conversations would be randomly monitored, forcing citizens to "assume the risk" of government intrusion. *Smith*, 442 U.S. at 750.

life, such as education, medical care, and travel.[58] Together these factors obviate the fictions of voluntary third-party disclosure and assumed risk, and should cause governmentally mandated collection and transmission of personal data to be considered a search under the Fourth Amendment.

If the Court did conclude that such compelled data collection was a search, that finding would not end the inquiry. It would then need to address the reasonableness of the practice. Given what could be characterized as the non-criminal purpose of such data collection, the analysis used for other administrative searches would probably be applied. As with regulatory seizures, this methodology balances "the government's need to search against the invasion the search entails."[59]

On the one hand, the Court has never upheld the kind of blanket invasion of personal privacy that the transmission of registration information into national databases would involve. While it is true that such personal information could be used to further important government interests, this practice would appear to lack many other attributes of the administrative searches sustained by the Court. Even though amassing personal information in comprehensive databases might ultimately increase their usefulness, the marginal utility is uncertain if not completely speculative. An enormous amount of data about virtually every person in the United States would have to be collected and retained in the hope that some of it might give some hint of terrorist or other illegal activity. The effectiveness of terrorist profiling using a wide array of personal data has yet to be demonstrated. There can thus be no claim that no "other technique would achieve acceptable results."[60]

With respect to the degree of intrusion, much would turn on how long the information was retained, what parties had access to it, and the purposes for which it would be used. These factors figured in the Court's assessment in *Whalen v. Roe* of New York's centralized filing system for controlled substances prescriptions, identifiable by patient name.[61] Challenged by

[58] As pointed out by Justice Marshall in dissent in *Smith*, 442 U.S. at 749, this distinguishes the consensual monitoring cases on which risk assumption analysis is built, because one can certainly live a full life without speaking to a given individual. In other words, talking to an unreliable auditor is truly a risk one "assumes," while engaging in these other activities is not. On the other hand, the Court has characterized the disclosures of private medical information to doctors, hospitals, insurance companies and public health agencies as "an essential part of modern medical practice." Whalen v. Roe, 429 U.S 589, 602 (1977).

[59] Camara v. Municipal Court of the City and County of San Francisco, 387 U.S. 523, 537 (1967).

[60] *Id.*

[61] 429 U.S. 589 (1977).

doctors and patients on right to privacy grounds,[62] this mandatory reporting program was upheld by a unanimous Court. The program contained limitations on who could access the data and made unauthorized disclosure a crime; no instances of unauthorized use had occurred in the first twenty months of the program's operation.[63] These and other features of the New York drug prescription library in *Whalen* point the way for designing a mandatory data collection system that could pass as a reasonable search. Comprehensive restrictions, particularly on disclosure and use, as well as on the duration of information retention, would help this practice pass constitutional muster. On the other hand, before the United States came close to becoming a "total information society" one would hope the Supreme Court would call a halt to its previous acceptance of mandated reporting.

Recording public encounters, including those in normal investigative stops and arrests, as well as those at checkpoints, raises its own set of issues. Keeping records of arrests is unexceptionable. Adding information about investigative stops to a database linked to a national identity card would probably be upheld also, despite considerable grounds for finding that the practice would upset the carefully constructed balance sustaining the constitutionality of seizures based upon reasonable suspicion. Such data collection and retention would clearly add to the imposition of a *Terry* stop. Now instead of the "brief intrusion" described by the Court, the individual would have a "brief intrusion" plus an endless record, not only of having been in a particular place at a particular time, but also perhaps of having generated reasonable suspicion of criminal activity.

Identification checkpoints present another opportunity for collecting, and then storing, information about the location of a particular individual at a particular time. Ordinarily, official surveillance of a person in public, even targeted surveillance, does not amount to a search under the Fourth Amendment.[64] Responding to the argument that planting a transmitter in an automobile was a Fourth Amendment search in *United States v. Knotts*, the Court stated, "A person traveling in an automobile on public thoroughfares has no reasonable expectation of privacy in his movements from one place to another."[65] If the government can follow (or stop) people traveling in public, presumably it can also record what it learns. With enough checkpoints,

[62] While the Court did not have before it pure Fourth Amendment claim, it rejected a Fourth Amendment-based right to privacy argument. *Id.* at 604, n. 32.

[63] *Id.* at 593-94, 600-01.

[64] The classic statement of this point comes from the majority opinion in Katz v. United States, 389 U.S. 347, 351 (1967): "What a person knowingly exposes to the public, even in his own home or office, is not a subject of Fourth Amendment protection."

[65] 460 U.S. 276, 281 (1980).

though, the government could virtually track everyone's movements. Knotts himself raised the specter of "twenty-four hour surveillance of any citizen," eliciting this response from the Court: "[I]f such dragnet type law enforcement practices . . . should eventually occur, there will be time enough then to determine whether different constitutional principles may be applicable."[66] The use of identification checkpoints to generate a database of individual activities would certainly bring that day closer.

4. CONCLUSION

The United States may never have or use a national identity card. Since September 11, however, the possibility has become real enough that it is not too soon to attempt to evaluate the constitutionality under the Fourth Amendment of the kinds of practices a national identity system might employ. Given the expense and effort entailed in creating any national identity system, costs and benefits should be evaluated at an early stage of its consideration. To the extent that the Constitution would stand in the way of particular national identity card attributes or uses, the projected benefits would be correspondingly reduced, decreasing the card's overall desirability. In addition, if there is going to be a national identity system, it is advisable to consider the Fourth Amendment issues in advance of its design so as to minimize civil liberties intrusions and maximize the prospects for judicial acceptance.

Clearly, the Fourth Amendment stands in the way of the kind of total surveillance and anytime identification demands that would allow such a system to operate at maximum efficiency. On the other hand, there is still a fair amount the government could do in both areas that would withstand Fourth Amendment challenge. Indeed, the brief analysis here reveals again the truth of *Katz v. United States'* famous epigram that "the Fourth Amendment cannot be translated into a general constitutional 'right to privacy.'"[67]

This review of Fourth Amendment issues should serve to demonstrate to proponents of an identity card that there are both limits and dangers to its use. It should also make clear to those who see a national identity system as an Orwellian nightmare that while the Constitution stands somewhat in its path, it does not make such a system impossible. Whether the kind of

[66] *Knotts*, 460 U.S. at 283-84.

[67] *Katz,* 389 U.S. at 350 ("[T]he protection of a person's general right to privacy – his right to be let alone by other people" is left to nonconstitutional sources of law.).

national identity system that could operate lawfully is worth the financial, administrative, and social costs, would ultimately be a policy, not a legal, judgment. Much of the policy assessment involves a balancing of the government's need for a national identity card, its effectiveness, and its interference with privacy and free movement. To a large degree, these are the same factors on which the constitutionality of a national identity system turns. The legal analysis thus provides a useful perspective on the desirability as well as the constitutionality of adopting a national identity card.

Chapter 18

PRIVACY ISSUES IN AN ELECTRONIC VOTING MACHINE

Arthur M. Keller[1], David Mertz[2], Joseph Lorenzo Hall[3], and Arnold Urken[4]
[1]UC Santa Cruz and Open Voting Consortium, ark@soe.ucsc.edu; [2]Gnosis Software, Inc., mertz@gnosis.cx; [3]UC Berkeley, School of Information Management and Systems, jhall@sims.berkeley.edu; [4]Stevens Institute of Technology, aurken@stevens.edu

Abstract: The Open Voting Consortium has a developed a prototype voting system that includes an open source, PC-based voting machine that prints an accessible, voter-verified paper ballot along with an electronic audit trail. This system was designed for reliability, security, privacy, accessibility and auditability. This paper describes some of the privacy considerations for the system.

Key words: electronic voting, privacy, secret ballot, Open Voting Consortium, Electronic Ballot Printer, paper ballot, barcodes, accessible, reading impaired interface, multiple languages, accessible voter-verified paper ballot

1. INTRODUCTION – WHY A SECRET BALLOT?

The requirements for secrecy in elections depend upon the values and goals of the political culture where voting takes place. Gradations of partial and complete privacy can be found in different cultural settings. For instance, in some cantons in Switzerland, voters traditionally communicate their choices orally in front of a panel of election officials.[1] In contrast, in most modern polities, the ideal of complete privacy is institutionalized by relying on anonymous balloting.[2]

[1] Benjamin Barber, Strong Democracy (Twentieth Anniversary Edition, University of California Press, 2004).

[2] Alvin Rabushka and Kenneth Shepsle, POLITICS IN PLURAL SOCIETIES: A THEORY OF DEMOCRATIC INSTABILITY (1972).

The use of secret balloting in elections – where a ballot's contents are disconnected from the identity of the voter – can be traced back to the earliest use of ballots themselves. The public policy rationales for instituting anonymous balloting are typically to minimize bribery and intimidation of the voter. For example, in Athens, Greece during the sixth century B.C.E., Athenians voted by raising their hands "except on the question of exiling someone considered dangerous to the state, in which case a secret vote was taken on clay ballots."[3] In this case, presumably it was deemed necessary to vote via secret ballot to avoid bodily harm to the voter.

Secret ballots, although not always required, have been in use in America since colonial times.[4] The Australian ballot,[5] designed to be uniform in appearance because it is printed and distributed by the government, was adopted throughout most of the U.S. in the late 1800's. Today, approximately one hundred years after most states in the U.S. passed legal provisions for anonymous balloting, a strong sense of voter privacy has emerged as a third rationale. All fifty states have provisions in their constitutions for either election by "secret ballot" or elections in which

[3] Spencer Albrecht, THE AMERICAN BALLOT (1942) at 9.

[4] In 1682, the Province of Pennsylvania in its Frame of the Government required "THAT all the elections of Members or Representatives of the People, to serve in the Provincial Council and General Assembly ... shall be resolved and determined by ballot." (Votes and Proceedings of the House of Representatives of the Province of Pennsylvania. Printed and sold by B. Franklin and D. Hall, at The New Printing Office, near the Market. Philadelphia, Pennsylvania MDCCLII, at xxxi.) In 1782, the legislature of the Colony/State of New Jersey tried to intimidate Tories by requiring viva voce voting. (At that time, about half of New Jersey voted with ballots and the other half viva voce.) They rescinded this in their next session. (Richard P. McCormick, THE HISTORY OF VOTING IN NEW JERSEY 74 (1953). In 1796, the State of New Jersey required federal elections to be by ballot and extended that to state elections the following year. (*Id.* at 106.) In the 1853 pamphlet SECRET SUFFRAGE, Edward L. Pierce recounted Massachusetts' battle to make the secret ballot truly secret. The Massachusetts Constitution in 1820 required elections for representatives to have "written" votes. In 1839, the legislature attacked the secrecy of the written ballot by requiring the ballot to be presented for deposit in the ballot box open and unfolded. In 1851, the legislature passed the "Act for the better security of the Ballot," which provided that the ballots are to be deposited in the ballot box in sealed envelopes of uniform size and appearance furnished by the secretary of the Commonwealth (State of Massachusetts). The battle waged until a provision in the State Constitution made the secret ballot mandatory. (Edward L. Pierce, SECRET SUFFRAGE 7 (1853)(published by the Ballot Society, No. 140 Strand, London, England).

[5] The more general "Australian ballot" is a term used for anonymous balloting using official non-partisan ballots distributed by the government. See Albright 1942 at 26. "The very notion of exercising coercion and improper influence absolutely died out of the country." *See supra* note 3, at 24, quoting Francis S. Dutton of South Australia in J. H. Wigmore's THE AUSTRALIAN BALLOT SYSTEM (2nd ed., Boston, 1889) at 15-23.

"secrecy shall be preserved," which has been interpreted by the courts as an implied requirement for secret balloting.[6] West Virginia does not *require* a secret ballot and leaves that to the discretion of the voter.[7] Fourteen states'[8] constitutions do not list "secret" balloting or "secrecy" of elections and/or ballots explicitly. These states have either state laws (election code) or case law (decided legal cases in that state) that mandate secret balloting or interpret the phrase "election shall be by ballot" to mean a "secret ballot."

These cultural values and practices contribute to the sets of user requirements that define the expectations of voters in computer-mediated elections[9] and determine alternative sets of specifications that can be considered in developing open source software systems for elections. The Open Voting Consortium (OVC)[10] has developed a model election system that aims as one of its goals to meet these requirements. This paper describes how the OVC model ensures ballot privacy.

The OVC has developed its model for an electronic voting system largely in response to reliability, usability, security, trustworthiness, and accessibility concerns about other voting systems. Privacy was kept in mind throughout the process of designing this system. Section 2 of this paper discusses the requirements for a secret ballot in more detail. Section 3 considers how secrecy could be compromised in some systems. Section 4 describes the architecture of the polling place components of the OVC system. Section 5 describes how the OVC handles privacy concerns. While this paper focuses mostly on privacy issues for U.S.-based elections, and how they are addressed in the OVC system, many of the issues raised are relevant elsewhere as well.

[6] For example, The Delaware Supreme Court recognized that the Delaware's constitutional language amounts to an "implied constitutional requirement of a secret ballot." *Brennan v. Black,* 34 Del. Ch. 380 at 402. (1954).

[7] *See* W. Va. Const. Art. IV, §2.

[8] "In all elections by the people, the mode of voting shall be by ballot; but the voter shall be left free to vote by either open, sealed or secret ballot, as he may elect." (W. VA. CONST. ART. IV, § 2 (2003).

[9] Arthur B, Urken, *Voting in A Computer-Networked Environment, in* THE INFORMATION WEB: ETHICAL AND SOCIAL IMPLICATIONS OF COMPUTER NETWORKING (Carol Gould, ed., 1989).

[10] The Open Voting Consortium (OVC) is a non-profit organization dedicated to the development, maintenance, and delivery of open voting systems for use in public elections. *See* http://www.openvotingconsortium.org/.

2. SECRET BALLOT REQUIREMENTS

The public policy goals of secret balloting[11] – to protect the privacy of the elector and minimize undue intimidation and influence – are supported by federal election laws and regulations. The Help America Vote Act of 2002[12] codifies this policy as "anonymity" and "independence" of all voters, and "privacy" and "confidentiality" of ballots. It requires that the Federal Election Commission create standards that "[preserve] the privacy of the voter and the confidentiality of the ballot."[13]

The Federal Election Commission has issued a set of Voting System Standards (VSS)[14] that serve as a model of functional requirements that elections systems must meet before they can be certified for use in an election. The VSS state explicitly:

> To facilitate casting a ballot, all systems shall:
>
> [...] Protect the secrecy of the vote such that the system cannot reveal any information about how a particular voter voted, except as otherwise required by individual State law;[15]

and:

> All systems shall provide voting booths [that shall] provide privacy for the voter, and be designed in such a way as to prevent observation of the ballot by any person other than the voter;[16]

as well as a lengthy list of specific requirements that Direct Recording Electronic voting systems must meet.[17] The basic, high-level requirement not to expose any information about how an individual voted is required of all

[11] There are two aspects to anonymous voting. The first is ballot privacy – the ability for someone to vote without having to disclose his or her vote to the public. The second is secrecy – someone should not be able to prove that they voted one way or another. The desire for the latter is rooted in eliminating intimidation while the former is to curb vote buying. The history of these two concepts is beyond the scope of this paper.

[12] The Help America Vote Act of 2002, 42 U.S.C.A. §§ 15301 – 15545 (West, 2004).

[13] *Id.*, § 301(a)(1)(C). (*Also see* §§ 242(a)(2)(B), 245(a)(2)(C), 261(b)(1), 271(b)(1), 281 (b)(1), 301(a)(3)(A)).

[14] Federal Election Commission, Voting System Standards, Vols. 1 & 2 (2002), *available at* http://www.fec.gov/pages/vsfinal (Microsoft Word .doc format) *or* http://sims.berkeley.edu/~jhall/fec_vss_2002_pdf/ (Adobe PDF format)

[15] *Id.* at Vol. 1, §2.4.3.1(b).

[16] *Id.* at Vol. 1, §3.2.4.1.

[17] *Id.* at Vol. 1, §3.2.4.3.2(a)-(e) and §4.5.

voting systems before certification and is the most important. The second requirement listed above is a corollary.

It is not sufficient for electronic voting systems merely to anonymize the voting process from the perspective of the voting machine. Every time a ballot is cast, the voting system adds an entry to one or more software or firmware logs that consists of a timestamp and an indication that a ballot was cast. If the timestamp log is combined with the contents of the ballot, this information becomes much more sensitive. For example, it can be combined with information about the order in which voters voted to compromise the confidentiality of the ballot. Such information can be collected at the polling place using overt or covert surveillance equipment – such as cell phone cameras or security cameras common at public schools. As described below, system information collected by the voting system should be kept separated from the content of cast ballots and used in conjunction only by authorized, informed election officials.

3. HOW SECRECY COULD BE COMPROMISED

3.1 A voter's secret identity

When a voter enters a polling place, she enters with a valuable secret: her identity. A secret ballot is not really "secret" in a general sense – it is possible, and even required, for certain recipients to disclose ballots. A secret ballot is "secret" only in the sense that it is blind as to the identity of the voter who cast it. The anonymity of ballots must apply even to most statistical properties of the voters who cast them; a notable exception, however, is in the disclosure of the geographic distribution of voters who vote certain ways in the aggregate. We all know there are "Republican precincts" and "Democratic precincts," and anyone can easily and legally find out which are which.

Complicating matters is the fact that a voter's secret, her identity, *must* be disclosed at a certain stage in the voting process. To be allowed to vote at all, a voter must authenticate her right to vote using her identity, if only by a declaration of purported identity to elections workers. Depending on jurisdiction, different standards of identity authentication apply – some require identification cards and/or revelation of personal information outside the public domain – but in all cases, identity acts as a kind of key for entry to voting. However, legally this key must be removed from all subsequent communication steps in the voting process.

The act of voting, and the acts of aggregating those votes at subsequently higher levels (called "canvassing" in voting parlance) can be thought of as involving a series of information channels. At a first step, a voter is given a token to allow her vote to pass through later stages; depending on the system model, this token may be a pre-printed ballot form, a PIN-style code, a temporary ballot-type marker, an electronic smart card, or at a minimum simply permission to proceed. Although the OVC has not yet settled on a particular token, we will focus on smart cards in this paper, because they have the most serious implications for privacy. Outside the US, tokens such as hand stamps in indelible ink are also used, particularly to preclude duplicate votes being cast.

Once at a voting station, a voter must perform some voting actions using either pen-and-paper, a mechanical device like a lever machine or a punch card guide, or an electronic interface, such as a touchscreen or headphones-with-keypad. After performing the required voting actions, some sort of record of the voter's selections is created, either on paper, in the state of gears, on electronic/magnetic storage media, or using some combination of those. That record of selections becomes the "cast ballot." Under the Open Voting Consortium system, the paper ballot produced at a voting station undergoes final voter inspection before being *cast* into a physical ballot box.

After votes are cast, they are canvassed at several levels: first by precinct; then by county, district, or city; then perhaps statewide. At each level of canvassing, either the literal initial vote records or some representation or aggregation of them must be transmitted.

3.2 Understanding covert channels

At every stage of information transmission, from voter entry, through vote casting, through canvassing, a voter's identity must remain hidden. It is relatively simple to describe the overt communication channels in terms of the information that actually *should* be transmitted at each stage. But within the actual transmission mechanism it is possible that a *covert* channel also transmits improper identity information.

Covert channels in a voting system can take a number of forms. Some covert channels require the cooperation of collaborators, such as voters themselves or poll workers. Other covert channels can result from (accidental) poor design in the communication channels; while others can be created by malicious code that takes advantage of incomplete channel specification. A final type of covert channel is what we might call a "sideband attack" – that is, there may be methods of transmitting improper information that are not encoded directly in the overt channel, but result indirectly from particular implementations.

For illustration, let us briefly suggest examples of several types of covert channels. One rather straightforward attack on voter ballot anonymity is repeatedly missed by almost every new developer approaching design from a databases-and-log-files background. If the voting channels contain information about the times when particular ballots are cast and/or the sequence of ballots, this information can be correlated with an under-protected record of the sequence of times when voters enter a polling place. We sometimes call this a "covert videotape" attack. In part, this attack uses a sideband: the covert videotaping of voters as they enter; but it also relies on a design flaw in which ballots themselves are timestamped, perhaps as a means to aid debugging.

A pure sideband attack might use Tempest[18] equipment to monitor electro-magnetic emissions of electronic voting stations. In principle, it might be possible for an attacker to sit across the street from a polling place with a van full of electronics, watch each voter enter, then detect each vote she selects on a touchscreen voting station.

Cooperative attacks require the voter or poll worker to do something special to disclose identity. As with other attacks, these covert channels need not rely on electronics and computers. For example, a malicious poll worker might mark a pre-printed blank paper ballot using ultraviolet ink before handing it to a targeted voter. The covert channel is revealed only with an UV lamp, something voters are unlikely to carry to inspect their ballots. A voter herself might cooperate in a covert channel in order to facilitate vote buying or under threat of vote coercion. One such covert channel is to instruct a bought or coerced voter to cast "marked votes" to prove she cast the votes desired by her collaborator. Unique write-in names and unusual patterns in ranked preference or judicial confirmations are ways to "mark" a ballot as belonging to a particular voter.

3.3 Links between registration data and ballots

Since a voter must identify herself when signing in at the polling place, there is the potential for her identity to be tied to her vote. The token given to the voter to allow her to vote may contain her identity. For example, the voter's registration number could be entered into the smart-card writer and then encoded on the smart card that is given to the voter to enable use of a Direct Recording Electronic voting machine. When the voter registration list is given to the polling place on paper, this channel appears less of an issue. However, if the voter registration list is handled electronically, then the smart card could easily contain the voter's identity. Diebold's stated intent makes this issue a potentially serious privacy risk.

[18] *See* http://www.cryptome.org/nsa-tempest.htm (Last visited February 13, 2005).

Diebold already has purchased Data Information Management Systems, one of two firms that have a dominant role in managing voter-registration lists in California and other states. "The long-term goal here is to introduce a seamless voting solution, all the way from voter registration to (vote) tabulation," said Tom Swidarski, Diebold senior vice president for strategic development.[19]

4. OVC SYSTEM OVERVIEW

The Open Voting Consortium is developing a PC-based open source voting system based on an accessible voter-verified paper ballot. We mostly describe the components of the system that operate in the polling place.[20] In addition, we briefly discuss the components at the county canvassing site.

4.1 Voter sign-in station

The Voter Sign-In Station is used by the poll worker when the voter signs in and involves giving the voter a "token." It is a requirement that each voter cast only one vote and that the vote cast be of the right precinct and party for the voter. The "token" authorizes the voter to cast a ballot using one of these techniques.

- Pre-printed ballot stock
    - Option for scanning ballot type by Electronic Voting Machine
    - Poll worker activation
- Per-voter PIN (including party/precinct identifier)
- Per-party/precinct token
- Smart cards

The token is then used by the Electronic Voting Machine or an Electronic Voting Machine with a Reading Impaired Interface to ensure that each voter votes only once and only using the correct ballot type.

If the voter spoils a ballot, the ballot is marked spoiled and kept for reconciliation at the Ballot Reconciliation Station, and the voter is given a new token for voting.

[19] Ian Hoffman, *With e-voting, Diebold treads where IBM wouldn't*, OAKLAND TRIB., May 30, 2004, *available at* http://www.oaklandtribune.com/Stories/0,1413,82~1865~2182212,00.html.

[20] *See* Arthur M. Keller, *et al.*, *A PC-Based Open Source Voting Machine with an Accessible Voter-Verifiable Paper Ballot*, 2005 USENIX ANNUAL TECHNICAL CONF., FREENIX/OPEN SOURCE TRACK, April 10-15, 2005, at 163–174, and *available at* http://www-db.stanford.edu/pub/keller/2004/electronic-voting-machine.pdf.

4.2 Electronic voting machine

The Electronic Voting Machine (EVM) includes a touch-screen interface for the voter to view the available choices for each contest and select among them. The EVM then prints a paper ballot, which the voter verifies (possibly using the Ballot Verification Station) and places in the ballot box. The EVM is activated by a token, such as a smart card, obtained at the sign-in station. The EVM maintains an electronic ballot image as an audit trail and to reconcile with the paper ballots at the Ballot Reconciliation Station.

4.3 Electronic voting machine with reading impaired interface

The Electronic Voting Machine with Reading Impaired Interface is a PC similar to the Electronic Voting Machine described above which provides auditory output of the ballot choices and selections made and also supports additional modes of making selections suitable for the blind or reading impaired. Whether these features are integrated into a common voting machine with all functionality, or whether there is a separate configuration for the disabled, is an open question. For example, additional modes of input may be useful for those who can read printed materials, but have physical limitations. The idea is to have a universal design that accommodates all voters.

4.4 Ballot verification station

The Ballot Verification Station reads the ballot produced by the Electronic Voting Machine or the Electronic Voting Machine with Reading Impaired Interface and speaks (auditorily) the selections on the voter's ballot. A count is kept of usage, including counts of consecutive usage for the same ballot, but no permanent record is kept of which ballots are verified.

The Ballot Verification Station could also have a screen for displaying the selections. Such an option, enabled by the voter upon her request, would enable a voter who can read to verify that her ballot will be read correctly for automated tallying.

4.5 Ballot reconciliation station

The Ballot Reconciliation Station reads the paper ballots, both cast and spoiled, and reconciles them against the Electronic Ballot Images from the Electronic Voting Machine or the Electronic Voting Machine with Reading Impaired Interface.

4.6 Paper ballot

The paper ballot is printed by the Electronic Voting Machine or the Electronic Voting Machine with Reading Impaired Interface. It must be "cast" in order to be tallied during canvassing, testing, or a manual recount.

The paper ballot is intended to be easily read by the voter so that the voter may verify that his or her choices have been properly marked. It also contains security markings and a bar code. The bar code encodes the voter's choices, as expressed in the human readable portion of the ballot. The human readable text should be in an OCR-friendly font so it is computer-readable as well. Voters may use the Ballot Verification Station to verify that the bar code accurately reflects their choices. The Ballot Verification Station not only assists sight-impaired and reading-impaired voters in verifying their ballots, but will also give all voters the assurance that the bar-code on the ballot properly mirrors their choices, as represented in the human-readable text on the ballot.

4.7 Privacy folder

The paper ballot contains the voter's choices in two forms: a form that can be read by people and a bar code that expresses those choices in a machine-readable form.

Poll workers may come in contact with the ballot should they be asked to assist a voter or to cast the ballot into the ballot box. In order to protect voter privacy it is desirable to minimize the chance that a voting place worker might observe the voter's ballot choices. A privacy folder is just a standard file folder with an edge trimmed back so that it reveals only the bar code part of a ballot. The voter is expected to take his/her ballot from the printer of the Electronic Voting Machine or the Electronic Voting Machine with Reading Impaired Interface and place it into a privacy folder before leaving the voting booth.

The privacy folder is designed so that the voter may place the ballot, still in its folder, against the scanning station of the Ballot Verification Station to hear the choices on the voter's ballot spoken.

When handed the ballot by the voter, the poll worker casts the ballot by turning the privacy folder so the ballot is face down, and then sliding the paper ballot into the ballot box.

4.8 Ballot box

The ballot box is a physically secure container, into which voters have their paper ballots placed, in order to "cast" their votes. The mechanical aspects of the ballot box will vary among jurisdictions, depending on local laws and customs. Optionally, a perforated tab is removed from the ballot before placing the ballot into the ballot box, and the tab is handed to the voter. The removal of the tab ensures that the ballot cannot be marked "spoiled."

4.9 Box for spoiled ballots

When a voter spoils a ballot, perhaps because the ballot does not accurately reflect her preferences, the ballot is marked spoiled and placed in a box for spoiled ballots for later reconciliation.

5. OVC BALANCES SECURITY, RELIABILITY AND PRIVACY

This section discusses how the Open Voting Consortium is balancing security, reliability and privacy in its electronic voting system.

5.1 Free and open source software

Opening the source code to a voting system – all stages of it, not only the voting station – is a necessary, though not sufficient, condition for ensuring trustworthiness, including the absence of trapdoors and covert channels. For practical purposes, no system that functions as a black box, in which the implementing source code is maintained as a trade secret, can be known to lack covert channels. Any channel with non-optimal utilization includes non-utilized content that is potentially malicious rather than merely accidental – behavior analysis, in principle, cannot distinguish the two.

Of course, free and open source code is not sufficient to prevent covert channels. Sideband channels, in particular, are never exposed by direct examination of source code in isolation; it is necessary to perform additional threat modeling. But even direct encoding of extra information *within* an overt channel can sometimes be masked by subtle programming tricks. More eyes always reduce the risk of tricks hidden in code. Parallel implementation to open specifications and message canonicalization also help restrict channels to overt content.

A frequent criticism of free and open source software is that, while the code is available for inspection, no coordinated inspection is actually conducted.[21] The absence of Non-Disclosure Agreements and restrictive intellectual property agreements makes it possible for a large body of open source developers to inspect the code. Furthermore, in the realm of elections systems, which are mission-critical for a democratic government, open source software could benefit from a specific group of developers who are tasked with recognizing and repairing vulnerabilities. This is a common need in many open source software projects, and in this sense, it might be an appropriate role for a non-profit institution that has delivered such services to other important projects like GNU/Linux, BIND, the Mozilla tool suite and the Apache web server.

5.2 Privacy in the voting token (e.g., smart card)

The token given to the voter to enable her to use the electronic voting machine might contain information that could compromise her anonymity. Indeed, it is not possible to demonstrate the absence of covert channels through black box testing. Thus, analysis of the software is important to show how the data for the smart card is assembled. Above, we considered the benefits of open source software in that numerous people, both inside and outside the process, have the ability to inspect and test the software to reduce the likelihood of covert channels. The hardware that enables smart-card use also includes an interface used by the poll worker (the Voter Sign-In Station). The nature of that interface limits the type of information that can be encoded. Encoding the time of day in the smart card, either intentionally or as a side effect of the process of writing files to the smart card, is a potential avenue for attack. However, the electronic voting machine receiving the smart card knows the time as well, so the smart card is not needed to convey this information.

We propose to encode in the voting token the ballot type and (particularly for multiple precincts at the same polling place) the precinct. The smart card should also be digitally signed by the smart card enabling hardware, so as to help reduce forgeries.

5.3 Printed ballot

The printed ballot contains a human readable version of the voter's selections. After all, that is how it is a voter-verifiable paper ballot. However, the secrecy of the voter's selections is at risk while the voter

[21] Fred Cohen, *Is Open Source More or Less Secure?* MANAGING NETWORK SECURITY, (July 2002).

carries the paper ballot from the electronic voting machine, optionally to the ballot validation station, and on to the poll worker to cast her ballot.

Our approach is to use a privacy folder to contain the ballot. When the voter signs in, she receives the token plus an empty privacy folder. When the EVM prints the ballot, the voter takes the ballot and places it in the privacy folder, so that only the barcode shows. The barcode can be scanned by the Ballot Validation Station without exposing the human readable portion of the ballot. When the privacy folder containing the ballot is given to the poll worker to be cast, the poll worker turns the privacy folder so the ballot is face down and then slides the ballot out of the privacy folder and into the official ballot box. The poll worker thus does not see the text of the ballot, with the possible exception of precinct and (for primaries) party identifiers that may be printed in the margin.

The privacy folder is an ordinary manila folder trimmed along the long edge so that the barcode sticks out.

5.4 Reading impaired interface

The reading impaired interface is used both by voters who cannot read and by voters who cannot see. Having a segregated electronic voting machine used only by the reading and visually impaired can compromise privacy. It is therefore desirable for the electronic voting machines with the reading impaired interface to be used also by those who can read. For example, if all electronic voting machines incorporated the reading impaired interface, then reading impaired voters would not be segregated onto a subset of the voting machines.

It is important that the ballot not record the fact that a particular ballot was produced using the reading impaired interface. Nor should the electronic voting machine record that information for specific ballots. Using a separate voting station for the reading impaired means that the audit trail is segregated by whether the voter is reading impaired.

Nonetheless, it is useful for the electronic voting machine to maintain some statistics on the use of the reading impaired interface, provided that these statistics cannot identify specific ballots or voters. These statistics could be used to improve the user interface, for example.

5.5 Privacy issues with barcodes

The Open Voting Consortium system design uses a barcode to automate the scanning of paper ballots. Such barcodes raise several possibilities for introducing covert channels.

The prototype/demo system presented by OVC, for example, used a 1-D barcode, specifically Code128. For vote encoding, selections were first converted to a decimal number in a reasonably, but not optimally, efficient manner; specifically, under the encoding particular digit positions have a direct relationship to corresponding vote selections. These digits, in turn, are encoded using the decimal symbology mode of Code128.

Co-author David Mertz identified the problem that even though barcodes are not per-se human readable, identical patterns in barcodes – especially near their start and end positions – could be recognized by observers. This recognition would likely even be unconscious after poll workers saw hundred of exposed barcodes during a day. For example, perhaps after a while, a poll worker would notice that known Bush supporters always have three narrow bars followed by a wide bar at the left of their barcode, while known Kerry supporters have two wide bars and two narrow bars. To prevent an attack based on this kind of human bar code recognition, 1-D barcodes undergo a simple obfuscation of rotating digits by amounts keyed to a repetition of the random ballot-id. This "keying" is not even weak encryption – it resembles a Caesar cipher,[22] but with a known key; it merely makes the same vote look different on different ballots.

In the future, OVC anticipates needing to use 2-D barcodes to accommodate the information space of complex ballots and ancillary anonymity-preserving information such as globally unique ballot-IDs and cryptographic signatures. At this point, we anticipate that patterns in 2-D barcodes will not be vulnerable to visual recognition; if they are, the same kind of obfuscation discussed above is straightforward. But the greatly expanded information space of 2-D barcodes is a vulnerability as well as a benefit. More bit space quite simply provides room to encode more improper information. For example, if a given style of barcode encodes 2000 bits of information, and a particular ballot requires 500 bits to encode, those unused 1500 bits can potentially contain improper information about the voter who cast the ballot.

Just because a barcode has *room* for anonymity-compromising information does not mean that information is actually encoded there, of course. Preventing misuse of an available channel requires complementary steps. Moreover, even a narrow pipe can disclose quite a lot; it only takes about 10 bits to encode a specific address within a precinct using a lookup table. Even a relatively impoverished channel might well have room for a malicious ten bits. For example, if a non-optimal vote encoding is used to

[22] *See* http://www.fact-index.com/c/ca/caesar_cipher.html (Last visited February 13, 2005).

represent votes, it is quite possible that multiple bit-patterns will correspond to the same votes. The choice among "equivalent" bit patterns might leak information.

Eliminating barcodes, it should be noted, does not necessarily eliminate covert channels in a paper ballot. It might, however, increase voter confidence as average voters become less *concerned* about covert channels (which is both good and bad). For example, even a barcode-free printed ballot could use steganography[23] to encode information in the micro-spacing between words, or within security watermarks on the page.

5.6 Ballot validation station

The Ballot Validation Station allows reading impaired voters – or anyone – to hear and therefore validate their paper ballots. Since only the barcode of the ballot (and possibly the ballot type – the precinct and party for primaries) is viewable (and as mentioned above, the barcode is obscured), it is best to keep the paper ballot in the privacy folder. So the Ballot Validation Station should be able to read the barcode without removing the paper ballot from the privacy folder. The back of the ballot should have a barcode (possibly preprinted) saying "please turn over," so a Ballot Validation Station will know to tell the blind voter that the ballot is upside down. So that others will not hear the Ballot Validation Station speak the choices on the ballot, the voter should hear these choices through headphones.

It is useful to know how many times the Ballot Validation Station is used, and how many consecutive times the same ballot is spoken. It is important to assure that ballot-IDs are not persistently stored by the Ballot Validation Station. In particular, to tell how many consecutive times the same ballot was spoken, the Ballot Validation Station must store the previous ballot-ID. However, once another ballot with a different ballot-ID is read, then that new ballot-ID should replace the previous ballot-ID. And the ballot-ID field should be cleared during the end-of-day closeout. The counts of consecutive reads of the same ballot should be a vector of counts, and no other ordering information should be maintained. Inspection of the code together with clear interfaces of persistently maintained records can help assure privacy.

[23] Neil F. Johnson and Sushil Jajodia, *Steganography: Seeing the Unseen*, IEEE COMPUTER (February 1998) at 26-34.

5.7 Languages

Steve Chessin has identified a problem with ballots for non-English speakers. For the voter, the ballot must be printed in her own language. However, for canvassing and manual counts, the ballot and its choices must also be printed in English. However, this approach makes bilingual ballots easy to identify, and that can compromise ballot anonymity if only a small number of voters in a given precinct choose a particular language. Steve Chessin's solution is to have all ballots contain both English and another language, where the other language is randomly chosen for English speakers.[24]

It is important that the Ballot Validation Station handle multiple languages so the voter can choose the language for validating the ballot. To simplify this process, the ballot barcode can include a notation of the second language, but only if that information does not compromise anonymity. Always choosing a second language at random where none is specifically requested reduces the risk. When the ballot's barcode is scanned by the Ballot Validation Station, the voter is given a choice of these two languages for the spoken review of choices listed on the ballot.

5.8 Randomization of ballot-IDs

Under the OVC design, ballots carry ballot-IDs. In our prototype, these IDs are four digit numbers, which provides enough space for ten thousand

[24] It is important to note that the procedure for randomizing the second, non-English language printed on a ballot would have to be quite good. Flaws in the randomization or maliciously planted code could result in the "marking" of certain ballots leading to a compromise of ballot privacy. A simple solution would be to have all ballots printed only in English, and requiring non-English literate voters to use the BVA to verify their vote auditorily. As an alternative for ballots printed only in English, ballot overlays could by provided for each language needed for each ballot type. The overlay could either be in heavy stock paper printed with the contest names with holes for the selections to show through, or it could be a translation sheet showing all the contest names and selections translated into non-English language. In the former case, the ballots would have to be have the layout of each contest fixed, so it would be necessary to have extra spaces when the length of the results vary, such as for pick up to 3 candidates when only 2 were selected. These overlays could be tethered to every voting machine so that voters who read only a specific language could simply place the overlay over their ballot so that she could read their selections as if the ballot was printed in their native language. The overlay approach reduces confusion for English speakers and it also reduces the length of the printed ballot.

ballots to be cast at a polling place. We anticipate this ballot-ID length to remain sufficient in production. The main purpose of ballot-IDs is simply to enable auditing of official paper ballots against unofficial electronic ballot images.

The crucial feature of ballot-IDs is that they must not reveal any information about the sequence of votes cast. The prototype and current reference implementation use Python's 'random' module to randomize the order of ballot-IDs. The module uses the well-tested Mersenne Twister algorithm, with a periodicity of $2^{19937}-1$. Seeding the algorithm with a good source of truly random data – such as the first few bytes of /dev/random on modern Linux systems – prevents playback attacks to duplicate ballot-ID sequences.

Because the ballot-IDs are generated at random by each of the electronic voting machines, it is important that two machines do not use the same random ballot-ID. As a result, the first digit (or character) of the ballot-ID in the reference platform will represent the voting machine ID for that polling place.

The remaining 3 digits of the ballot-ID are randomly selected from the range of 000 to 999. A list is maintained of already used ballot-IDs for this electronic voting machine for this election. (One way to obtain such a list is to scan the stored electronic ballot images for the ballot numbers used.) If the random number generated matches an already used ballot-ID, then that number is skipped and a new random number is generated.

5.9 Information hidden in electronic ballot images and their files

The electronic ballot images (EBIs) are stored on the electronic voting machine where the ballot was created. One purpose of maintaining these EBIs is to reconcile them against the paper ballots, to help preclude paper ballot stuffing. The EBIs are in XML format, which can be interpreted when printed in "raw" form.

We prefer not to store the EBIs in a database on the electronic voting machine. A database management system incurs additional complexity, potential for error, and can contain sequence information that can be used to identify voters. On the other hand, flat files in XML format would include the date and time in the file directory, and that is also a potential privacy risk. We can mitigate this risk by periodically "touching" EBI files electronically during voting station operation, in order to update the date and time of all files to the latest time. The placement order of the files on the disk, however, may still disclose the order of balloting.

Another approach is to store all the EBIs in a single file as if it were an array. Suppose that it is determined that the largest XML-format EBI is 10K bytes. Since there are 1000 possible ballot-IDs for this electronic voting machine, it is possible to create a file with 1000 slots, each of which is 10K in length. When the ballot is to be printed, the random ballot-ID is chosen, and the EBI is placed in that slot in the file, padded to the full 10K in length with spaces (which would be removed during canonicalization). The file can be updated in place, thereby having only the latest date and time. Alternatively, two files can be used, and the electronic voting machine can write to one, wait for completion, and then write to the other. The benefit of this approach is increased reliability of persistent storage of the EBI file.

A similar technique can be used to maintain copies of the Postscript versions of the ballots.

When the polling place closes, the electronic voting machine is changed to close out the day's voting. At this time, the EBIs are written as individual flat files in ascending ballot-ID order to a new session of the CD-R that already contains the electronic voting machine software and personalization. Because the EBIs are written all at once, and in order by ascending random ballot-ID, anonymity is preserved.

5.10 Public vote tallying

It is important that the ballots be shuffled before publicly visible scanning occurs using the Ballot Reconciliation System. The ballots will naturally be ordered based on the time they were placed in the ballot box. As described above, the time or sequence of voting is a potential risk for privacy violations.

An illustration of this problem was reported privately to co-author Arthur Keller about a supposedly secret tenure vote at a university. Each professor wrote his or her decision to grant or deny tenure on a piece of paper. The pieces of paper were collected and placed on top of a pile one-by-one in a sequence determined by where each person was sitting. The pile was then turned over and the votes were then read off the ballots in the reverse of that sequence as they were tallied. One observer noted how each of the faculty members voted in this supposedly secret vote.

5.11 Results by precinct

A key approach to ensuring the integrity of county (or other district) canvassing (i.e., vote tallying) is to canvass the votes at the precinct and post the vote totals by contest at the precinct before sending on the data to the county. As a crosscheck, the county should make available the vote totals by contest for each precinct. However, because the county totals include absentee votes, it is difficult to reconcile the posted numbers at the precinct against the county's totals by precinct, unless the county separates out absentee votes (plus hand-done polling place votes). However providing these separations may reduce the aggregation size to impair anonymity. An even worse threat to anonymity arises when provisional ballots are incrementally approved and added to the tally one-by-one.

We propose to exclude provisional ballots from the results posted at the precinct. The county tallies by precinct should be separated into a group of votes included in the precinct-posted tally and a group of votes not included in the precinct-posted tally. As long as there is a publicly viewable canvassing of the votes not included in the precinct-posted tally, the issue of voter confidence in the system will be addressed. If that canvassing process involves ballots that have already been separated from the envelope containing the voter's identity, privacy is enhanced.

The totals by precinct are aggregate counts for each candidate. There is no correlation among specific ballots, an important factor to help assure privacy. However, ranked preference voting schemes, such as instant runoff voting, require that the ordering of the candidates must be separately maintained for each ballot. Vote totals are useful to help assure that each vote was counted, but they do not contain enough information to produce an absolute majority winner. Therefore, vote totals can be posted at the precinct – independent of ranking – and those totals can also be posted at the county. A voter who specifies a write-in candidate for a ranked preference voting race might in principle be doing so as a marker for observation during the canvassing process. To ensure anonymity, write-in candidates whose vote totals are below a certain threshold could be eliminated from the canvassing process. This threshold must be set to avoid distortions of aggregate scores at the county level.

5.12 Privacy in the face of voter collusion

Complex cast ballots, taken as a whole, inevitably contain potential covert channels. We reach a hard limit in the elimination of improper identifying information once voter collusion is considered. In an ideal case, voters cooperate in the protection of their own anonymity; but threats of vote coercion or vote buying can lead voters to collaborate in disclosing – or rather, proving – their own identity. It is, of course, the right of every voter to disclose her own votes to whomever she likes; but such disclosure must not be subject to independent verifications that attack voter anonymity as a whole.

Elections with many contests, with write-ins allowed, or with information-rich ranked preference contests, implicitly contain extra fields in which to encode voter identity. For example, if an election contains eight judicial retention questions, there are at least 6561 possible ways to complete a ballot, assuming Yes, No, and No Preference are all options for each question. Very few precincts will have over 6561 votes cast within them, so a systematic vote buyer could demand that every voter cast a uniquely identifying vote pattern on judicial retentions. That unique pattern, plus the precinct marked on a ballot, in turn, could be correlated with a desired vote for a contested office.

Ballots may not generally be completely separated into records by each individual contest. For recounts or other legal challenges to elections, it is generally necessary to preserve full original ballots, complete with correlated votes. Of course it is physically possible to cut apart the contest regions on a paper ballot, or to perform a similar separation of contests within an EBI. However, doing so is not generally permissible legally.

The best we can do is to control the disclosure of full ballots to mandated authorities, and maintain the chain of custody over the ballots, including the EBIs. A full ballot must be maintained, but only aggregations of votes, per contest, are disclosed to the general public. The number of people who have access to full ballots should be as limited as is feasible, and even people with access to some full ballots should not necessarily be granted general access to all full ballots.

5.13 Privacy in electronic voting machines with voter-verifiable paper audit trails

This section discusses other approaches to voter-verifiable paper audit trails. These issues do *not* apply to the design described in this paper — the voter-verifiable paper *ballot*.[25]

Rebecca Mercuri has proposed that Direct Recording Electronic voting machines have a paper audit trail that is maintained under glass, so the voter does not have the opportunity to touch it or change it.[26] Some vendors are proposing that paper from a spool be shown to the voter, and if the ballot is verified, a cutter will release the paper audit trail piece to drop into the box for safekeeping.[27] The challenge with this approach is to make sure that all of the paper audit trail is readable by the voter and does not curl away out of

[25] See http://evm2003.sourceforge.net/security.html for the difference between a paper receipt and a paper ballot, and between a paper audit trail and an electronically generated paper ballot.

[26] Rebecca Mercuri, *A Better Ballot Box?*, IEEE SPECTRUM ONLINE (October 2002), *available at* http://spectrum.ieee.org/WEBONLY/publicfeature/oct02/evot.html.

[27] For reference, *see* Avanti VOTE-TRAKKER™EVC308, *available at* http://aitechnology.com/votetrakker2/evc308.html

view, and yet that paper audit trails from previous voters are obscured from view. Furthermore, there is the problem that the paper audit trail would fall in a more-or-less chronologically ordered pile. It is also difficult to reconcile the paper audit trail with the electronic ballot images in an automated manner if the paper audit trail cannot be sheet-fed.

Another approach is to keep the paper audit trail on a continuous spool.[28] While this approach has the potential to allow the audit trail to be more easily scanned in an automated fashion for reconciliation, privacy is compromised by maintaining an audit trail of the cast ballots in chronological order. We described above why maintaining order information is a problem for privacy.

6. CONCLUSION

We have described the Open Voting Consortium's voting system that includes a PC-based open-source voting machine with a voter-verifiable accessible paper ballot, and discussed the privacy issues inherent in this system. By extension, many of the privacy issues in this paper also apply to other electronic voting machines, such as Direct Recording Electronic voting machines. The discussion illustrates why careful and thorough design is required for voter privacy. Even more work would be required to ensure that such systems are secure and reliable.

ACKNOWLEDGEMENTS

We acknowledge the work of the volunteers of the Open Voting Consortium who contributed to the design and implementation we describe. In particular, Alan Dechert developed much of the design and Doug Jones provided significant insights into voting issues. The demonstration software was largely developed by Jan Kärrman, John-Paul Gignac, Anand Pillai, Eron Lloyd, David Mertz, Laird Popkin, and Fred McLain. Karl Auerbach wrote an FAQ on which the OVC system description is based. Amy Pearl also contributed to the system description. Kurt Hyde and David Jefferson gave valuable feedback. David Dill referred some of the volunteers.

[28] Press Release, Sequoia Voting Systems, *Sequoia Voting Systems Announces Plan to Market Optional Voter Verifiable Paper Record Printers for Touch Screens in 2004, available at* http://www.sequoiavote.com/article.php?id=54

An extended abstract of this paper appeared at the Workshop on Privacy in the Electronic Society on October 28, 2004 in Washington DC, part of ACM CCS 2004 (Conference on Computer and Communications Security). Other papers on this topic are at http://www-db.stanford.edu/pub/keller under electronic voting. More information on the Open Voting Consortium may be found at http://www.openvotingconsortium.org.

Chapter 19

HIDDEN-WEB PRIVACY PRESERVATION SURFING (HI-WEPPS) MODEL

Yuval Elovici[1], Bracha Shapira[2] and Yael Spanglet[3]
[1]*Department of Information Systems Engineering, Ben-Gurion University of the Negev, Beer-Sheva, Israel, Tel- 972-8-6477138 Fax – 972-8-6477527, E-mail: elovici@bgu.ac.il;* [2]*Department of Information Systems Engineering, Ben-Gurion University of the Negev, Beer-Sheva, Israel, Tel- 972-8-6477551 Fax – 972-8-6477527, E-mail: bshapira@bgu.ac.il;* [3]*Department of Information Systems Engineering, Ben-Gurion University of the Negev, Beer-Sheva, Israel, Tel- 972-8-6477003 Fax – 972-8-6477527, E-mail: yaels@bgu.ac.il.*

Abstract: A new model for privacy preservation named Hidden-web Privacy Preservation Surfing (Hi-WePPS) is proposed. A hidden-web site often requires a subscription in order to access information stored in the site's database. The basic assumption motivating this proposal was that such websites cannot be trusted to preserve their surfers' privacy since site owners know the identities of their users and can monitor their activities. The new privacy preservation model includes an agent installed in the user computer and generates "intelligent" noise when a user accesses a hidden-web site in order to conceal the user's interests (profile). The noise is generated by submitting fake requests providing wrong data to the automatic programs collecting data about the users. A prototype of Hi-WePPS is being developed for preserving a surfer's privacy while accessing the U.S. patent office site (www.uspto.gov). This prototype enables industrial companies to search for patents without exposing their interests to any eavesdroppers on the path between their computers and the site, or to the patent site itself.

Key words: privacy, hidden-web, user profile, anonymity, web browsing, user modeling

1. INTRODUCTION

Protecting personal information privacy is an important challenge facing the growth and prosperity of the Internet.[1] Many eavesdroppers on the Web violate the privacy of users for their own commercial benefit and, as a result, users concerned about their privacy refrain from using useful web services in order to prevent exposure.[2] Web users leave identifiable tracks while surfing websites.[3] The tracks can be differentiated between those related to information rooted in the communication infrastructure and those related to users' explicit or implicit information. Explicit information is provided directly by users, while implicit information is inferred from users' behavior and actions.[4] In the current study we deal only with situations in which the information is acquired implicitly by a potential privacy attacker. Various eavesdroppers observe the tracks users leave and use them for their own benefit. These eavesdroppers may include the websites, the users' Internet Service Providers (ISPs), or some other adversaries interested in listening to communications between a user and the website. The eavesdroppers are able to record users' transactions and derive significant data about the content of pages a user visited, time spent, frequency and duration of exchanges, and other navigational behavior. This data could be used to infer users' personal information, interests and needs, *i.e.*, profiles, by employing various machine learning and user modeling techniques.[5] The profile constructed by eavesdroppers might be used without the authorization or awareness of the user. Examples of such uses range from industrial espionage to advertisements and junk-mail targeting, and even to assessing a user's medical condition for insurance companies.

Existing solutions provided by many studies performed by the computer security community aim at improving users' anonymity on the Web and thereby improving privacy. These solutions hide identifiable tracks originating from the communication infrastructure involved and create an anonymous channel between the user and the accessed site. However,

[1] S. Brier, How to Keep Your Privacy: Battle Lines Get Clearer, N.Y. TIMES, Jan. 13, 1997.

[2] P. Harris, *It Is Time for Rules in Wonderland*, BUSINESSWEEK, Mar. 20, 2000.

[3] *See supra* note 1.

[4] M. Claypool et al., Implicit Interest Indicators, 6 PROC. OF THE INT'L CONF. ON INTELLIGENT USER INTERFACES (2001); M.Q. Luz & M. Javed, Empirical Evaluation of Explicit versus Implicit Acquisition of User Profiles in Information Filtering Systems, 4 PROC. OF THE ACM CONF. ON DIGITAL LIBR. (1999); D. Oard & J. Kim, Implicit Feedback for Recommender Systems, 1998 PROC. OF THE AAAI WORKSHOP ON RECOMMENDER SYSTEMS.

[5] *Id.*

concealing users' identities is not an adequate solution in many situations on the Web in which the user is requested to identify himself to the end server. Some examples of web transactions requiring user identification include services provided only to subscribers, such as digital libraries (*e.g.*, the IEEE digital library available at http://ieeexplore.ieee.org). In these cases the users are identified to the operating site but still would not want eavesdroppers on the path between their computers and the site, or at the operating site itself, to infer their interests. Derivation of user interests is sometimes not only a matter of the user's privacy but might cause tangible harm to a user. For example, industrial espionage might be involved when stock or patent sites are spied upon, or denial of service might be caused when insurance companies spy on potential customers by monitoring health-related sites.

Hi-WePPS is an innovative model aimed at effectively concealing users' interests while browsing hidden-web (invisible) sites requiring identification. The model is based on generating intelligent fake transactions that would confuse an eavesdropper's automated programs that are attempting to derive the user's interests. The Hi-WePPS model learns the hidden site domain and the actual user interests and accordingly generates fake transactions that cannot be identified as fake by sophisticated data mining techniques. The remainder of the paper is organized as follows: Section 2 discusses related background on web anonymity and user profiling; Section 3 introduces Hi-WePPS while explaining its objectives, goals and significance; Section 4 describes the evaluation plan to examine the feasibility of the model, and Section 5 concludes with a summary and discussion of future research issues.

2. BACKGROUND

The background for understanding this model encompasses two topics. The first is a review of previous models of anonymity and existing Privacy Enhancing Technologies (PETs) for the Internet. The second is a review of widely accepted methods for user profile representation that are being used in the model.

2.1 Anonymity and privacy on the Internet

Much effort has been put into protecting the privacy of Internet surfers. This effort includes legislation and a new breed of technologies, so-called

Privacy Enhancing Technologies. In America and Europe various activities aim[6] to define the relevant players and goals for maintaining proper privacy while using the Internet.[7] Privacy may be preserved by Anonymity Enhancing Technology and Privacy Enhancing Technology, as described in the following sections.

2.1.1 Anonymity Enhancing Technologies

Anonymity Enhancing Technologies focus on ensuring that requests to websites cannot be linked to an IP address identifying the user.

Some of the better-known tools are:

- **Anonymizer**[8] (available at http://www.anonymizer.com): This application submits HTTP requests to the Internet on behalf of its users so that the only IP address revealed to the websites is that of **Anonymizer**. However, users have to trust the **Anonymizer** and their own ISPs which can still observe their activities.
- **Crowds:** This anonymity agent developed at Bell Labs[9] is based on the idea of "blending into a crowd", i.e., concealing one's actions among the actions of others. To execute Web transactions using this approach, a user first joins a crowd of users. A user's request to a Web server is passed to a random member of the crowd who can either submit the request directly to the end server or forward it to another randomly chosen member. Neither the end server nor any of the crowd members can determine where the requests originated.
- **Onion-Routing:**[10] In this approach users submit layered encrypted data specifying the cryptographic algorithms and keys. One layer of encryption is removed at each pass through each onion-router on the way to the recipient.

[6] Council Directive 2002/58/EC on Privacy and Electronic Comm., 2002 O.J. (L 201) 37.

[7] V. Senicar *et al.*, *Privacy-Enhancing Technologies – Approaches and Development*, COMPUTER STANDARDS & INTERFACES, May 2003.

[8] J. Claessens *et al.*, *Solutions for Anonymous Communication on the Internet,* 487 PROC. OF THE IEEE INT'L CARNAHAN CONF. ON SECURITY TECH. 298-303 (1999).

[9] M.K. Reiter and A.D. Rubin, *Crowds: Anonymity for Web Transactions*, 1 ACM TRANSACTIONS ON INFO. AND SYSTEM SECURITY 66-92 (1998).

[10] P.F. Syverson *et al.*, *Anonymous Connections and Onion Routing*, 18 PROC. OF THE ANNUAL SYMP. ON SECURITY AND PRIVACY 44-54 (1997); D.M. Goldschlag *et al.*, *Hiding Routing Information, Information Hiding,* 1174 SPRINGER-VERLAG LLNCS 137-150 (R. Anderson ed., 1996); D.M. Goldschlag *et al.*, *Onion Routing for Anonymous and Private Internet Connections*, 42 COMM. OF THE ACM 39-41 (1999).

- **LPWA:**[11] The Lucent Personalized Web Assistant generates pseudonymous aliases, user names and email addresses for Web users that enable users to securely browse the Web and send and receive email messages. LPWA facilitates the use of identified and personalized services but still maintains their security. LPWA requires an intermediate server to assign and maintain the aliases.

All the above-mentioned anonymity tools assume that users have no need or desire to be identified. However, this assumption is too strong for the Web today where many services require users' identification. Hi-WePPS targets Internet users who do not wish to remain anonymous but still wish to preserve their privacy. Hi-WePPS is designed to send the end server additional dummy messages to ensure that the real users' interests cannot be derived.

2.1.2 Privacy Enhancing Technologies

Privacy Enhancing Technologies include tools that try to build users' trust in the Internet, and tools that provide users with information about their current levels of privacy.

Existing trust centers aim at building users' trust in the Internet.[12] A trust center acts like a notary for both parties - users on the one hand, and any communication and business partners on the other hand. Its tasks include distributing public keys and issuing certificates for them, and acting as a trustee for holding data. Similar ideas for building users' trust are reflected in initiatives such as *TRUSTe*[13] that license websites to display a "trust-mark" on their sites indicating that users can safely surf without having to be concerned about their privacy. However, these trust solutions depend on the good will of websites and limit users' access to licensed sites when privacy is desired. Additionally, these trust centers provide privacy solutions only at sites (at the end servers) and do not protect users from eavesdroppers located on the path from a user's computer to the end server.

[11] E. Gabber *et al.*, *On Secure and Pseudonymous Client-Relationships with Multiple Severs*, 2 ACM TRANSACTIONS ON INFO. AND SYSTEMS SECURITY 390-415 (1999); E. Gabber et al., *Consistent, yet Anonymous, Web Access with LPWA*, 42 COMM. OF THE ACM 42-47 (1999).

[12] *See supra* note 7.

[13] P. Benassi, *TRUSTe: An Online Privacy Seal Program*, 42 COMM. OF THE ACM 56-59 (1999).

Projects such as P3P, the Platform for Privacy Preferences,[14] provide Internet users with information about the privacy level maintained by a specific site. When applied, P3P presents users with information such as whether their personal data is collected by the site, and how it will be used. This privacy information is taken from the site's privacy policy. P3P treats the site as a trusted entity but this is not always accurate.

2.2 User Profiling

While a user browses the Net, eavesdroppers may intercept the communication and derive a user profile from it. A profile in the information retrieval (IR) domain refers to a representation of users' interests and behaviors.[15] Such profiles are usually generated to enable personalized services such as filtering relevant documents or tailoring content to the user's interest.[16] In a typical personalization process, the similarity between information items and user profiles is measured to determine the relevancy of an item to a user. Personalization systems acquire data for deriving the user profiles using explicit or implicit methods. Explicit methods consist of interrogation of users and are usually employed by on-line or off-line forms which users are required to complete, as is carried out in WebWatcher,[17] or by asking users explicitly to rate information items they observed, as is carried out in AntWorld.[18] Implicit acquisition methods consist of derivation of user profiles from recorded user activities such as browsed sites, time spent reading a page, addition of data items to a "favorites" list, printing

[14] WC3, Platform for Privacy Preferences Project (May 27, 2003), *available at* http://www.w3.org/P3P.

[15] M. Eirinaki and M. Vazirgiannis, *Web Mining for Web Personalization*, 3 ACM TRANSACTIONS ON INTERNET TECH. 1-28 (2003).

[16] U. Hanani *et al.*, *Information Filtering: Overview of Issues, Research and Systems*, 11 USER MODELING AND USER-ADAPTED INTERACTION 203-259 (2001); Oard & Kim, *supra* note 4.

[17] D. Malden, *Machine Learning Used by Personal WebWatcher*, 1999 PROC. OF ACAI-99 WORKSHOP ON MACHINE LEARNING AND INTELLIGENT AGENTS.

[18] P.B. Kantor *et al.*, *Capturing Human Intelligence on the Net*, 43 COMM. OF THE ACM 112-115 (2000).

[19] H. R. Kim & P. K. Chan, *Learning Implicit User Interest Hierarchy for Context in Personalization*, FLORIDA INST. OF TECH. TECHNICAL REPORT CS-2002-15 (2002).

information items, and so on.[19] Implicit user profile acquisition is performed without any user intervention or awareness. Eavesdroppers on the Web use implicit acquisition methods to infer the interests and needs of users (i.e., their profiles) without their knowledge or consent, while violating users' privacy, mainly for commercial purposes. Many users are concerned about privacy violation while browsing the Net,[20] and some even forgo services on the Web in order to protect their privacy.[21]

2.2.1 Profile Presentation

Most user profiles are represented as sets of weighted keywords, usually derived from the documents that the user observed. The weighted list (a vector) has been proven effective in many systems.[22] Some profile representation systems consider additional sources of information such as the user's bookmark[23] or temporal information.[24] Other representations of profiles include semantic networks such as SiteIF,[25] rule based systems,[26] and neural networks.[27]

[20] *See supra* note 15.

[21] *See supra* note 1.

[22] S. Gauch *et al.*, Ontology-Based User Profiles for Search and Browsing (forthcoming).

[23] M. Montebello *et al.*, *A Personable Evolvable Advisor for WWW Knowledge-Based Systems*, 1998 PROC. OF THE INT'L DATABASE ENGINEERING AND APPLICATION SYMP 224-233.

[24] D.H. Widyantoro *et al.*, *Learning User Interest Dynamics with a Three-Descriptor Representation*, 52 J. OF THE AM. SOC'Y OF INFO. SCI. AND TECH. 212-225 (2001).

[25] A. Stefani and C. Strappavara, *Personalizing Access to Websites: The SiteIF Project*, 2 PROC. OF THE WORKSHOP ON ADAPTIVE HYPERTEXT AND HYPERMEDIA (1998).

[26] G. Adomavicius and A. Tuzhilin, *User Profiling in Personalization Applications through Rule Discovery and Validation*, 5 PROC. OF THE ACM SIGKDD INT'L CONF. ON KNOWLEDGE DISCOVERY AND DATA MINING (1999).

[27] Z. Boger *et al.*, *Automatic Keyword Identification by Artificial Neural Networks Compared to Manual Identification by Users of Filtering Systems*, 37 INFO. PROCESSING & MGMT. 187-198 (2001).

In the Hi-WePPS model user profiles are represented using the vector space model,[28] which is commonly used in IR applications to represent profiles. In the vector space model each data item, such as a document, a query, or a profile, is represented as a vector of its significant weighted terms. The weight of a term in a vector represents how indicative the term is for the specific document it represents, i.e., to what extent the term if included in a query will cause the document to be retrieved. A weight of a term in a vector representing a user profile should indicate the interest of the user in the subject corresponding to the term. A data item d is represented by a vector in an n-dimensional space:

$$d = (w_1, w_2, \ldots, w_n) \tag{1}$$

where the weight of the *i*th term in the *d*th document is represented by w_i. We use the common IDF weighting scheme.[29] TF, the frequency of a term in the document, denotes the importance of the term to the document. For profile applications it is possible to predict the relevancy of new items to a user by measuring the similarity between the user profile represented as a vector and the vector of the new item.

To evaluate the privacy obtained by using Hi-WePPS we consider the similarity between the real user profile (represented as a vector) and the most accurate vector that an eavesdropper might approximate. Similarity between these two vectors is computed by the commonly used cosine distance measuring method,[30] reflecting the cosine of the angle between two vectors in a vector space. The similarity between the two vectors is in the interval of (0,1), where lower angles are closer to 1 and denote higher similarity.

[28] G. Salton, INTRODUCTION TO MODERN INFO. RETRIEVAL (W.J. McGill, ed., McGraw-Hill 1983).

[29] *Id.*

[30] *Id.*

The following is the cosine similarity equation:

$$S(u_j,u_k)=\frac{\sum_{i=1}^{n}\left(tu_{ij}\cdot tu_{ik}\right)}{\sqrt{\sum_{i=1}^{n}tu_{ij}^2\cdot\sum_{i=1}^{n}tu_{ik}^2}} \tag{2}$$

Where

u_j, u_k= vectors
tu_{ij} = the ith term in the vector u_j.
tu_{ik} = the ith term in the vector u_k.
n = the number of unique terms in each vector.

One limitation of the vector space model is that it assumes independence of the terms in a document. This is not a realistic assumption since it ignores semantic and syntactic relationships between terms. There are newer methods for conceptual representation of documents and user profiles other than representation by terms (such as Latent Semantic Indexing[31]). In our model we will first develop the prototype with the common vector space model to prove the feasibility of the privacy model. At a second stage we will incorporate concept-based representation in order to enhance the model.

3. HI-WEPPS MODEL

Lack of privacy on hidden-web sites affects a surfer's sense of security and might, in some situations, also cause financial or personal harm to the surfer. For example, a company willing to apply for a patent needs to search for similar patented inventions in the formal patent site, and utilize the site's dedicated search engine. While searching in the domain of the invention, the company might expose its interests and intentions and the situation might result in industrial espionage. Another example is a user performing a search in one of the many health-related sites (such as health on the Net-www.hon.ch/) for information about diseases or health services. Such a search might be tracked by medical insurance companies and result in refusal of service.

[31] S. Deerwester *et al.*, *Indexing by Latent Semantic Indexing*, 41 J. OF THE AM. SOC'Y FOR INFO. SCI. (1990).

The main goal of this study is to develop a new method for privacy preservation while surfing or searching the hidden (invisible) Web. The new model should protect users' privacy from eavesdroppers on the path between the user computer and the operating site and also from eavesdroppers on the site. It should not require user anonymity, as hidden-web sites may require identification on many subscribed sites. In order to achieve this goal we propose to develop HI-WePPS - a Hidden-Web Privacy Preservation System - to protect users' privacy while surfing hidden-web sites.

HI-WePPS generates "intelligent" noise *i.e.*, performs fake searches, while a user is surfing (or searching) in a subscribed hidden-web site. The noise is aimed at concealing a user's profile by providing incorrect data to any automatic programs collecting data on the user. The model "learns" the user's interests and the site's domain in order to automatically complete the forms on the site and generate fake transactions relevant to the site's domain to create a fuzzy cloud around the user's actual information needs.

HI-WePPS was designed to protect users from two types of eavesdroppers:

1. An eavesdropper intercepting communication between a website and the user.
2. An eavesdropper that accesses the website's log file or database.

In order to formulate the model we use the following definitions, similar to those used in PRAW,[32] as Hi-WePPS and PRAW are based on a similar concept of concealing privacy by fake transaction generation. (PRAW is designed for the surface Web and cannot be used to protect the privacy of users accessing hidden-web sites.)

Internal user profile (IUP) – A user profile constructed inside the user's computer. It is based on the terms the user uses for his queries and the content of the pages the user accesses.

External user profile (EUP) – A user profile based on the information that might be collected about the user in the subscribed Website or on the path between the user and the Website. It is based on user and fake transactions, queries, and result content.

An eavesdropper on the target website is able to compute the EUP by observing the website's log and recording every user's actions on the website. We assume that eavesdroppers do not have any smart agents, such as computer viruses, inserted in the user's computer and are therefore unable

[32] B. Shapira *et al.*, *PRAW – The Model for the Private Web*, J. OF THE AM. SOC'Y FOR INFO. SCI. AND TECH. (forthcoming); Y. Elovici *et al.*, *A New Privacy Model for Hiding Group Interests while Accessing the Web*, WORKSHOP ON PRIVACY IN THE ELECTRONIC SOC'Y IN ASS'N WITH 9TH ACM CONF. ON COMPUTER AND COMM. SECURITY (2002).

to compute the IUP. Under ordinary circumstances, the IUP and the EUP are identical. The goal of Hi-WePPS is to assure that the EUP is not identical to the IUP but differs in an intelligent way so that the IUP cannot be derived from the observed EUP even if data-mining techniques are applied. The EUP is actually an expansion of the IUP with the addition of fake transactions located in the larger domain of the IUP. For example, we expect that while a user looks for patents related to "tennis," the model will generate faked transactions looking for patents in the general sports domain. If the EUP contained fake information related to totally different subjects than the subjects in the IUP, it might be possible to differentiate the fake data from the actual data.

The model architecture consists of four main components: *System Initializer, Profile Builder, Wrapper and Transaction Generator,* as illustrated in Figure 19-1.

The system is configured and initialized by the System Initializer. The Profile Builder component computes the internal user profile (IUP) based on the user's query and the contents of the pages the user downloads.

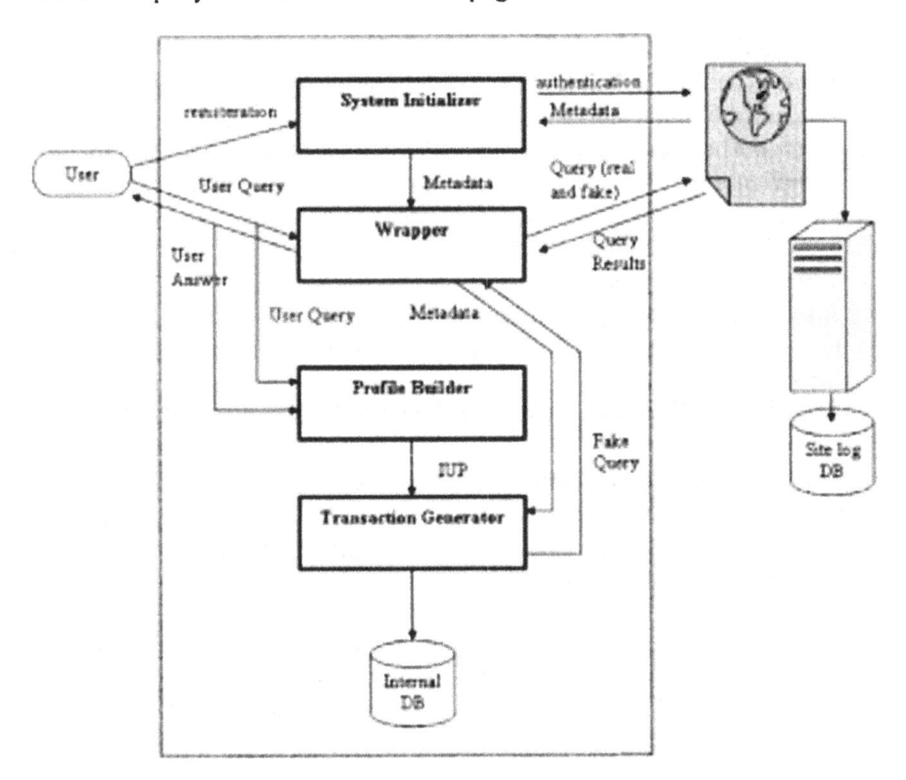

Figure 19-1. Hi-WePPS Architecture

The model's components are:

System Initializer

Input: The System Initializer receives the target website URL from the user and login information (username and password) for the authentication process.

The System Initializer obtains from the target website a metadata file that contains the website's structure and hierarchical content.

Output: The System Initializer passes the username and password to the target website during the authentication phase.

The System Initializer sends the metadata to the wrapper component.

Functionality: The System Initializer receives a request to initiate a connection with a specific site requiring authentication using username and password (it is assumed that the user has already registered with the site and that the site cooperates with the privacy model and provides a metadata file that the model can interpret). The System Initializer downloads the metadata file from the website and forwards it to the wrapper component.

Profile Builder

Input: The Profile Builder receives the user query and the pages that the user accessed from the query results.

Output: The Profile Builder outputs the IUP to the transaction generator.

Functionality: The Profile Builder computes the IUP based on the user's queries and pages that the user accessed based on the query results.

Transaction Generator

Input: The Transaction Generator receives the IUP from the profile builder, the metadata file from the wrapper, and fake transaction results from the wrapper.

Output: The Transaction Generator outputs to the wrapper fake queries that extend the IUP to the EUP.

Functionality: The Transaction Generator is responsible for the generation of fake transactions based on the IUP, the metadata and an internal database of terms. The fake query content is based on the metadata describing the surfed Website's forms and the required query format. It is also based on the IUP in order to generate fake transactions that are in the same general domain as those of the user. The transaction generator computes the EUP based on the IUP and the fake transaction results. The dissimilarity between the IUP and the EUP will indicate the system's success.

Wrapper

Input: The Wrapper receives the user queries and fake queries from the Transaction Generator. In addition, the Wrapper receives the metadata file from the website initializer. The Wrapper sends the queries to the website (the real and fake queries) and receives the query results from the website.

Output: The Wrapper output to the site is a mixture of fake and real queries. The Wrapper sends the real query results to the user and the fake query results to the Transaction Generator.

Functionality: The Wrapper component is acting as a mediator between the model's components and the site.

4. EVALUATION OF HI-WEPPS FEASIBILITY

The Hi-WePPS model is aimed at enabling users to access websites privately even when their identities are known to the target website.

Hi-WePPS will be empirically evaluated by developing a prototype system implementing the model and running experiments. The empirical study will examine the following research questions:

1. Hi-WePPS feasibility;
2. The effect of the number of average fake transactions generated for each actual transaction (denoted by Tr) on the privacy level obtained. The privacy level is measured as the similarity between the actual user profile and the observed user profile that would be inferred by eavesdroppers;
3. Hi-WePPS ability to withstand auto-clustering attacks aimed at differentiating between real and fake user transactions.

During the experiments, ten different users will be instructed to search for patents in the official U.S. patent site (www.uspto.gov) as a case study. Each user will search within a different domain, simulating an inventor seeking information about the domain of his or her invention before registering a patent. During the users' searches, Hi-WePPS will operate in the background and generate fake transactions aimed at concealing the users' searches. Hi-WePPs will generate the maximal number of fake transactions for each actual transaction (maximal Tr). All users' actual and fake transactions will be recorded for later analysis.

The feasibility of Hi-WePPS will be proven once the prototype is operated by users without interfering with their regular usage of the hidden-web sites.

The effect of the average number of fake transactions for each actual transaction (Tr) will be evaluated by calculating the similarity between the real internal user profile (IUP) and the external profile (EUP) when different number of Trs are considered in the EUP.

In addition, auto-clustering attacks will be simulated on the EUP, i.e., the transaction as observed by eavesdroppers which includes real and fake transactions. These simulations will evaluate the model's ability to withstand attacks that attempt to differentiate between real and fake transactions.

5. SUMMARY

In this paper we describe a novel method of preserving privacy while surfing or searching the hidden-web. The model is important for services on the Web that require identification, form completion or the use of a dedicated search engine. As the Internet transforms into a non-free service, the number of such websites will grow. Most existing solutions suggest anonymous routing and are based on the concept of concealing users' identities. However, for those cases in which the identity of the user is required, anonymity cannot be a solution to protect privacy. Hi-WePPS focuses on preserving privacy without anonymity, and thus appears to be an adequate and novel solution for preserving privacy while surfing or searching the hidden-web.

In addition, a new measure of privacy is suggested that is based on the distance between the original user profile (IUP) and the profile that is observed by eavesdroppers (EUP). It quantifies the degree of privacy that a system can guarantee its users. Privacy quantification is important as it enables users to define their preferences on a quantifiable scale.

Chapter 20

GLOBAL DISCLOSURE RISK FOR MICRODATA WITH CONTINUOUS ATTRIBUTES

Traian Marius Truta[1], Farshad Fotouhi[2] and Daniel Barth-Jones[3]

[1]Department of Mathematics and Computer Science, Northern Kentucky University, Highland Heights, KY 41076, USA; [2]Department of Computer Science, Wayne State University, Detroit, MI 48202, USA; [3]Center for Healthcare Effectiveness, Wayne State University, Detroit, MI 48202, USA

Abstract: In this paper, we introduce three global disclosure risk measures (minimal, maximal and weighted) for microdata with continuous attributes. We classify the attributes of a given set of microdata in two different ways: based on its potential identification utility and based on the order relation that exists in its domain of value. We define inversion factors that allow data users to quantify the magnitude of masking modification incurred for values of a key attribute. We create vicinity sets from microdata for each record based on distance functions or interval vicinity for each key attribute value. The disclosure risk measures are based on inversion factors and the vicinity sets' cardinality computed for both initial and masked microdata.

Key words: disclosure risk, microdata, information loss, inversion factor, disclosure control, random noise, microaggregation

1. INTRODUCTION

"Privacy is dead, deal with it," Sun Microsystems CEO Scott McNealy is widely reported to have declared some time ago. Privacy in the digital age may not be as dead and buried as McNealy believes, but it's certainly on life support[1]. While releasing information is one of their foremost reasons to exist, data owners must protect the privacy of individuals. Privacy concerns are being fueled by an ever-increasing list of privacy violations, ranging from accidents to illegal actions.[2] Privacy issues are increasingly important in today's society, and, accordingly a number of privacy regulations have been recently enacted in various fields. In the U.S., for example, privacy regulations promulgated by the Department of Health and Human Services as part of the *Health Insurance Portability and Accountability Act (HIPAA)* went into effect in April 2003 to protect the confidentiality of electronic healthcare information.[3] Other countries have promulgated similar privacy regulations (for example, the *Canadian Standard Association's Model Code for the Protection of Personal Information*[4] and the *Australian Privacy Amendment Act of 2000*[5]).

In this paper, we introduce global disclosure risk measures for microdata. *Microdata* consists of a series of records, each record containing information on an individual unit such as a person, a firm, or an institution.[6] Microdata can be represented as a single data matrix in which the rows correspond to the individual units and the columns to the attributes (such as name, address, income, sex, etc.). Typically, microdata is released for use by a third party after the data owner has masked the data to limit the possibility of disclosure of personally identifiable information. We will call the final microdata *masked* or *released microdata.*[7] We will use the term *initial microdata* for

[1] B.N. Meeks, *Is Privacy Possible in Digital Age*, MSNBC NEWS, Dec. 8, 2000, *available at* http://msnbc.msn.com/id/3078854.

[2] R. Agrawal et al., *Hippocratic Databases*, 28 Proc. of the Int'l Conf. on Very Large Databases (2002).

[3] Health Insurance Portability and Accountability Act, Pub. L. No. 104-191, 110 Stat. 1936 (1996), *available at* http://www.hhs.gov/ocr/hipaa.

[4] THE PRIVACY LAW SOURCEBOOK 2000: UNITED STATES LAW, INTERNATIONAL LAW, AND RECENT DEVELOPMENTS (M. Rotenberg ed., 2000).

[5] The Australian Privacy Amendment (Private Sector) Act, No. 155 (2000), No. 119 (1988) (Austl.), *at* http://www.privacy.gov.au/publications/npps01.html.

[6] L. Willemborg, *Elements of Statistical Disclosure Control*, SPRINGER VERLAG (T. Waal ed., 2001).

[7] T. Dalenius and S.P. Reiss, *Data-Swapping: A Technique for Disclosure Control*, 6 J. OF STATISTICAL PLANNING AND INFERENCE 73-85 (1982).

[8] G. Chen and S. Keller-McNulty, *Estimation of Deidentification Disclosure Risk in Microdata*, 14 J. OF OFFICIAL STATISTICS 79-95 (1998).

[9] *See supra* note 6.

[10] I.P. Fellegi, *On the Question of Statistical Confidentiality*, 67 J. OF THE AM. STATISTICAL ASS'N 7-18 (1972).

microdata to which no masking methods (also called disclosure control methods) have been applied. *Disclosure risk* is the risk that a given form of disclosure will be encountered if masked microdata are released.[8] *Information loss* is the quantity of information which existed in the initial microdata but which is not present in the masked microdata.[9] When protecting the confidentiality of individuals, the owner of the data must satisfy two conflicting requirements: protecting the confidentiality of the records from the initial microdata and maintaining analytic properties (*statistical integrity*[10]) in the masked microdata.[11]

Recent work in disclosure risk assessment can be categorized into two approaches: individual and global disclosure risk. Benedetti and Franconi introduced *individual risk methodology.*[12] In this approach, the risk is computed for every released record from masked microdata. Domingo-Ferer, Mateo-Sanz and Torra[13] describe three different disclosure risk measures. The first measure is *distance-based record linkage,*[14] in which the Euclidian distance from each record in masked microdata to each record in the initial data is computed. The *nearest* and the *second nearest* records are considered. The percent of correct matching with the nearest and second nearest records from initial microdata is a measure of disclosure risk. The second measure is *probabilistic record linkage.*[15] This matching algorithm uses the linear sum assignment model to match records in initial and masked microdata. The percent of correctly paired records is a measure of the disclosure risk. The last metric presented is *interval disclosure*. The measure of disclosure risk is the proportion of original values that fall into an interval centered around their corresponding masked value. *Global disclosure*

[11] J.J. Kim & W.E. Winkler, *Multiplicative Noise for Masking Continuous Data*, ASA PROC. OF THE SEC. ON SURV. RES. METHODS (2001).

[12] R. Benedetti & L. Franconi, *Statistical and Technological Solutions for Controlled Data Dissemination*, 1 PRE-PROCEEDINGS OF NEW TECHNIQUES AND TECHNOLOGIES FOR STATISTICS 225-32 (1998).

[13] J. Domingo-Ferrer & J. Mateo-Sanz, *Practical Data-Oriented Microaggregation for Statistical Disclosure Control*, 14 IEEE TRANSACTIONS ON KNOWLEDGE AND DATA ENGINEERING 189-201 (2002).

[14] D. Pagliuca & G. Seri, *Some Results of Individual Ranking Method on the System of Enterprise Accounts Annual Survey, in* ESPRIT SDC PROJECT, DELIVERABLE MI3/D2 (1999).

[15] M.A. Jaro, *Advances in Record-Linkage Methodology as Applied to Matching the 1985 Census of Tampa, Florida*, 84 J. OF AM. STATISTICS ASS'N, 414-20 (1989).

risk is defined in terms of the expected number of identifications in the released microdata. Eliot[16] and Skinner[17] define a new measure of disclosure risk as the proportion of correct matches amongst those records in the population which match a sample unique masked microdata record.

We call actions taken by the owner of the data to protect the initial microdata with one or more disclosure control methods the *masking process.* The masking process can alter the initial microdata in three different ways: *changing the number of records*, *changing the number of attributes* and *changing the values of specific attributes.* The change in the number of attributes is always used, since the removal of identifier attributes is the first step in data protection. We call this first mandatory step in the masking process the *remove identifiers method.* The other two types of changes may or may not be applied to the initial microdata. The most general scenario is when all three changes are applied to the given initial microdata. Changing the number of records is used by two techniques: *simulation*[18] and *sampling.*[19] Changing attribute values is part of a large number of disclosure methods (*microaggregation,*[20] *data swapping,*[21] and *adding noise,*[22] for example).

The initial microdata consists of a set of *n* records with values for three types of attributes: identifier (*I*), confidential (*S*), and key (*K*) attributes. Identifier attributes, such as *Address* and *SSN,* can be used directly to identify a record. Such attributes are commonly deleted from

[16] M.J. Elliot, DIS: A New Approach to the Measurement of Statistical Disclosure Risk, INT'L. J. OF RISK MGMT. 39-48 (2000).

[17] C.J. Skinner and M.J. Elliot, *A Measure of Disclosure Risk for Microdata*, 64 J. OF THE ROYAL STATISTICAL SOC'Y 855-67 (2002).

[18] N.R. Adam & J.C. Wortmann, *Security Control Methods for Statistical Databases Comparative Study*, 21 ACM COMPUTING SURVEYS NO. 4 (1989).

[19] C.J. Skinner et al., *Disclosure Control for Census Microdata*, J. OF OFFICIAL STATISTICS 31-51 (1994).

[20] *See supra* note 13.

[21] *See supra* note 7.

[22] J.J. Kim, *A Method for Limiting Disclosure in Microdata Based on Random Noise and Transformation*, ASA PROC. OF THE SEC. ON SURV. RES. METHODS 79-95 (1998).

[23] T.M. Truta *et al.*, *Disclosure Risk Measures for Sampling Disclosure Control Method*, ANN. ACM SYMP. ON APPLIED COMPUTING 301-306 (2004).

[24] T.M. Truta *et al.*, *Privacy and Confidentiality Management for the Microaggregation Disclosure Control Method*, PROC. OF THE WORKSHOP ON PRIVACY AND ELECTRONIC SOC'Y 21-30 (2003).

the initial microdata in order to prevent direct identification. Key attributes, such as *Sex* and *Age,* may be known by an intruder. Key attributes are often fields that would ideally be retained in masked microdata for their analytic utility, but which, unfortunately, often pose potential disclosure risks. Confidential attributes, such as *Principal Diagnosis* and *Annual Income,* are rarely known by an intruder. Confidential attributes are present in masked microdata as well as in the initial microdata.

The attributes from a microdata set can also be classified in several categories (continuous, discrete, ordered, partially ordered, etc.) using their domain of value properties as the main classification criteria. We consider an attribute to be continuous if the domain of values theoretically ranges over an infinitely divisible continuum of values. The attributes *Distance* and *Length,* as well as many financial attributes, fit into this category.

In this paper, we extend disclosure risk measures that have been previously presented for discrete attributes and specific disclosure control methods (disclosure risk measures for sampling,[23] microaggregation[24] and combinations of those[25]) to make them suitable when all key attributes are continuous and the masking process changes the values of those attributes. We assume that the number of records does not change during the masking process. Many financial datasets have only continuous key attributes, and in order to understand how to protect this data from presumptive intruders, new disclosure risk measures that are applicable to continuous attributes are needed.

Our formulations for disclosure risk measures compute overall disclosure risks for given datasets and, thus, are not linked to specific individuals. We define an inversion factor that allows data users to quantify the magnitude of masking modification incurred for values of a key attribute. The disclosure risk measures are based on these inversion factors and on distance functions computed between pairs of records in the initial and masked microdata.

[25] T.M. Truta *et al.*, *Disclosure Risk Measures for Microdata*, INT'L CONF. ON SCI. AND STATISTICAL DATABASE MGMT. 15-22 (2003).

2. GLOBAL DISCLOSURE RISK FOR MICRODATA

In order to describe the masking process, a few assumptions are needed. The first assumption we make is that the intruder does not have specific knowledge of any confidential information. The second assumption is that an intruder knows all the key and identifier values from the initial microdata, usually through access to an external dataset. In order to identify individuals from masked microdata, the intruder will execute a record linkage operation between the external information dataset and the masked microdata. This conservative assumption maximizes the amount of external information available to an intruder. Since disclosure risk increases when the quantity of external information increases, this second assumption guarantees that any disclosure risk value computed by one of the proposed measures is an upper bound to the disclosure risk value when the amount of external information available to an intruder is not maximal. Based on the above assumptions, only key attributes are subject to change in the masking process.

For any continuous (or discrete) ordered attribute we define the notion of an inversion. The pair (x_{ik}, x_{jk}) is called an *inversion for attribute* O_k if $x_{ik} < x_{jk}$ and $x'_{ik} > x'_{jk}$ for i, j between 1 and n where n is the number of records in both sets of microdata, x_{ik} is the value of attribute k for the record j in the initial microdata, and x'_{ik} is the value of attribute k for the record j in the masked microdata. Without simplifying our assumptions we assume that the records preserve their order during the masking process, and consequently, x'_{ik} is the masked value for x_{ik}. We label the total number of inversions for the attribute O_k, inv_k. By definition:

$$inv_k = |\{(x_{ik}, x_{jk}) | \, x_{ik} < x_{jk} \text{ and } x'_{ik} > x'_{jk}, \; 1 \leq i, j \leq n\}| \tag{1}$$

Next, an *inversion factor for attribute* O_k is defined as the minimum between 1 and the ratio of the number of inversions for attribute O_k to the average number of inversions.

$$if_k = \min\left(1, \frac{4 \cdot inv_k}{n \cdot (n-1)}\right) \tag{2}$$

The inversion factor will help in the assessment of disclosure risk, since the intruder may try to link an external dataset to the masked microdata based on how the records are ordered for a specific key attribute.

The other tool we use for assessing disclosure risk is based on the creation of a vicinity set of records for each record from both initial and masked microdata. There are two possible approaches for this construction. One approach is to use a distance function between records. Another approach is to use interval vicinities for each of the key attributes individually, and to consider in the corresponding vicinity only the records that are in the vicinity of each key attribute. The width of an interval is based on the rank of the attribute or on its standard deviation. Using this method we will be able to compute for any pair (x_k, x'_k) its corresponding vicinity sets, labeled V_k and V'_k (x_k is the record j in initial microdata, x'_k is the record j in masked microdata, and x'_k represents the same record as x_k). To represent the cardinality of the vicinity sets, V_k and V'_k, we use the notations $n(V_k)$ and $n(V'_k)$ respectively. The probability of correct linking for the pair (x_k, x'_k) is a function of $n(V_k)$ and $n(V'_k)$, and, therefore, we will classify each pair (x_k, x'_k) based on the cardinality of the vicinity sets. To capture this property we define the *classification matrix C* with size $n \times n$. Each element of C, c_{ij}, is equal to the total number of pairs (x_k, x'_k) that have a vicinity set of size i for the masked microdata record x'_k, and, have a vicinity set of size j for the initial microdata record v_j. Mathematically, this definition can be expressed in the following form: for all $i = 1, .., n$ and for all $j = 1, .. n$; $c_{ij} = | \{(x_k, x'_k) \mid (x_k, x'_k)$ is a pair of corresponding records from initial and masked microdata, $n(V'_k) = i$, and $n(V_k) = j$, where $k = 1, .., n \} |$. The matrix C has several properties; the most important of which is that the sum of all elements is equal to the number of records in the masked microdata. Based on this property, we developed an algorithm that for each pair of records (x_k, x'_k) computes the corresponding vicinity sizes $n(V_k)$ and $n(V'_k)$, and increments by one the matrix element on the row $n(V'_k)$ and column $n(V_k)$. We present this algorithm below:

Algorithm 1 (Classification matrix construction)

Create a square matrix of size n (the number of records in both microdata sets).
Initialize each element from C with 0.
For k =1 to n do
  Find the kth pair (x_k, x'_k) of corresponding records from
    initial and masked microdata.
  For the record x'_k from masked microdata find its vicinity set V'_k.
    Let $i = n(V'_k)$.
  For the record x_k from initial microdata find its vicinity set V_k.
    Let $j = n(V_k)$.
  Increment c_{ij} by 1.
End for.

The minimal, weighted and maximal disclosure risk measures proposed[26] can be generalized to apply to this extended classification matrix that analyses continuous key attributes.

There are many disclosure control methods used to mask continuous attributes. Some methods, such as microaggregation,[27] preserve the order of records for the microaggregated attribute from the initial microdata to the masked microdata; others alter this ordering (providing random noise[28]). We call the first category of methods *order preserving*, and the second category *non-order preserving* methods.

When only order preserving methods are used in the disclosure control process, our formulations for disclosure risk generalize the results presented for microaggregation.[29] In this scenario, all inversion factors are 0, and they are not present in the disclosure risk measures. The major improvement is that the classification matrix is defined using vicinity sets. As stated before, there are different methods the data owner can use to compute vicinity sets (intervals, ranks, etc.), and the property of the classification matrix presented for discrete attributes[30] that all values above the main diagonal are 0 do not stand.

Due to the above differences, the disclosure risk formulations for order preserving methods are:

$$DR_{min} = \frac{c_{11}}{n} \tag{3}$$

$$DR_{max} = \frac{1}{n} \cdot \sum_{k=1}^{n} \frac{1}{k} \left(\sum_{i=1}^{k} c_{ik} + \sum_{j=1}^{k-1} c_{kj} \right) \tag{4}$$

$$DR_{W} = \frac{1}{n \cdot w_{11}} \left(\sum_{k=1}^{n} \frac{1}{k} \left(\sum_{i=1}^{k} w_{ik} \cdot c_{ik} + \sum_{j=1}^{k-1} w_{kj} \cdot c_{kj} \right) \right) \tag{5}$$

[26] T.M. Truta et al., *Assessing Global Disclosure Risk Measures in Masked Microdata*, PROC. OF THE WORKSHOP ON PRIVACY AND ELECTRONIC SOC'Y 85-93 (2004).
[27] *See supra* note 13.
[28] Kim, *supra* note 22, at 303-308.
[29] *See supra* note 24.
[30] *Id.*

DR_{min} represents the percentage of records from the population that the intruders can de-identify because c_{11} represents the number of records in vicinity sets with cardinality 1 for both initial and masked microdata. This is the minimal disclosure risk value. DR_{max} takes into consideration the probability of correct linking for records that have vicinity sets with higher cardinality values.

For the third measure, we define the *disclosure risk weight matrix, W*, as:

$$W = \begin{pmatrix} w_{11} & w_{12} & \dots & w_{1n} \\ w_{21} & w_{22} & \dots & w_{2n} \\ \dots & \dots & \dots & \dots \\ w_{n1} & w_{n2} & \dots & w_{nn} \end{pmatrix} \quad (6)$$

The disclosure risk weight matrix gives greater importance to records with vicinity sets of cardinality 1 relative to the rest of records, and likewise, attributes a greater importance to records with vicinity sets of cardinality 2 relative to records vicinity sets of greater cardinality, and so on. Detailed properties of the disclosure risk weight matrix have been presented in previous work.[31] The combined effect of these weighting constraints allows the data owner considerable flexibility in addressing probabilistic record linkage risks, while also accounting for the potential disclosure risks posed by the data intruder's prior beliefs about confidential variables which could possibly be used to distinguish individuals. The data owner defines the weight matrix, and this matrix captures particularities of that specific initial microdata.

To include non-order-preserving methods, we consider inversion factors for all key attributes. In the following formulations, the number of key attributes is p, and if_k represents the inversion factor for the continuous attribute O_k.

$$DR_{min}^{int} = \left(\prod_{k=1}^{p} (1 - if_k) \right) \cdot \frac{c_{11}}{n} \quad (7)$$

[31] *See supra* note 25.

$$DR_{max}^{int} = \left(\prod_{k=1}^{p}(1-if_k)\right)\cdot\frac{1}{n}\cdot\sum_{k=1}^{n}\frac{1}{k}\left(\sum_{i=1}^{k}c_{ik}+\sum_{j=1}^{k-1}c_{kj}\right) \tag{8}$$

$$DR_{W}^{int} = \left(\prod_{k=1}^{p}(1-if_k)\right)\cdot\frac{1}{n\cdot w_{11}}\left(\sum_{k=1}^{n}\frac{1}{k}\left(\sum_{i=1}^{k}w_{ik}\cdot c_{ik}+\sum_{j=1}^{k-1}w_{kj}\cdot c_{kj}\right)\right) \tag{9}$$

We call the above formulations *intermediary disclosure risk measures* because they are not always accurate. Let us assume that the inversion factor for a continuous key attribute is 1. This rare situation occurs when, for instance, the attribute is ordered and all values are modified such that the highest value became the lowest value, the second highest value became the second lowest, and so on. In that case the intermediary disclosure risk (minimal, maximal or weighted) will be equal to 0. In reality the intruder may notice that is something wrong with values of that key attribute and, therefore, he may not consider this key attribute in the process of deriving confidential information from masked microdata. In that case, the set of useful key attributes should not include that specific attribute. Similar situations may exist for other attributes with non-null inversion factors. In order to find the correct value for disclosure risk, we have to consider all possible subsets of key attributes, compute the disclosure risk for each subset and select the maximum value. Based on this procedure, we compute the minimal, maximal, and weighted disclosure risk for each subset of key attributes, and the final disclosure risk values are the maximum values of all disclosure risks considered for each subset of key attributes.

The disclosure risk formulations for each subset of key values are:

$$DR_{min}^{v} = \left(\prod_{k=1}^{p}(1-v_k\cdot if_k)\right)\cdot\frac{c_{11}^{v}}{n} \tag{10}$$

$$DR_{max}^{v} = \left(\prod_{k=1}^{p}(1-v_k\cdot if_k)\right)\cdot\frac{1}{n}\cdot\sum_{k=1}^{n}\frac{1}{k}\left(\sum_{i=1}^{k}c_{ik}^{v}+\sum_{j=1}^{k-1}c_{kj}^{v}\right) \tag{11}$$

$$DR_{W}^{v} = \left(\prod_{k=1}^{p}(1-v_k\cdot if_k)\right)\cdot\frac{1}{n\cdot w_{11}}\left(\sum_{k=1}^{n}\frac{1}{k}\left(\sum_{i=1}^{k}w_{ik}\cdot c_{ik}^{v}+\sum_{j=1}^{k-1}w_{kj}\cdot c_{kj}^{v}\right)\right) \tag{12}$$

The vector v is a binary vector with p elements where the value zero on the position k means that the corresponding key attribute is not considered in disclosure risk computation. For each vector v we compute the classification matrix with C^v that considers continuous key attributes with value 1 for their corresponding position in the vector v.

The final formulations for disclosure risk are:

$$DR_{min} = \max_{v}\{DR^{v}_{min}\} \tag{13}$$

$$DR_{max} = \max_{v}\{DR^{v}_{max}\} \tag{14}$$

$$DR_{W} = \max_{v}\{DR^{v}_{W}\} \tag{15}$$

As noted above, in order to find the correct value for disclosure risk, we have to consider all possible subsets of key attributes, compute the disclosure risk for each subset and select the maximum value. The algorithm that generates all possible subsets of key attributes to compute the disclosure risk has exponential complexity $O(2^p)$. Fortunately, in a real initial microdata set, the number of key attributes is low (usually less than 5). Moreover, the owner of the data can reduce the number of subsets to check by determining if the inversion factor is either 0 or 1. It is easy to show that key attributes corresponding to the inversion factor 0 will be included in the search. Also, when the inversion factor is 1, the corresponding key attributes will be excluded from the search.

3. RESULTS

To illustrate the concepts from the previous section, in Table 20-1 we consider the following initial microdata set with two corresponding masked microdata sets (we show only key attribute values). The order of records remains unchanged from the initial microdata to both masked microdata.

Increase Rate and *Annual Income* are considered continuous key attributes. In the first masked microdata set, labeled masked microdata A (MM_A), univariate microaggregation is applied for both attributes with size 3. In the second masked microdata set, (MM_B), random noise is added to each attribute such that the mean value is unchanged.

Table 20-1. Initial microdata with two possible masked microdata sets

Initial Microdata		Masked Microdata A		Masked Microdata B	
Increase Rate	Annual Income	Increase Rate	Annual Income	Increase Rate	Annual Income
0.19	56,000	1.27	53,400	2.19	52,000
1.8	68,100	1.27	60,300	2.1	68,100
1.82	68,300	1.27	70,925	1.92	68,300
1.89	56,200	1.91	60,300	2.99	56,200
1.91	56,100	1.91	53.400	1.91	56,100
1.93	48,100	1.91	53,400	2.93	52,100
1.94	56,600	4.85	60,300	1.94	56,600
2.7	71,400	4.85	70,925	1.7	72,400
4.91	71,800	4.85	70,925	3.91	70,800
5.0	72,200	4.85	70,925	2.50	72,200

We computed inversion factors for *Increase Rate* and *Annual Income* for MM_B and the results are 0.978, and 0.133 respectively. As presented in the previous section, all subsets of key attributes must be considered when the inversion factor is not 0 or 1. For the remainder of this example, the superscript notation is used to show what subset of key attributes is used. We use the number 1 for the set {*Increase Rate, Annual Income*}, 2 for {*Increase Rate*}, and 3 for {*Annual Income*}.

To compute vicinity sets, we use the interval vicinity method. The choice for each interval range for each continuous key attribute is made by the data owner. For this example, we consider that x_j is in the vicinity set V_i if and only if the distance between the increase rate of x_i and x_j is less than or equal to 0.1 and the distance between the annual income of x_i and x_j is less than or equal to 500. This procedure is applied for each piece of microdata individually. Table 20-2 summarizes the vicinity set cardinality for each record.

Table 20-2. Vicinity set cardinality for each individual record

The kth record	$n(V_k(IM))$	$n(V_k(MM_A))$	$n(V_k^1(MM_B))$	$n(V_k^2(MM_B))$	$n(V_k^3(MM_B))$
1	1	1	1	2	2
2	2	1	1	2	2
3	2	1	1	3	2
4	2	1	1	2	3
5	2	2	2	3	3
6	1	2	1	2	2
7	1	1	2	3	3
8	1	3	1	1	2
9	2	3	1	1	1
10	2	3	1	1	2

Based on the vicinity set cardinality, the classification matrices are as follows:

$$C_{IM} = \begin{pmatrix} 4 & 0 & 0 & \dots & 0 \\ 0 & 6 & 0 & \dots & 0 \\ 0 & 0 & 0 & \dots & 0 \\ \dots & \dots & \dots & \dots & \dots \\ 0 & 0 & 0 & \dots & 0 \end{pmatrix} \quad C_{MMA} = \begin{pmatrix} 2 & 3 & 0 & \dots & 0 \\ 1 & 1 & 0 & \dots & 0 \\ 1 & 2 & 0 & \dots & 0 \\ \dots & \dots & \dots & \dots & \dots \\ 0 & 0 & 0 & \dots & 0 \end{pmatrix}$$

$$C^1_{MMB} = \begin{pmatrix} 3 & 5 & 0 & \dots & 0 \\ 1 & 1 & 0 & \dots & 0 \\ 0 & 0 & 0 & \dots & 0 \\ \dots & \dots & \dots & \dots & \dots \\ 0 & 0 & 0 & \dots & 0 \end{pmatrix}$$

$$C^2_{MMB} = \begin{pmatrix} 1 & 2 & 0 & \dots & 0 \\ 2 & 2 & 0 & \dots & 0 \\ 1 & 2 & 0 & \dots & 0 \\ \dots & \dots & \dots & \dots & \dots \\ 0 & 0 & 0 & \dots & 0 \end{pmatrix} \quad C^3_{MMB} = \begin{pmatrix} 0 & 1 & 0 & \dots & 0 \\ 3 & 3 & 0 & \dots & 0 \\ 1 & 2 & 0 & \dots & 0 \\ \dots & \dots & \dots & \dots & \dots \\ 0 & 0 & 0 & \dots & 0 \end{pmatrix} \qquad (16)$$

To compute the weighted disclosure risk, data owners may choose various weight matrices based on the initial microdata characteristics and the level of protection desired.[32] For this illustration, we consider the following weight matrix:

[32] *See supra* note 24.

$$W = \begin{pmatrix} 6 & 1 & 0 & \ldots & 0 \\ 1 & 2 & 0 & \ldots & 0 \\ 0 & 0 & 0 & \ldots & 0 \\ \ldots & \ldots & \ldots & \ldots & \ldots \\ 0 & 0 & 0 & \ldots & 0 \end{pmatrix} \tag{17}$$

In Table 20-3, we compute the minimal, maximal and weighted disclosure risk for all microdata sets presented in this illustration. The column MM_B shows the final disclosure risk values for the second masked microdata and the last three columns show the disclosure risk values for each subset of key attributes.

Table 20-3. Disclosure risk values

	IM	MM_A	MM_B	MM_B^1	MM_B^2	MM_B^3
DR_{min}	0.4	0.2	0.00866	0.00866	0.00333	0
DR_W	0.5	0.25	0.07222	0.01059	0.00555	0.07222
DR_{max}	0.7	0.55	0.39	0.01877	0.01666	0.39

As expected, the disclosure risk values are lower in both masked microdata sets compared with the initial microdata values. We notice that records with vicinity sets of size 1 in both the initial and masked microdata sets are at the highest risk of disclosure. We conclude that a major goal of all masking processes is to eliminate all such situations, if possible. If no record fits into this category the minimal disclosure risk will be 0.

When using the second masked microdata set, an intruder has a better chance to disclose individual information if he or she uses *Annual Income* as the only key attribute. This is another danger that the data owner should avoid in any masking process - overprotection of one specific key attribute with loss of information and no benefit in terms of decreasing disclosure risk values.

4. CONCLUSIONS AND FUTURE WORK

The disclosure risk measures presented in this paper can be computed when all key attributes are continuous and the masking process does not change the number of records. They may become an important decision factor for the data owners in selecting which disclosure control methods they should apply to given initial microdata sets. The illustrations presented show that usually the proposed disclosure risk measures decrease as the data is modified more and more from its original form.

By establishing the vicinity set's cardinality, data owners can identify at- risk records and apply disclosure control techniques that target those records. In the masked microdata set all records with vicinity sets of size 1 should have vicinity sets of higher cardinality in the initial microdata.

Data owners can use disclosure risk measures to determine when additional data modification will produce minimal disclosure control benefits. There are situations when data is modified with little or no effect in lowering disclosure risk.

In future work, we will test our disclosure risk measures on real financial data, and will analyze the correlation of disclosure risk with information loss. Of particular interest is automating the process of defining vicinity sets based on the data characteristics. The final goal is to derive patterns of applying more than one disclosure control method to a specific initial microdata set to minimize both disclosure risk and information loss.

References

Ackerman, M., L. Cranor and J. Reagle, *Privacy in E-Commerce: Examining User Scenarios and Privacy Preferences*, PROC. ACM CONFERENCE ON ELECTRONIC COMMERCE (1998)

Acquisti, A., *Privacy in Electronic Commerce and the Economics of Immediate Gratification, in* PROC. ACM CONF. ON ELECTRONIC COMMERCE (EC 04) (2004).

Adomavicius, G. and A. Tuzhilin, *User Profiling in Personalization Applications through Rule Discovery and Validation*, 5 PROC. OF THE ACM SIGKDD INT'L CONF. ON KNOWLEDGE DISCOVERY AND DATA MINING (1999).

Agrawal, R. and R. Srikant, *Privacy-Preserving Data* Mining, in 2000 PROC. ACM SIGMOD CONF. ON MANAGEMENT OF DATA.

Ainslie, George, BREAKDOWN OF WILL (2001).

Albrecht, Spencer, THE AMERICAN BALLOT (1942).

Allen, Anita L., *Coercing Privacy*, 40 WM. AND MARY L. REV. 723 (1999).

Allen, Anita L., T*he Wanted Gaze: Accountability for Interpersonal Conduct at Work*, 89 GEO. L. REV. 2013 (2001).

Ayres, Ian & Matthew Funk, *Marketing Privacy*, 20 YALE J. ON REG. 77 (2003).

Balkin, J.M., CULTURAL SOFTWARE: A THEORY OF IDEOLOGY 141 (1998).

Bell, Christopher L., *The ISO 14001 Environmental Management Systems Standard: A Modest Perspective*, 27 ENVTL. L. REP. 10 (Dec. 1997).

Benassi, P., *TRUSTe: An Online Privacy Seal Program*, 42 COMM. OF THE ACM (1999).

Benham, Lee, *The Effect of Advertising on the Price of Eyeglasses*, 15 J.L. & ECON. 337 (1972).

Berry, M. and G. Linoff, DATA MINING TECHNIQUES FOR MARKETING, SALES AND CUSTOMER SUPPORT (1997).

Blackstone, William, COMMENTARIES ON THE LAWS OF ENGLAND 2 (facsimile ed. 1979) (1766).

Boger, Z. *et al.*, *Automatic Keyword Identification by Artificial Neural Networks Compared to Manual Identification by Users of Filtering Systems*, 37 INFO. PROCESSING & MGMT. 187 (2001).

Boyle, James, *A Politics of Intellectual Property: Environmentalism for the Net?*, 47 DUKE L.J. 87 (1997).

Boyle, James, *Foucault in Cyberspace: Surveillance, Sovereignty, and Hardwired Censors*, 66 U. CIN. L. REV. 177 (1997).

Brand, Stewart, THE MEDIA LAB: INVENTING THE FUTURE AT MIT (1987).

Brandeis, Louis & Samuel Warren, *The Right to Privacy*, 4 HARV. L. REV. 193 (1890).

Branscomb, Anne W., WHO OWNS INFORMATION? (1994).

Calabresi, Guido & A. Douglas Melamed, *Property Rules, Liability Rules, and Inalienability: One View of the Cathedral*, 85 HARV. L. REV. 1089 (1972).

Cameron, Alex, *Digital Rights Management: Where Copyright and Privacy Collide*, 2 CANADIAN PRIVACY LAW REV. 14 (2004).

Chase, Adam, *The Efficiency Benefits of "Green Taxes," A Tribute to Senator John Heinz*, 11 UCLA J. ENVTL. L. & POL'Y 1 (1992).

Chen, G. and S. Keller-McNulty, *Estimation of Deidentification Disclosure Risk in Microdata*, 14 J. OF OFFICIAL STATISTICS 79 (1998).

Claessens, J. *et al.*, *Solutions for Anonymous Communication on the Internet,* 487 PROC. OF THE IEEE INT'L CARNAHAN CONF. ON SECURITY TECH. 298 (1999).

Claypool, M. et al., *Implicit Interest Indicators*, 6 PROC. OF THE INT'L CONF. ON INTELLIGENT USER INTERFACES (2001).

Coase, Ronald, *The Problem of Social Cost*, 3 J.L. & ECON. 1 (1960).

Cohen, Fred, *Is Open Source More or Less Secure?* MANAGING NETWORK SECURITY, (July 2002).

Cohen, Julie E., *Examined Lives: Informational Privacy and the Subject as Object*, 52 STAN. L. REV. 1373 (2000).

Conway, R. and D. Strip, *Selective Partial Access to a Database, in* 1976 PROC. ANN. ACM CONF.

Dagan, Hanoch & Michael A. Heller, *The Liberal Commons*, 110 YALE L.J. 549 (2001).

Dalenius, T. and S.P. Reiss, *Data-Swapping: A Technique for Disclosure Control*, 6 J. OF STATISTICAL PLANNING AND INFERENCE 73 (1982).

Deerwester, S. *et al.*, *Indexing by Latent Semantic Indexing*, 41 J. OF THE AM. SOC'Y FOR INFO. SCI. (1990).

Derrida, J., *Plato's Pharmacy, in* DISSEMINATION (B. Johnson trans., 1981).

Derry, T.K. & Trevor I. Williams, A SHORT HISTORY OF TECHNOLOGY: FROM THE EARLIEST TIMES TO A.D. 1900 (1960).

Dershowitz, Alan M., WHY TERRORISM WORKS: UNDERSTANDING THE THREAT, RESPONDING TO THE CHALLENGE (2002).

Dripps, Donald A., *Terror and Tolerance: Criminal Justice for the New Age of Anxiety*, 1 OHIO ST. J. OF CRIM. L. 9 (2003).

du Pont, George F., *The Time has Come for Limited Liability for Operators of True Anonymity Remailers in Cyberspace: An Examination of the Possibilities and Perils*, 6 FALL J. TECH. L. & POL'Y 3 (2001).

Duda, R. and P. Hart, PATTERN CLASSIFICATION AND SCENE ANALYSIS (1973).

Dworkin, G., THE THEORY AND PRACTICE OF AUTONOMY (1988).

Elliot, M.J., DIS: *A New Approach to the Measurement of Statistica Disclosure Risk*, INT'L. J. OF RISK MGMT. 39 (2000).

Ellmann, Stephen J., *Racial Profiling and Terrorism*, 22 N.Y.L.S. L. REV. 675 (2003).

Elovici, Y. *et al.*, *A New Privacy Model for Hiding Group Interests while Accessing the Web*, WORKSHOP ON PRIVACY IN THE ELECTRONIC SOC'Y IN ASS'N WITH 9TH ACM CONF. ON COMPUTER AND COMM. SECURITY (2002).

Emmerink, Richard H. M., INFORMATION AND PRICING IN ROAD TRANSPORTATION (1998).

Epstein, Richard A., *The Allocation of the Commons: Parking on Public Roads*, 31 J. LEGAL STUD. S515 (2002).

Fellegi, I.P., *On the Question of Statistical Confidentiality*, 67 J. OF THE AM. STATISTICAL ASS'N 7 (1972).

Fishbein, M. and I. Azjen, BELIEF, ATTITUDE, INTENTION AND BEHAVIOR: AN INTRODUCTION TO THEORY AND RESEARCH (1975).

Flaherty, David H., PROTECTING PRIVACY IN SURVEILLANCE SOCIETIES: THE FEDERAL REPUBLIC OF GERMANY, SWEDEN, FRANCE, CANADA, AND THE UNITED STATES (1989).

Fox, Susannah, *Trust and Privacy Online: Why Americans Want to Rewrite the Rule* (2000).

Froomkin, A. Michael, *Anonymity in the Balance*, *in* DIGITAL ANONYMITY AND THE LAW (C. Nicoll et al. eds., 2003).

Froomkin, A. Michael, *Symposium: Cyberspace and Privacy: A New Legal Paradigm?: The Death of Privacy?*, 52 STAN. L. REV. 1461 (2000).

Frum, David & Richard Perle, AN END TO EVIL: HOW TO WIN THE WAR ON TERROR 69-73 (2003).

Fulda, Joseph, *Data Mining and Privacy,* 11 ALB. L.J. SCI. & TECH. 105 (2000).

Gabber, E. *et al.*, *On Secure and Pseudonymous Client-Relationships with Multiple Severs*, 2 ACM TRANSACTIONS ON INFO. AND SYSTEMS SECURITY (1999)

Gabber, E., *et al.*, *Consistent, yet Anonymous, Web Access with LPWA*, 42 COMM. OF THE ACM 42-47 (1999).

Gandy, Oscar H., Jr., THE PANOPTIC SORT: A POLITICAL ECONOMY OF PERSONAL INFORMATION 10 (1993).

Gandy, Oscar H., Jr., THE PANOPTIC SORT: A POLITICAL ECONOMY OF PERSONAL INFORMATION (1993).

Garfinkel, S., DATABASE NATION: THE DEATH OF PRIVACY IN THE 21ST CENTURY, (2000).

Glancy, Dorothy, *At the Intersection of Visible and Invisible Worlds: United States Privacy Law and the Internet*, 16 SANTA CLARA COMPUTER & HIGH TECH. L.J. 357 (2000).

Goldschlag, D.M. *et al.*, *Hiding Routing Information*, *Information Hiding,* 1174 SPRINGER-VERLAG LLNCS 137 (R. Anderson ed., 1996).

Goldschlag, D.M. *et al.*, *Onion Routing for Anonymous and Private Internet Connections*, 42 COMM. OF THE ACM 39 (1999).

Gurstein, Rochelle, THE REPEAL OF RETICENCE: A HISTORY OF AMERICA'S CULTURAL AND LEGAL STRUGGLES OVER FREE SPEECH, OBSCENITY, SEXUAL LIBERATION, AND MODERN ART (1996).

Hagel, John III & Marc Singer, NET WORTH: SHAPING MARKETS WHEN CUSTOMERS MAKE THE RULES (1999).

Han, Jiawei and Micheline Kamber, DATA MINING: CONCEPTS AND TECHNIQUES (2001).

Hand, D. J., H. Mannila & P. Smyth, PRINCIPLES OF DATA MINING (2001).

Harris, David A., *Car Wars: The Fourth Amendment's Death on the Highway*, 66 GEO. WASH. L. REV. 556 (1998).

Harris, David A., PROFILES IN INJUSTICE: WHY RACIAL PROFILING CANNOT WORK 73 (2002).

Harris, David A., *The Stories, the Statistics, and the Law: Why "Driving While Black" Matters*, 84 MINN. L. REV. 265 (1999).

Hohfeld, Wesley Newcomb, Fundamental Legal Conceptions as Applied in Judicial Reasoning, 26 YALE L.J. 710 (1917).

Hughes, Arthur M., THE COMPLETE DATABASE MARKETER 354 (2d ed. 1996).

Hunter, Richard, WORLD WITHOUT SECRETS: BUSINESS, CRIME, AND PRIVACY IN THE AGE OF UBIQUITOUS COMPUTING 7 (2002).

Jamieson, Kathleen Hall and Karlyn Kohrs Campbell, THE INTERPLAY OF INFLUENCE: NEWS ADVERTISING, POLITICS, AND THE MASS MEDIA 195 (4th ed., 1997).

Jange, Edward J. and Paul M. Schwartz, *The Gramm-Leach-Bliley Act, Information Privacy, and the Limits of Default Rules*, 86 MINN. L. REV. 1219 (2002).

Jaro, M.A., *Advances in Record-Linkage Methodology as Applied to Matching the 1985 Census of Tampa, Florida*, 84 J. OF AM. STATISTICS ASS'N, 414 (1989).

Johnson, Neil F. and Sushil Jajodia, *Steganography: Seeing the Unseen*, IEEE COMPUTER (February 1998) at 26.

Johnson, Stephen M., ECONOMICS, EQUITY AND THE ENVIRONMENT (2004).

Juels, A. and J. Brainard, *Soft Blocking: Flexible Blocker Tags on the Cheap,* PROC. 2004 ACM WORKSHOP ON PRIVACY IN THE ELECTRONIC SOC'Y 1 (V. Atluri, P. F. Syverson, S. De Capitani di Vimercati, eds., 2004).

Juels, A., R.L. Rivest, M. Szydlo, *The Blocker Tag: Selective Blocking of RFID Tags for Consumer Privacy*, 8TH ACM CONF. ON COMPUTER AND COMM. SECURITY 103 (V. Atluri, ed., 2003).

Jurafsky, D. and J. H. Martin, SPEECH AND LANGUAGE PROCESSING (2000).

Kafka, Frank, THE TRIAL (Willa & Edwin Muir trans., 1937).

Kahneman, Daniel and Amos Tversky, eds., CHOICES, VALUES, AND FRAMES (2000).

Kang, Jerry, *Information Privacy in Cyberspace Transactions*, 50 STAN. L. REV. 1193 (1998).

Kao, Alice, *RIAA V. Verizon: Applying the Subpoena Provision of the DMCA*, 19 BERKELEY TECH. L.J. 405 (2004)

Karas, Stan, *Privacy, Identity, and Databases*, 52 AM. U. L. REV. 393 (2002).

Kennedy, Michael, THE GLOBAL POSITIONING SYSTEM AND GIS: AN INTRODUCTION (2nd ed., 2002).

Kerr, Ian R. & Daphne Gilbert, *The Changing Role of ISPs in the Investigation of Cybercrime, in* INFORMATION ETHICS IN AN ELECTRONIC AGE: CURRENT ISSUES IN AFRICA AND THE WORLD (Thomas Mendina & Johannes Brtiz eds., 2004).

Kerr, Ian, *The Legal Relationship Between Online Service Providers and Users*, 35 CAN. BUS. L.J. 40 (2001).

Korobkin, Russell, *The Endowment Effect and Legal Analysis, Symposium on Empirical Legal Realism: A New Social Scientific Assessment of Law and Human Behavior*, 97 NW. U.L. REV. 1227 (2003).

Lafave, Wayne R., *The "Routine Traffic Stop" from Start to Finish: Too Much "Routine," Not Enough Fourth Amendment,* 102 MICH. L. REV. 1843 (2004).

Lear, Rick S. & Jefferson D. Reynolds, *Your Social Security Number or Your Life: Disclosure of Personal Identification Information by Military Personnel and the Compromise of Privacy and National Security*, 21 B. U. INTL. L. J. 1 (2003).

Lessig, L., *The Path of Cyberlaw*, 104 YALE L.J. 1743 (1995).

Lessig, L., *The Law of the Horse: What Cyberlaw Might Teach*, 113 HARV. L. REV. 501 (1999).

Lipton, Jacqueline, *Information Property: Rights and Responsibilities,* 56 FLA. L. REV. 135 (2004).

Litan, Robert E., *Law and Policy in the Age of the Internet*, 50 DUKE L. J. 1045 (2001).

Litman, Jessica, *Information Privacy/Information Property*, 52 STAN. L. REV. 1283 (2000).

LoPucki, Lynn M., *Human Identification Theory and the Identity Theft Problem*, 80 TEX. L. REV. 89 (2001).

Luz, M.Q. & M. Javed, *Empirical Evaluation of Explicit versus Implicit Acquisition of User Profiles in Information Filtering Systems*, 4 PROC. OF THE ACM CONF. ON DIGITAL LIBR. (1999).

Lyon, David, ed.,SURVEILLANCE AS SOCIAL SORTING: PRIVACY, RISK AND AUTOMATED DISCRIMINATION (2002).

Malden, D., *Machine Learning Used by Personal WebWatcher*, 1999 PROC. OF ACAI-99 WORKSHOP ON MACHINE LEARNING AND INTELLIGENT AGENTS.

Mallman, Otto, § 6, in KOMMENTAR ZUM BUNDESDATENSCHUTZGESETZ 545 (Spiros Simitis ed., 5th ed. 2003).

Maltoni, D., D. Maio, A.K. Jain, and S. Prabhakar, HANDBOOK OF FINGERPRINT RECOGNITION (2003).

Mank, Brad, *The Environmental Protection Agency's Project XL and Other Regulatory Reform Initiatives: The Need for Legislative Authorization,* 1998 ECOLOGY L. Q. 1.

McAdams, Richard H., *The Origin, Development and Regulation of Norms*, 96 MICH. L. REV. 338 (1997).

McClurg, Andrew J., *A Thousand Words Are Worth a Picture: A Privacy Tort Response to Consumer Data Profiling*, 98 NW. U. L. REV. 63 (2003).

McCormick, Richard P., THE HISTORY OF VOTING IN NEW JERSEY (1953).

McLauglin, Marsha Morrow & Suzanne Vaupel, *Constitutional Right of Privacy and Investigative Consumer Reports: Little Brother Is Watching You*, 2 HASTING CONST. L.Q. 773 (1975).

Mill, John Stuart, ON LIBERTY (Gateway 1955) (1859).

Monahan, Kevin and Don Douglas, GPS INSTANT NAVIGATION: FROM BASIC TECHNIQUES TO ELECTRONIC CHARTING (2nd ed., 2000).

Monmonier, Mark, SPYING WITH MAPS: SURVEILLANCE TECHNOLOGIES AND THE FUTURE OF PRIVACY (2002).

Murphy, Richard S., *Property Rights in Personal Information: An Economic Defense of Privacy*, 84 GEO. L.J. 2381 (1996).

Nagel, R., *Unraveling in Guessing Games: An Experimental Study*, 85 AMER. ECON. REV. 1313 (1995).

Netanel, Neil Weinstock, *Cyberspace Self-Governance: A Skeptical View from Liberal Democratic Theory*, 88 CAL. L. REV. 395 (2000).

Orts, Eric, *Reflexive Environmental Law*, 89 NW. U. L. REV. 1227 (1995).

Orwell, George, 1984 (1949).

Overton, Hon. Ben F. & Katherine E. Giddings, *The Right of Privacy in Florida in the Age of Technology and the Twenty-First Century: A Need for Protection from Private and Commercial Intrusion*, 25 FLA. ST. U. L REV. 25 (1997).

Pearson, Eric, ENVIRONMENTAL AND NATURAL RESOURCES LAW (2002).

Posner, R.A., *The Economics of Privacy*, 71 AM. ECON. REV. 405 (1981).

Posner, Richard A., *The Right of Privacy*, 12 GA. L. REV. 393 (1977-78).

Post, Robert C., *The Unwanted Gaze, by Jeffrey Rosen: Three Concepts of Privacy*, 89 GEO. L.J. 2087 (2001).

Rabushka, Alvin and Kenneth Shepsle, POLITICS IN PLURAL SOCIETIES: A THEORY OF DEMOCRATIC INSTABILITY (1972).

Radin, Margaret Jane, CONTESTED COMMODITIES (1996).

Ramsay, Iain, ADVERTISING, CULTURE AND THE LAW – BEYOND LIES, IGNORANCE AND MANIPULATION (1996).

Raz, J., THE MORALITY OF FREEDOM (1986).

Richards, Jef, DECEPTIVE ADVERTISING (1990, Lawrence Erlbaum).

Rose, Carol M., *Canons of Property Talk, or, Blackstone's Anxiety*, 108 YALE L.J. 601 (1998).

Rose-Ackerman, Susan, *Inalienability and the Theory of Property Rights*, 85 COLUM. L. REV. 931 (1985).

Salton, G., INTRODUCTION TO MODERN INFO. RETRIEVAL (W.J. McGill, ed., McGraw-Hill 1983).

Salzman, James & Barton H. Thompson Jr., ENVIRONMENTAL LAW AND POLICY (2003).

Samarati, Pierangela and Latanya Sweeney, *Protecting Privacy when Disclosing Information: k-Anonymity and Its Enforcement through Generalization and Suppression*, SRI TECH. REP. (1998).

Samuelson, Pamela, *Privacy as Intellectual Property*, 52 STAN L. REV. 1125 (2000).

Schultz, Brian S., *Electronic Money, Internet Commerce, and the Right to Financial Privacy: A Call for New Federal Guidelines*, 67 U. CIN. L. REV. 779 (1999).

Schwartz, Paul M., *Privacy and Democracy in Cyberspace*, 52 VAND. L. REV. 1609 (1999).

Schwartz, Paul M., *Privacy and Participation: Personal Information and Public Sector Regulation in the United States*, 80 IOWA L. REV. 553 (1995).

Schwartz, Paul M., *Privacy and the Economics of Personal Health Care Information*, 76 TEX. L. REV. 1 (1997).

Schwartz, Paul M., *Property, Privacy, and Personal Data*, 117 HARV. L. REV. 2055 (2004).

Schwartz, Paul, *The Computer in German and American Constitutional Law: Towards an American Right of Informational Self-Determination*, 37 AM. J. COMP. L. 675 (1989).

Selten, R., *What is Bounded Rationality? in* BOUNDED RATIONALITY: THE ADAPTIVE TOOLBOX (G. Gigerenzer and R. Selten, eds., 2001).

Shabecoff, Philip, A FIERCE GREEN FIRE: THE AMERICAN ENVIRONMENTAL MOVEMENT (1993).

Shapira, B. *et al.*, *PRAW – The Model for the Private Web*, J. OF THE AM. SOC'Y FOR INFO. SCI. AND TECH. (forthcoming).

Shenk, David, DATA SMOG 114 (1997).

Simitis, Spiros, *Das Volkszählungsurteil oder der Lane Weg zur Informationsaskese*, 83 KRITISCHE VIERTELJAHRESSCHRIFT FÜR GESETZGEBUNG UND RECHTSWISSENSCHAFT 359 (2000).

Simitis, Spiros *Einleitung*, in KOMMENTAR ZUM BUNDESDATENSCHUTZGESETZ 545 (Spiros Simitis ed., 5th ed. 2003).

Simon, H.A., MODELS OF BOUNDED RATIONALITY (1982).

Skinner, C.J. and M.J. Elliot, *A Measure of Disclosure Risk for Microdata*, 64 J. OF THE ROYAL STATISTICAL SOC'Y 855 (2002).

Skinner, C.J. et al., *Disclosure Control for Census Microdata*, J. OF OFFICIAL STATISTICS 31 (1994).

Slobogin, Christopher, *Public Privacy: Camera Surveillance of Public Places and the Right to Anonymity*, 72 MISS. L.J. 213 (2002).

Solove, Daniel J. & Marc Rotenberg, INFORMATION PRIVACY LAW (2003).

Solove, Daniel J., THE DIGITAL PERSON (2004).

Solove, Daniel J., *The Virtues of Knowing Less: Justifying Privacy Protections against Disclosure*, 53 DUKE L. J. 967 (2003).

Spiekermann, Sarah, Jens Grossklags, and Bertina Berendt, *E-Privacy in 2nd Generation E-Commerce: Privacy Preferences v. Actual Behavior*, *in* 3RD ACM CONFERENCE ON ELECTRONIC COMMERCE - EC '01 (2002).

Staples, William G., THE CULTURE OF SURVEILLANCE: DISCIPLINE AND SOCIAL CONTROL IN THE UNITED STATES (1997).

Stefani, A. and C. Strappavara, *Personalizing Access to Websites: The SiteIF Project*, 2 PROC. OF THE WORKSHOP ON ADAPTIVE HYPERTEXT AND HYPERMEDIA (1998).

Sterne, Jim, WHAT MAKES PEOPLE CLICK: ADVERTISING ON THE WEB (1997).

Stewart, Richard, *A New Generation of Environmental Regulation?*, 29 CAPITAL L. REV. 21 (2001).

Stigler, G.J., *An Introduction to Privacy in Economics and Politics*, 9 J. LEGAL STUDIES 623 (1980).

Strandburg, Katherine J., Privacy, Rationality, and Temptation: A Theory of Willpower Norms, RUTGERS L. REV. (forthcoming 2005).

Syverson, P. and A. Shostack, *What Price Privacy? in* THE ECONOMICS OF INFORMATION SECURITY (L.J. Camp and S. Lewis, eds., 2004).

Syverson, P.F. *et al.*, *Anonymous Connections and Onion Routing*, 18 PROC. OF THE ANNUAL SYMP. ON SECURITY AND PRIVACY 44 (1997).

Truta, T.M. *et al.*, *Disclosure Risk Measures for Microdata*, INT'L CONF. ON SCI. AND STATISTICAL DATABASE MGMT. 15 (2003).

Truta, T.M. *et al.*, *Disclosure Risk Measures for Sampling Disclosure Control Method*, ANN. ACM SYMP. ON APPLIED COMPUTING 301 (2004).

Trute, Hans-Heinrich, *Verfassungsrechtliche Grundlagen*, in HANDBUCH DATENSCHUTZ: DIE NEUEN GRUNDLAGEN FÜR WIRTSCHAFT UND VERWALTUNG 156 (Alexander Rossnagel ed., 2003).

Turk, M. and A. Pentland, *Eigenfaces for Recognition*, 3 J. COGNITIVE NEUROSCIENCE 71 (1991).

Turkington, Richard C. & Anita L. Allen, PRIVACY LAW: CASES AND MATERIALS (2nd ed. 2002).

Urken, Arthur B, *Voting in A Computer-Networked Environment*, *in* THE INFORMATION WEB: ETHICAL AND SOCIAL IMPLICATIONS OF COMPUTER NETWORKING (Carol Gould, ed., 1989).

Volokh, Eugene, *Freedom of Speech and Information Privacy: The Troubling Implications of a Right to Stop People From Speaking About You,* 52 STAN. L. REV. 1049 (2000).

Wayman, J.L., FUNDAMENTALS OF BIOMETRIC TECHNOLOGIES, *available at* http://www.engr.sjsu.edu/biometrics/publications_tech.html.

Weber, Max, ECONOMY AND SOCIETY (Guenther Roth & Claus Wittich eds., 1978).

Westin, Alan F., PRIVACY AND FREEDOM (1967).

Westin, Alan F., *Privacy in the Workplace: How Well Does American Law Reflect American Values?*, 72 CHI.-KENT L. REV. 271 (1996).

Westin, Alan, PRIVACY AND FREEDOM (1967).

Widyantoro, D.H. *et al.*, *Learning User Interest Dynamics with a Three-Descriptor Representation*, 52 J. OF THE AM. SOC'Y OF INFO. SCI. AND TECH. 212 (2001).

Wigmore, J. H., THE AUSTRALIAN BALLOT SYSTEM (2nd ed., Boston, 1889).

Winter, Steven L, A CLEARING IN THE FOREST: LAW, LIFE, AND MIND (2001).

Woodward, John *et al.*, IDENTITY ASSURANCE IN THE INFORMATION AGE (2003).

Wuchek, Wendy, *Conspiracy Theory: Big Brother Enters the Brave New World of Health Care Reform*, 3 DEPAUL J. HEALTH CARE L. 293 (2000).

Yao, Andrew, *Protocols for Secure Computations*, PROC. OF THE IEEE SYMP. ON THE FOUNDS. OF COMPUTER SCI., 160 (1982).

Zarsky, Tal Z., *"Mine Your Own Business!": Making the Case for the Implications of the Mining of Personal Information in the Forum of Public Opinion*, 5 YALE J.L. & TECH. 4 (2003).

Zarsky, Tal Z., *Desperately Seeking Solutions – Using Implementation-Based Solutions for the Troubles of Information Privacy in the Age of Data Mining and the Internet Society*, 56 ME. L. REV.13 (2004).

Zarsky, Tal Z., *Thinking Outside the Box: Considering Transparency, Anonymity, and Pseudonymity as Overall Solutions to the Problems of Information Privacy in the Internet Society*, 58 U. MIAMI L. REV. 991 (2004).

Zarsky, Tal, *Cookie Viewers and the Undermining of Data-Mining: A Critical Review of the DoubleClick Settlement*, 2002 STAN. TECH. L. REV. P1 (2002).

Zimmerman, Diane L., 68 CORNELL L. REV. 291 (1983).

Index